Medicinal and Aromatic Plants of the World

Volume 2

Series Editor
Ákos Máthé
University of West Hungary
Faculty of Agriculture and Food Sciences
Mosonmagyaróvár, Hungary

More information about this series at http://www.springer.com/series/11192

Zohara Yaniv • Nativ Dudai
Editors

Medicinal and Aromatic Plants of the Middle-East

Springer

Editors
Zohara Yaniv
Department of Plant Introduction and Seed Research
Institute of Plant Sciences
Agricultural Research Organisation
Bet-Dagan, Israel

Nativ Dudai
The Aromatic, Medicinal and Spice Crops Unit
Newe Ya'ar Research Centre
Agricultural Research Organization
Ramat Yishay, Israel

ISSN 2352-6831 ISSN 2352-684X (electronic)
ISBN 978-94-017-9275-2 ISBN 978-94-017-9276-9 (eBook)
DOI 10.1007/978-94-017-9276-9
Springer Dordrecht Heidelberg New York London

Library of Congress Control Number: 2014949857

Printed on acid-free paper

Springer is part of Springer Science+Business Media (www.springer.com)

Preface

The use of medicinal plants as recognized therapies in modern health care has been increasing over the past three decades. Treatments that were considered "old wives tales" have been found to have underlying biochemical and even epigenetic foundations for their efficacy. The search for medically useful plants and the means to produce their active principles and ingredients now encompasses large research programs and even entire scientific institutes.

The series "Medicinal Plants of the World", under the capable and thoughtful leadership of Prof. Dr. Ákos Máthé, is intended to bring the latest research to the attention of the broad range of botanists, ethnopharmacists, biochemists, plant and animal physiologists and others who will benefit from the information gathered therein. Plants know no political boundaries, and bringing specific folklore to general medical awareness can only be for the benefit of all. The current volume brings together chapters on medicinal plants of the Mediterranean region from researchers in Israel, Jordan and Turkey, and covers both wild (non-cultivated) and domestic (cultivated) crops with therapeutic value. Some of these plants are well-known medicinally, such as opium poppy and khat, while others such as apharsemon and citron have both ritual and medicinal uses. All have specific and valuable uses in modern society.

Bet-Dagan, Israel | Zohara Yaniv

Acknowledgments

In commissioning, collecting, and editing these chapters I have been greatly assisted by Nativ Dudai in locating authors, by Roy Amram and Prof. Uriel Bachrach in editing, by the staff of Springer Publishing in numerous technical details, and by Ákos Máthé in many general tasks. My profound thanks go to all of them and to the authors of the various chapters, with best wishes for continued good health for them and for the readers of this volume.

Contents

Introduction: Medicinal Plants in Ancient Traditions

Zohara Yaniv

Abstract Plants have been used for medicinal purposes since time immemorial, and to this day, many of the important and familiar remedies originate in plants.

This chapter outlines the history and early traditions of medicinal plants in the Middle-east. The importance of the early "medicine-men" in ancient cultures, as collectors and healers, is emphasized. Archaeological findings in sites such as Iraq and Babylon, as well as clay tablets and ancient manuscripts from Egypt, Sumaria and Assyria, India and China reveal the immense body of knowledge that existed during these old times in history.

A special place is devoted to the Bible, as an ancient document describing the use of plants in this region during biblical times.

Since before the Common Era, great herbalists, such as Dioscorides, Hippocrates, Theophrastus and Galenius acted as scientists and therapists leaving us with prominent books, such as De Materia Medica of Dioscorides. This priceless document provided the world with vast knowledge regarding hundreds of medicinal plants which are found in the region of the Middle-East. Most of these plants are still used at the present time for therapy and some of them are rich sources of natural compounds with medicinal properties.

It was only by the mid-fifteenth century that the influence of Dioscorides, and that of the classic herbalists, began to fade within European botany and medicine. During this period, and until our times, the European herbalists began researching plants for pure research purposes, which resulted in great scientific discoveries.

Although there is no doubting the predominance of chemical research in modern medicine, there is a notably increasing interest, within both medical circles and the general public alike, in plant-oriented folk medicine. Further research into the biochemical mechanisms of herbal medicines will enable a synthesis of traditional and modern methods of health care, to the benefit of all.

Keywords History • Tradition • Dioscorides • Bible • Galen • Egypt • Babylon • Middle-ages • Avicenna • Doctrine of signatures • India • Papyrus of Ebers

---

Z. Yaniv (✉)
Department of Plant Introduction and Seed Research, Institute of Plant Sciences, Agricultural Research Organization, Bet-Dagan, Israel
e-mail: zoharayb@netvision.net.il

Z. Yaniv, N. Dudai (eds.), *Medicinal and Aromatic Plants of the Middle-East*,
Medicinal and Aromatic Plants of the World 2, DOI 10.1007/978-94-017-9276-9_1

The History of Medicinal Plants

The gathering of medicinal plants is one of the oldest branches of the medical profession. It is most likely that at first there was no distinction between the gathering of plants for food and collecting them for medicinal needs, especially since most human food sources originated in plants. However, mankind has always suffered from various diseases and disabilities, and through trial and error, has unravelled, the medicinal value of plants.

In ancient cultures, as well as in primitive societies of our own time, the shamans, or medicine men, have combined their practical experience, knowledge of botany and psychology – with knowledge of witchcraft and various mystical ceremonies, through which they are claimed to have contact with the world of powerful spirits and occult wisdom. This gave them a unique position within the tribe, and attributed to them fantastic healing powers, including the ability to cure diseases and injuries, as well as to expel evil spirits which were believed to have invaded human bodies (Gordon 1977).

Historical Findings

Archaeological findings at an ancient burial site in **Iraq** demonstrate the use of several plants that still serve in folk medicine today: yarrow, marshmallow, hyacinth, groundsel, thistle and ephedra.

In ancient **Babylon**, medical practice was one of the functions of priests. Evidence of the use of medicinal herbs has been found there in many archaeological excavations.

Clay tablets bearing medical prescriptions have been found in numerous **Sumerian** and **Babylonian** sites. In the laws of Hammurabi, king of Babylon in 1728 BCE, we find evidence of the use of medicinal plants such as cassia, henbane, liquorice, and varieties of mint, which continue to be used in folk medicine to the present day.

About 400 **Assyrian** remedies derived from plant and mineral sources are known today. One well-known Assyrian remedy is a resin derived from pine, which was used both externally as a muscle embrocation, and internally as a treatment for kidney and liver diseases. The Assyrians knew of the narcotic properties of poppy seeds and of the medicinal qualities of mandrake.

Medical prescriptions written on Egyptian papyri often include the various ingredients and their quantities. Ancient Egyptian medicine excelled in its "empirical-rational" approach, as part of the general Egyptian frame-of-mind.

The oldest medical document known is the papyrus of Ebers. It is dated to the sixteenth century BCE and was buried until its discovery, during the nineteenth century CE, in **Egypt**. The document contains 877 prescriptions and remedies, some of which include myrrh, frankincense, fennel and many other spice plants.

In **India** we find evidence of traditional folk medicine dating back to 1500 BCE, where it was passed on by word of mouth from father to son, for many generations, until it was included in the holy writings known as the Vedas. Many of the plants known in the Indian folklore were put to use by the Egyptians and by the Greeks after them, and, ultimately, found their way into European folk medicine as well (Savnur 1984).

The Chinese, as long as 5,000 years ago, were accustomed to drinking tea made from a leafless creeper known as "Ma-Huang" (Ephedra sinica), as a treatment for lung diseases. An ancient Chinese book, written in 2900 BCE, includes 365 remedies, and particularly specifies the value of this tea, which accelerates the circulation of the blood, reduces fever and relieves coughing; its main importance, however, was in clearing disturbances in the respiratory system. 'Ephedrine' was first extracted from this plant in the nineteenth century. The compound is still used as an important medication for asthma, various allergies and influenza to this day (Lee 2011).

The Chinese, and with them the Indians, contributed to the world the use of cannabis (hashish), which is known today mainly for its narcotic qualities, but which for thousands of years was famed for its medicinal properties, such as a cure for eye problems and pain relief.

We have only indirect evidence of the medicinal use of plants during the biblical period. Several plants which were known in the ancient Orient as medicinal plants are referred to in the Bible as sources of perfume; these include labdanum, tragacanth, myrrh and frankincense. Various poisonous plants mentioned in the Bible include mandrake, hemlock, wormwood and bitter apple, while spices mentioned include garlic, onion, cumin, hyssop and fennel. In spite of the fact that there is much evidence of the medicinal use of these plants in the region, there are no direct references in the Bible as to their usage for these purposes (Zohary 1982).

References to medicinal uses of plants in the Bible are usually episodic, with no discussion on the subject as such. For instance, the prophet Jeremiah mentions in three places the use of balm in the treatment of sores (Jeremiah, 8, 22; 46, 11; 51, 8). Various sources indicate that he was referring to a plant-resin which can be found in the Gilead; nowadays, it is considered to have been resin from the styrax. It could be that the lack of sustained attention to medicinal plants in the Bible stems from the belief that only the will of God cures mankind of its illnesses, and therefore, it does not necessarily indicate that there was no medicinal use of the plants at the time (Lev 2002; Palevitch and Yaniv 2000).

In contrast to the total of about 120 plants which are mentioned in the Bible, about 400 are introduced in the Mishnah and the Gemara, which together make up the Oral Law (Talmud). It is clear that until the period when the Talmud was codified (about 550 CE), the people of Israel had their own tradition of folk medicine, which they had not received from the neighboring peoples. Within the framework of this practice they must have used a large number of medicinal plants. In any case, the evidence of practical use of medicinal plants is much greater in the Talmud than in the Bible (Palevitch and Yaniv 2000; Duke 2007).

The ancient Greeks excelled in the field of herbal medicine and their heritage left its mark on European culture for hundreds of years to come.

Hippocrates (fourth century BCE) is considered to be the father of modern medicine. In his writings he lists about 300 medicinal plants. Some of these plants are used to this day in folk medicine, including: mint, poppy, sage, rosemary, rue and vervain.

Theophrastos (third century BCE), the Athenian botanist, studied the plants of Greece and the neighboring countries, and his book "Enquiry into Plants" was the pioneering work in both botanical research and in the history of medicinal plants. Some of his prescriptions are, in their essence, in use up to this day. In his book, he introduced details of the plants' appearances, their manner of cultivation, their classifications and their uses. Theophrastos is therefor known as "the Father of Botany" and his book has been used as a basic textbook for more than 2,000 years.

Dioscorides (first century BCE) was a Greek doctor who served in the Roman army in the days of the Caesar Nero. In his book, "De Materia Medica" he describes nearly 800 remedies derived from plant, animal and mineral sources, as well as directions of how to find them, the proper harvesting time, preparation procedures and medical designation. The book proved to be extremely influential in the West and laid the foundations for herbal medicine up to the Renaissance period (Gunther 1959). The Romans did not introduce many innovations to the field of herbal medicine and, as in other issues, mainly derived their knowledge from the Ancient East and especially from the Greeks. Both Greeks and Romans used medicinally active plants for production of cosmetics, extraction of perfumes, and as edible spices. Much of this information was gathered by Pliny the Elder (79–23 BCE), in his book, "Natural History", a compilation of information from about 2,000 Greek and Roman manuscripts. Pliny's writings, although not exclusively medicinal, do contain some information on the use of plant-derived medicine.

The influence of Galen (131 BCE) persisted in the medical world for many centuries. Many of his treatments included the use of medicinal plants. Plant-based remedies to this day bear the heading "Galenic medicine". After the fall of the Roman Empire, the active development of medicine, including the medicinal use of plants, ceased in the West; early Christianity believed that diseases were divine punishment for sins, and that their cure was a matter for Heaven and not for mankind to tamper with.

In sharp contrast to the "all in God's hands" approach, Arabic/Muslim medicine flourished during the seventh century CE on, up to its peak at the eleventh century CE, influenced mainly by the ancient medicinal practices of Mesopotamia, Greece and India. Many principles of Arabic medicine, including the use of medicinal plants, are widespread today among many millions of people across North Africa, Europe and Asia. Although ancient Arab medicine had inherited traditions from Greek and Roman practices, it extended the use of plants, and introduced many which had not been known previously in the West.

Among the outstanding figures of that period was Avicenna (980–1037 CE), whose book "Canon of Medicine" would later provide the basis for medical schools in Europe. For instance, Avicenna recommended the use of groundsel as a purgative for clearing the gall bladder, and to this day this plant is widely used in medicine, as a remedy for stomach aches and for improving digestion.

Ibn el Beithar (1197–1248), a famous Arab doctor, wrote a wide-ranging book on medicinal plants; in it he describes the medical practices of Dioscorides and Galen, and introduces additional information from his own research. His book was written with extreme clarity, and was ordered alphabetically for ease of use. It specifies the names of the remedies and of the plants in various languages, thus providing a first class tool for comparative research of medicinal plants.

Maimonides (Moses ben Maimon, known by the Hebrew acronym Rambam 1134–1240), in addition to his vast publications in the fields of Torah and Jewish Law (Halachah), published numerous medical books such as "The Book on Asthma", "Poisonous Drugs" and "Names of Medicines", which were considered the epitome of advanced medicine at that time. These books attribute great importance to the use of medicinal plants, several hundred of which are mentioned in Maimonides' writings (Lev and Amar 2000).

In medieval Europe, knowledge of medicinal plants was preserved mainly in the monasteries. The monks grew medicinal plants with great dedication in the monastery gardens, and also knew of their natural habitats. They gathered and dried the plants, extracted essential oils from them, and used them for medicinal treatment according to the traditions of the Greek and Roman manuscripts, which they had copied and translated with great diligence, drawings included (Lev 2008).

The moratorium on medicine in Europe was total in the early to mid-Middle Ages, and even the "Black Plague", which spread across Europe during the mid-fourteenth century, did not encourage fresh research into the causes of the disease. As mentioned previously, Christian belief was that disease was a punishment ministered directly by God. However, many people did hang bunches of bay leaves, dill, mint, roses and other aromatic plants above entrances to hospitals and houses, because these were believed to fend off the plague by power of their fragrance. It would later be known that some of these plants do possess disinfecting properties. In Europe of the Middle-Ages, grave importance was attributed to plants of the Solanaceae family: mandrake, henbane, stinkweed and belladonna. These plants owe their medicinal characteristics to the high concentration of tropane alkaloids in them. These were mainly used during the Middle Ages for witchcraft and magic, due to their hallucinogenic, anesthetic and poisonous properties. For this reason, later medical publications would refuse to include them in an authorized list of medicinal plants, in spite of their medicinal importance. It should be noted that several plants of the Solanaceae family, especially belladonna, are used up to this day in folk medicine, while in modern medicine they are sources of indispensable drugs such as atropine and scopolamine.

In medieval Europe, an interesting philosophy of herbal medicine was developed, called the "Doctrine of Signatures". It was first introduced by herbalists many years before, but was documented in writing by Paracelsus, a Swiss fifteenth century doctor. He believed that plants were created to serve mankind, and that they had been given their forms by the Divine Creator in accordance with the objective that they were to serve. Thus, heart-shaped leaves and flowers were believed to be designated for treatment of heart diseases; kidney-shaped fruits, such as pulses, were to treat diseases of the kidneys. Red flowers and plants with red juice, such as pomegranate, were considered excellent for treatment of problems in blood circulation; yellow plants or those with yellow juice were to be beneficial for curing jaundice; walnuts, which resemble the skull, were used for treatment of illnesses related to the head; hairy plants would benefit the hair; plants which produce white milky juice would promote the production of milk in the breasts of nursing mothers; long-lived plants would bring long life to those who ate them, and the list goes on and on.

It was only by the mid-fifteenth century that the influence of Dioscorides, and that of the classic herbalists, began to fade within European botany and medicine. During this period, the European herbalists began researching plants for pure research purposes, which resulted in descriptions and drawings of plant life worldwide, as advances in ship-building technology allowed Europeans to begin exploring the globe. Although there is no doubting the predominance of chemical research in modern medicine, there is a notably increasing interest, within both medical circles and the general public alike, in plant-oriented folk medicine. Further research into the biochemical mechanisms of herbal medicines will enable a synthesis of traditional and modern methods of health care, to the benefit of all (Yaniv and Bachrach 2005).

No-one expects to turn-back time nor to ignore the achievements of modern science, or to restore folk-medicine dominancy into society, just because its use has persisted throughout the long history of mankind. Nevertheless, in spite of the mystery and magic associated with the practice of herbal medicine, the centuries-old traditions and knowledge of herbal practitioners has been passed down to us; practices and methods that has been tested throughout time and tradition hold an invaluable source for further developments, in both herbal and conventional medicine.

These days there is an ongoing growth in the worldwide market of plant-based cosmetics and of over-the-counter, non-prescription herbal medicine. However, the marketing and administration of such products are subjected to strict supervision of the appropriate authorities (Palevitch and Yaniv 2000).

It is important to carry on the knowledge of traditional practices in order to develop new remedies for modern use. The process is not easy. It involves, first, the proper cultivation of selected plants with a commercial and medicinal potential, using controlled production methods. The next step involves modern extracting methods. The purpose is to come up with an active fraction or, if possible, a purified active compound. At this stage tests aimed at standardizing the compound and testing for toxicity and validity of function should follow.

Last but not least is the need to get a health permit from health authorities in the different countries.

This long and rather expensive process can explain the slow release of new herbal supplements into today's markets. However, in view of the immense potential contribution of this research we hope that in spite of the difficulties the search for new herbal remedies will continue.

References

Duke JA (2007) Medicinal plants of the Bible. CRC Press, Atlanta, 488 pp

Gordon L (ed) (1977) Green magic – flowers plants and herbs in lore and legend. Webb and Bower Limited, London

Gunther RT (1959) The Greek herbal of Dioscorides. Hafner Publishing Company, New York

Lee MR (2011) The history of Ephedra (ma-huang). J R Coll Physicians Edinb 41(1):78–84. doi:10.4997/JRCPE.2011.116

Lev E (2002) Medicinal substances of the Medieval Levant. Eretz, Tel Aviv

Lev E (2008) Ethnopharmacology in Israel and neighboring countries. In: Pretorius R (ed) Trends and developments in ethnopharmacology. Research Signpost, Kerala, pp 35–71

Lev E, Amar Z (2000) Ethnopharmacological survey of traditional drugs sold in Israel at the end of the 20th century. J Ethnopharmacol 72:191–205

Palevitch D, Yaniv Z (2000) Medicinal plants of the holy land. Modan, Tel-Aviv, 358 pp

Savnur HV (1984) Ayurvedic materia medica: principles of pharmacology and therapeutics. Sri Satguru Publications, Delhi

Yaniv Z, Bachrach U (eds) (2005) Handbook of medicinal plants. The Haworth Medical Press, New York

Zohary M (1982) Plants of the Bible. A complete handbook to all the plants with 200 full-color plates taken in the natural habitat. Cambridge University Press, Cambridge, pp 1–220

References

Wild Edible Plants in Israel Tradition Versus Cultivation

Uri Mayer-Chissick and Efraim Lev

Abstract Studying the traditional knowledge of gathering wild edible plants around the world might teach us a great deal about the patterns of domestication and the connections between wild edible plants and "domesticated" cultivated plants. It is even more important and relevant if we consider that until not long ago, gathering of wild edible plants was a core daily practice, alongside the practice of agriculture and that the source if all our cultivated plants are gathered wild plants.

Gathering edible plants is a well-known habit among both the local Arab and the Jewish population and folklore. In the Land of Israel this tradition is in rapid decline due to excessive commercial gathering that almost brought to extinction some of the plants. The continuous urbanization of traditional communities also contributes to the loss of foraging knowledge and traditions. The wild edible plants and the traditions connected to their use can teach us a lot about different issues and raise questions about the co-evolution between humans, agriculture and wild edible plants.

Keywords Israel • Traditional knowledge • Gathering • Wild edible plants • Foraging • Cultivated plants • Malva • Marjoram

Introduction

The study of traditional knowledge concerning the gathering of wild edible plants anywhere in the world, can teach us a great deal about the relationship between wild edible plants and "domesticated" cultivated plants. It is important and relevant if we consider that until not long ago, gathering of wild edible plants was a core daily practice, side by side with the practice of agriculture.

Babai and Molnár (2013) studied the traditional habitat knowledge in Gyimes, Eastern Carpathians, Romania. They found that the locals used a rich and sophisticated vocabulary to name and describe habitat categories. They distinguished altogether at least 142–148 habitat types, and named them by 242 habitat terms, which implies that their traditional ecological knowledge was wide. Their rich

U. Mayer-Chissick, B.A., M.A., Ph.D. (✉) • E. Lev, B.A., M.Sc., Ph.D.
Department of Eretz Israel Studies, University of Haifa, Haifa 31905, Israel
e-mail: uri@mayerchissick.com; elev@univ.haifa.ac.il

Z. Yaniv, N. Dudai (eds.), *Medicinal and Aromatic Plants of the Middle-East*,
Medicinal and Aromatic Plants of the World 2, DOI 10.1007/978-94-017-9276-9_2

vocabulary helped to explain the habitat preference of a particular wild plant species. Their diverse knowledge was a fundamental part in the tradition of gathering wild edible plants in their area.

All over the world, the decline in the traditional gathering of wild edible plants caused the loss of the vocabulary used to describe the acts of gathering and the natural habitats of wild edible plants. There are less and less places like Gyimes where this knowledge is preserved. In the Land of Israel this tradition is in rapid decline. However, some attempts are made to recover and preserve the local knowledge of gathering wild edible plants (Dafni 1985; Mayer-Chissick 2010), even though the Land of Israel is one of the best locations for the study of the co-evolution of human and edible plants in general. The land is situated in the center of the Fertile Crescent and there are many archeological sites that shed light over the beginning of agriculture, the inhabitants of the land had been collecting edible plants, along the history, for their nutrition at the diverse regions, as parallel processes.

Plant remains had been found in prehistoric sites with some evidences of being consumed by humans. The plant assemblage of the Acheulian site of Gesher Benot Ya'aqov, for example, includes nuts of Atlantic Pistacio (*Pistacia atlantica*), acorns of Mt. Tabor Oak (*Quercus ithaburensis*), and wild Almonds (*Amygdalus communis*) are thought to have been consumed by humans (Goren-Inbar et al. 2002).

Hence, the large Mousterian carbonized plant assemblage retrieved during the excavations at Kebara cave fills a major gap in our knowledge of Middle Paleolithic gathering. It also provides critical information about subsistence strategies, thus, a fuller picture of Middle Paleolithic foraging as practiced by the Kebara inhabitants is achieved due to the fortuitous preservation of plant remains. The plant remains found in the Kebara cave including few thousands seeds of legumes, half a dozen cereals, few nuts of Atlantic Pistacio, and few parts of acorns of Mt. Tabor Oak seem to indicate that the inhabitants had a sufficient supply of all necessary elements for a healthy diet, available mostly during spring, early summer and fall. Although legumes are potentially suited for starting fires (morphological characteristics such as small leaves, and narrow stems) we believe that most were brought into the cave when ripe or almost ripe, and the presence of thousands of charred seeds strongly suggests that they were used for human consumption (Lev 2008).

In addition to the prehistoric remains indicating wild plants consumption in the area in the far past, vegetal food gathering could be studied from vast historical sources describing the phenomena and from plants remains found in archeological sites all over the country.

The Land of Israel is especially diverse in its nature, mainly due to its geographic location at the meeting point of three continents (Asia, Africa, Europe). Its special climate, influenced by the temperate Mediterranean on the one hand and the arid Arabian and Asian deserts on the other, and its particular topographic structure, including the Rift Valley, have contributed to the area's richness of flora and fauna. Different climatic, phyto-geographic, and zoo-geographic zones – Mediterranean, Irano-Turanian, Saharo-Arabian and Sudanian – converge here, creating enormous biological diversity (Mendelssohn and Yom-Tov 1999). About 2,700 plant species are to be found, of which 150 (5.5 %) are recorded as endemic (Shmida 1982) and more than 185 as edible plants (Dafni 1985).

The ethnic groups which compose this impressive mosaic in the Holy Land retained to varying degrees their languages, religions, clothing, ceremonial customs, traditional medicine and traditional food including the consumption of wild plants (Lev 2006).

In the early spring one can not miss, while driving through the country, many Arab families mostly the women wondering around the roadsides of the fields and gathering many different kinds of wild edible plants. A few studies had proven the health benefits of diet consisting wild plants in preventing some of the health problems of the modern civilization (Abu-Rabia 2005). Gathering edible plants is well known habit both among the local Arab and Jewish population and folklore.

According to Jewish religious the habit of wild plants gathering is one of the ways to deal with the rabbinical law of the "Shmita" (Huchberg 2000a, b), the last year in a 7-year cycle during which lands in Israel must lie fallow and debts are canceled. But this traditional way is also declining in the modern world when vegetables can be grown in special technology not involving ground and bought from the Moslems around the country.

The tradition of gathering wild edible plant is disturbed by the modernization of the Arab village, the invention of the fridge for example, enable the family to pick more than the amount needed for the season and freeze it for the summer. This excessive gathering combined with the growth of the community brought almost to a disappearance of some of the plants, and some of them like the wild (Syrian) marjoram (*Origanum syriacum*) and the tumble thistle (*Gundelia tournefortii*) had to be declared protected by law and their gathering is forbidden.

Among the most important sources in Rabbinical literature we have used the *Mishnah* (the six orders of the *Mishnah* are an assembled legal code arranged by topics most relevant to Jewish life such as agriculture, festivals, civil law, etc.) written by the *tannaim*, namely third-century CE Jewish Sages. Then come the Talmud, the Talmud is a commentary on the *Mishnah* and the *Tosefta* (laws added to the Mishnaic corpus). The final dates for the editing of the *Yerushalmi* (i.e., Palestinian) reflects very little reality after the early to mid-fifth century CE. The Talmud was written by the *amoraim,* namely later Jewish Sages of the late Roman and the Byzantine periods (the latter in the Land of Israel dates from 324 CE to 638 CE, the year of the Muslim conquest).

The plants presented here were selected because they represent different issues and raise questions of the co-evolution between humans, agriculture and wild edible plants. Descriptions, origin and ecology of all plants presented here are mostly taken from the Flora Palaestina (Zohary 1966–1986).

The popular and thus important wild edible plant in the area according to the tradition is ***Origanum syriacum*** and ***Gundelia tournefortii,*** two plants that are still picked today and that are in danger of extinction due to commercial gathering. ***Asparagus aphyllu*** and ***Pisum fulvum***, plants that were cultivated through history but are not grown intensively anymore are only picked in the wild nowadays. These are examples of *plants that give us clues to the beginning of cultivation* ***Arum palaestinum***, ***Malva nicaeensis*** and ***Quercus ithaburensis*** are more examples of common wild edible plants in the area.

Specimen Monographs

Malva

Both in Hebrew and Arabic the names of Malva come from the same lingual root that the word 'bread' comes from, the root that represent in the Semitic languages the most important foods: bread or meat. This connection points to the significance of the plant as a central food in the local tradition of gathering wild edible plants.

Description Annual or biennial herb with upright and hairy stem (up to 60 cm). The leaves are big (up to 12 cm), wide and have several slight lobes along the edges. Flowers are with pink or purple. The fruits are disc-shaped and has several segments (Table 1).

Local Gastronomy The Leaves are eaten raw or cooked and according to the local tradition the seeds were picked when dry, and then grinded and used as flour.

Local Folklore The Malva is considered to be the most important wild edible plant in the local gathering tradition. A lot of tales are connected to its ability to help people in nutritional stress, as well as to its high nutritional value (Picture 1).

Syrian Marjoram

The Syrian marjoram (Picture 2) is one of the most popular seasoning plants in Israel and is the main ingredient in the local blend of the za'atar spice that is called after the plant name in Arabic. It has been declared a protected species in Israel as early as 1977, and that protection proved a key factor in its survival as a wild plant.

Description Woody based perennial, 30–50 cm tall, with soft-woolly, glandular hairs. Stems erect, rigid paniculately branched on upper part. Leaves short-petiolate, ovate, rather thick. Spikes oblong. Corolla white, 4 mm long, tube exerted from calyx (Table 2).

Table 1 Botanical features of Malva

Scientific name	Family	Origin	Ecology	Local folk name	Part(s) used in the local cuisine	Gathering period
Malva nicaeensis	Malvaceae	E. Mediterranean	Batha, garigue and fallow fields	Hebrew: helmit Arabic: *hubeza*	Leaves and seeds	December–May

Picture 1 *Malva nicaeensis*

Picture 2 Syrian marjoram

Table 2 Botanical features of Syrian marjoram

Scientific name	Family	Origin	Ecology	Local folk name	Part(s) used in the local cuisine	Gathering period
Origanum syriacum (=*Majoran syriaca*)	Labiatae	E. Mediterranean	Batha and garigue on rocky hills; rarely on walls	Hebrew: *ʾEzov* Arabic: *Zaʿtar*	Leaves	March–May

Sensorial Botany Brunches and leaves are hairy and scanted when touched, their test is bitter.

Local Gastronomy The leaves are dry grounded and blend with salt, olive oil, local sumac (*Rhus coriaria*) fruits and sesame (*Sesamum indicum*) seeds in to a very popular spies called za'atar that is served mostly with pita bread. The fresh leaves are also used in moderation (to much is bitter) as seasoning in different dishes.

Local Folklore Wild marjoram or Syrian hyssop are mentioned in the bible (Exodus 12:21–22; I Kings 4:33; Psalms 51:7) and in the New Testimony (John 19:28–30). Zohary (1982) argues it was used to treat leprosy (Leviticus 14:4) and for worship (Number 19:6). The wild marjoram traditionally used by the Samaritans to sprinkle the blood of the Passover sacrifice.

Marjoram was used in the past for heating, and the wild species is a main substance in the famous eastern spice called 'za'atar.' Ibn Masawayh (1932–1933) suggested using 'marzangosh' oil while staying in the hot bathhouse (hamam). Arab villagers ('fellahin') used to pad their shoes with wild marjoram leaves in the winter for warmth and disinfection of the feet. The seeds of the plant consist of ethereal oils used in the food and cosmetic industries (Alon 1982–1990).

Syrian marjoram is one of the typical plants of the Land of Israel. Its use as a traditional spice is very common among all the local inhabitants until the present day. Large quantities of the plant and its products are exported to neighboring Arab countries and for this reason it has been declared a protected species in Israel (Lev 2002).

Tumble Thistle

This plant that was gathered in the area throughout history has been declared a protected species in Israel due to an excessive commercial gathering in the beginning of the second millennium. After a complicated debate between traditional local gatherers and the authorities and a partial success of its cultivation, the limits on its gathering were reduced and enabled the restoration of its existence in some reserved areas.

Description Perennial herb, 30–50 cm tall, cobwebbed to almost hairless. Stem thick, breaking of near base at maturity, than the whole plant is rolled by wind (tumbleweed). Leaves are large, rigid, thick & oblong (Picture 3). Leaf lobes are spiny-toothed. Head compound, ovoid, 4–8 cm across. Head bracts are cobwebbed ending in a strong spine. Florets dull purple outside, yellow inside, anthers yellow (Table 3).

Picture 3 Thuble thistle

Table 3 Botanical features of tumble thistle

Scientific name	Family	Origin	Ecology	Local folk name	Part(s) used in the local cuisine	Gathering period
Gundelia tournefortii	Asteraceae	W. Irano-Turanian, invading E. Mediterranean	Batha, tragacanth vegetation, open steppe forests, fallow fields and steppes	Hebrew: *ʿAkuvit ha-galgal*	Young leaves, shoots, and inflorescence	Early winter (young leaves) Early spring (stems and inflorescence)
				Arabic: *ʿAkkub*[a]		

[a]Bailey and Danin (1981)

Sensorial Botany The green soft young leaves in the winter become hard brown spiny towards the end of the spring.

Local Gastronomy The leaves of the plant are eaten while they are still green, they are picked and after they had been cleaned they are fried with onion and spices. The stems and inflorescence are made into few different dishes, fried, boiled with meet or grilled with meet (Buchman 1993; Dafni 1985). The seeds are rich in fats and are edible (Lev Yadun and Abbo 1999). In Turkey they use it sometimes as a coffee substitute (Dafni 1984, 1985; Uphof 1968).

Local Folklore Most biblical researchers think that this is the plant referred to in the Bible: "... as a wheel, as the stubble before the wind" (Psalms 83:14) The Hebrew name meaning wheel and resembles the way of seed spreading of the plant when the plant rolls in the wind (Buchman 1993).

The Mishna recognizes and categorizes the plant as a special wild plant that is edible for humans (Mishna, Okatzim, 3). During the Byzantine period, the Tumble thistle seeds were an important component in the food of monks in the Judean desert (Herschfeld 1991). According to the Jerusalem geographer, al-Muqaddasi (1906), the plant was one of the plant species with which "Falastin" was favored. The plant has been declared a protected species in Israel duo to an excessive gathering.

Pricky Asparagus

Wild Asparagus is an expensive foodstuff because of its short shelf life. It is still gathered according to local tradition but not for commercial use and mostly for self-consumption. This allows its survival in the wild.

Description Climbing perennial herb (to 1 m tall) with a woody base and swollen roots. Stem spreadingly branched, brunches angular. Lower scale-like leaves and phylloclades spreading, unequal 5–10 mm (Picture 4). Flowers 1–2, berry 6–7 mm in diameter, blackish (Table 4).

Sensorial Botany Phylloclades are spiny.

Local Gastronomy The young shoots are eaten fresh, fried with an egg or pickled in brine.

Local Folklore The asparagus was an agricultural crop in the Land of Israel that was explicitly mentioned from the early Islamic period (Amar 2000). Various species of asparagus are climbing plants in the woody groves of the Land of Israel, and some of them are used for food until today (Dafni 1985).

The medical use of asparagus was mainly as a diuretic drug, to relieve pain, to increase sperm and also to fatten. These uses have been preserved in the framework of traditional medicine in the region until today (Lev 2002). The Hebrew name of the species is derived from the scientific name that originates from ancient Greek (Lev 2002).

Picture 4 Pricky Asparagus (Taken by: Ramon Casha and Flicker)

Table 4 Botanical features of Pricky Asparagus

Scientific name	Family	Origin	Ecology	Local folk name	Part(s) used in the local cuisine	Gathering period
Asparagus aphyllus	Liliaceae	S. and E. Mediterranean	Batha and maquis, among rocks	Hebrew: *Asparagus* Arabic: *Halyn̈*	Young shoots	Early spring

Yellow Wild Pea

Although wild pea and other wild legumes are still spread in the wild in the Levant, they are not widely picked by local gatherers any more. This is due to the high labor needed to pick a quantity that is satisfying nutritionally. Abbo et al. (2008) asked in their research, what guided the cultivation process of wild pea? They got the best yield gathered from the *Pisum fulvum*, 500 g. of dried in an hour of work, which is definitely not a lot. However, this was most probably not the reason for the cultivation of the pea because although *Pisum fulvum* was the best yield it was not the specie that was cultivated.

Description Hairless annual (15–30 cm tall) with ascending or creeping stems. Leaves 6–12 cm long, spreading; stipules 2–3.5 cm long, ovate. Flowers to 1 cm long, rusty-yellow or reddish-brown. Pods are short beaked, seeds black, 4 mm across (Table 5).

Sensorial Botany Seeds and pods are green, moisture and soft in winter time and become hard and brown towards the end of the spring.

Table 5 Botanical features of yellow wild pea

Scientific name	Family	Origin	Ecology	Local folk name	Part(s) used in the local cuisine	Gathering period
Pisum fulvum	Fabaceae	E. Mediterranean	Batha, garigue and fallow fields	Hebrew: Afun Arabic: *Bazilla*	Young leaves, seeds and pods	February–May

Picture 5 Yellow wild pea

Local Gastronomy The legumes are eaten raw or cooked (Dafni 1985). Leaves are eaten raw.

Local Folklore Seeds of the plant where found in the Kebara cave (Mt. Carmel, Israel) and where reconstructed as part of the vegetal gathering diet of the Middle Paleolithic Humans that lived in the area (Lev 2008). Seeds were also found in sites from the sixth and seventh millennium BC around Turkey, Iraq and Israel (Ledizinsky 1985–1986) (Picture 5).

Palestine Arum

Arum is known today as a poisonous plant, but was known through history as a highly valued edible plant nutritionally and medicinally. We learn from the local traditions to transform it from a poison to a healthy and delicious food.

Description Scapose perennial herb with tuber, 4–8 cm in diameter, long leaves, petiole 15–50 cm, longer than blade (Picture 6). Spathe (16–50 cm) with tube (3–6 cm) and 2.5–4.5 cm across, oblong, outer side green, inner side purple in its upper half and greenish in the lower. Appendage of spadix dark purple (Table 6).

Picture 6 Palestine arum

Table 6 Botanical features of yellow Palenstine arum

Scientific name	Family	Origin	Ecology	Local folk name	Part(s) used in the local cuisine	Gathering period
Arum palaestinum	Araceae	E. Mediterranean	Batha, garigue and maquis edges; alluvial soils, rocky places	Hebrew: Luf Arabic: ʾuḍn ilFil[a]	Leaf blades	Winter

[a]The Arabic name means "Elephant Ear" and is driven from the shape of the leaf (Dafni 1985)

Sensorial Botany Inflorescence mostly strong scented of decaying fruit and fermentation or even dung. In the tissues of the plant there are crystals of sodium acid (Sodium Oxalate) in the form of needle bunches. When the tissues disintegrate, the crystals remain in form of a white powder. There are also poisons in the foliage and the gourds that prevent it from being eaten, but roasting or cooking neutralizes these poisons (Alon 1982–1990; Uphof 1968; Dafni 1985; Crowfoot and Baldensperger 1932; Grieve 1994).

Local Gastronomy The stems are cut off from the leaves, leaf blades are cut into thin pieces, then cooked thoroughly with lemon or sorrel leaves (*Rumex* sp.) the acidity of the sorrel and the lemon helps to neutralize the leaf poisons. The cooked leaves are eaten with a flat bread and considered healthy and a delicacy.

Local Folklore Various species of arum growing in the Mediterranean Basin have served as food and medicine since early times in spite of their toxicity. In the temple of Thutmose III in Karnak there are engraved drawings of plants that were brought from Canaan in the year 1447 BCE, among which there were various species of arum (Krispil 1983–1989).

The plant is mentioned in Jewish rabbinical literature, and apparently they knew even then how to overcome its toxicity. Palestine arum was a cultivated crop and it was customary to eat the leaves and the gourds that appeared perennially (for example: Mishna, Kilayim, 2:5). The gourds of the plant are also mentioned in connection with the sabbatical year (Mishna, Shevi'it, 5:3). The seeds of the gourd were not used for human consumption but only for birds (Mishna, Ma'aserot, 5:8) (Lev 2002).

Cooked leaves are considered a delicacy by the local Arabs, and a treatment for many diseases and especially for cancer. It is a commonplace among the Arabs of Israel that the cooked leaves of the plant kill intestinal worms even among animals, and also that they strengthen the bones of the body. Crushed leaves are used to treat infection in open wounds, obstructions in the urinary tract, and kidney stones (Krispil 1983–1989). The Jews of Iraq use the plant to cure skin sores, syphilis, rheumatism, tuberculosis, diarrhea, and stomach worms (Ben-Ya'akov 1992).

Mt. Tabor Oak

Oaks are widely found in the region under research, and are very well known in the landscape. Oak was used in many different ways in local traditions. It was used for medicine, building, dying and more. The reason for it not being cultivated was researched and it was found that unlike the almond that has single dominant gene controlling its bitterness, the oak has many genes controlling its bitterness, which made it impossible to cultivate (Diamond 1997).

Description A tree that may reach a height of more than 10 m, and a trunk with a circumference of 6 m. The branches form a spherical crown. The bark of the branches and the trunk is gray-dark brown and it is hard with deep grooves. The leaves are 5–10 cm long. They are stiff, ovate, with dentate-thorny margins. The upper side of the leaf is shiny; its lower side is covered with felt-like hairs (Table 7).

Local Gastronomy The oak acorns are roasted and them grinded and used in various cooking technics (Picture 7).

Local Folklore Many traditional tales are told about the ability of the oak to provide food at times of scarcity. It is mentioned in the bible and forests of it covered large areas in the Galilee and on Coastal Plain in Israel, but these have been

Table 7 Botanical features of Mt. Tabor Oak

Scientific name	Family	Origin	Ecology	Local folk name	Part(s) used in the local cuisine	Gathering period
Quercus ithaburensis	Fagaceae	Mediterranean	Heavy soils	Hebrew: Alon	Acorns	October–December
				Arabic: Balut		

Picture 7 Mt. Tabor Oak

greatly reduced in the Othman era. The oldest tree in Israel, near the tomb of Rabbi Abba Halfeta in the Lower Galilee, was dated to a little over 500 years.

Dwarf Chicory

In the last few years, when the awareness of the traditional cuisine grew and the tradition of gathering wild edible plants has risen, the demand for wild chicory in local restaurants rose. Therefore farmers make an effort to grow wild species of it to meet the demand.

Description 5–100 cm tall annual, somewhat waxy, hairless, stems divaricately branched. Leaves oblong (Picture 8), often dry at flowering time, heads stalkless, peduncles thickened at maturity. Florets sky-blue (Picture 9, Table 8).

Sensorial Botany Leaves soft and bitter at winter time and get hard, dry and their bitterness is sharpened towards the end of the spring.

Local Gastronomy The green leaves are cooked in water in order to neutralize their bitterness and then fried with onion and spices, the sugar that comes out during the cooking of the onion is also neutralizing the bitterness. The young leaves are also eaten fresh in a salad with seasoned with lemon and olive oil. A coffee substitute is made from roasted roots and dry heads of the chicory, the heads are cooked for a long time until it is made bright and the cooking water with some sugar added is used as coffee (Dafni 1985).

Picture 8 Dward chicory (leaves)

Picture 9 Dwarf chicory (florets)

Table 8 Botanical features of dwarf chicory

Scientific name	Family	Origin	Ecology	Local folk name	Part(s) used in the local cuisine	Gathering period
Cichorium pumilum (=endivia)	Asteraceae	Mediterranean and W. Irano-Turanian	Fields, road sides and seasonally wet places	Hebrew: 'Olesh Arabic: 'Iet, Hindibe Barriyye	Mostly the leaves but also the roots and the heads	Winter and spring

Local Folklore In the Mishna the 'olashim' are mentioned regarding the laws of the sabbatical year and cross-fertilization. The reference is apparently to the common chicory (*Cichorium intybus*) which is one of the varieties which can be used as the bitter herbs required for the Passover meal (Feliks 1987; Low 1924–1934).

Dioscorides distinguished between the wild and the cultivated variety, which was called *endive* and was used as a remedy for the stomach, heart and eyes and for scorpion stings (Gunther 1959). The Talmud also describes the two varieties of chicory. Some of the Jewish medieval commentators identified this plant by its Arabic name 'hundaba' (Feliks 1980–1987). The plant 'hundaba' was listed among the crops grown in the al-Sham region (the Levant) during the Crusader and Mamluk period (Taqi al-Din al-Badri 1980). According to sixteenth century Jewish Rabbi, Hayim Vital, the juice of the chicory called 'hindaba' was a component in a medication to relieve stomach swellings (Benajahu 1952).

Arabs in Israel use the chicory juice to clear the blood, to stimulate the urinary system, to treat constipation, arthritic pains, rash and light injuries, to eliminate intestinal worms and to cure internal diseases such as malaria and jaundice (Dafni 1985). The roots of the cultivated chicory are considered as a strengthening substance and are used to cure diseases of the stomach and the gall bladder (Uphof 1968). In Iraq, the plant is used as a diuretic, a purgative and a cure for stomach problems. The roots are used among other things as a cooling medicine to treat the gall bladder and also as a cathartic drug, a diuretic, a strengthener and to reduce fever (Hooper 1937; Chizik 1952; Al-Rawi and Chaakravarty 1964). The cultivated chicory is widely used as a remedy and a food in Europe as well (Grieve 1994).

Field Eryngo

The Field eryngo is an example of a plant that remained a wild edible plant, and is known mostly to the traditional gatherers. It is very common in the wild.

Description Perennial or biennial herb, hairless, 20–50 cm tall, stems divaricately branched from below, basal leaves 5–10 cm long (Picture 10), herbaceous, bluish. Inflorescences forked repeatedly, heads 7–8 mm across. Flowers overtopped by bracteoles. Fruit scaly-bristly, obscurely ribbed (Picture 11, Table 9).

Picture 10 Field eryngo (leaves)

Picture 11 Field eryngo (fruits)

Table 9 Botanical features of field eryngo

Scientific name	Family	Origin	Ecology	Local folk name	Part(s) used in the local cuisine	Gathering period
Eryngium creticum	Umbelliferae	E. Mediterranean, with slight extension into Irano-Turanian territories	Fallow fields, batha and roadsides	Hebrew: Ḥarḥavina makḥila Arabic: Qurßuʿanne	Young leaves and thick roots	February–March

Sensorial Botany Soft bluish brunches during winter time become spiny towards the end of the spring.

Local Gastronomy The young leaves are eaten fresh in a salad and the thick roots are eaten raw or cooked.

Local Folklore The scientific name is derived from the Greek term for one of the varieties of this species, the Hebrew name for this species appears in the Mishna where it is listed with the type of vegetables permitted to be used as bitter herbs eaten at the Passover meal. The name is related to the root h-r-b (dry) (Feliks 1976; Kislev 1997). The Arabic name meaning "sting from me" came from the wide use of the plant for treatment of stings (Dafni 1985).

The plant is recognized as one the varieties which can be used as the bitter herbs required for the Passover meal (Mishna, Pesahim, 2:6). The Arabs of Israel are accustomed to eat various kinds of sea holly (Dafni 1985; Uphof 1968) and use the *Eryngium creticum* in their traditional medicine. They eat the leaves or smear the ground root to cure open wounds and scorpion stings. The seeds serve to cure stomach sores, cataract in the eyes, and to expel worms. The leaves are also eaten to strengthen the gums, and their juice is used to treat diabetes. The juice of the roots is drunk in order to treat kidney stones (Palevitch et al. 1985).

Discussion and Conclusions

Gathering of wild edible plants was an essential practice in prehistoric times, become common practice after the agricultural revolution mainly as supplement to the usual diet. Today it is a rare tradition among the traditional population and a modern trend among the well fed "Western" populations. Similar to the traditional medicine and the use of medicinal substances, edible plants unify together member of different cultures, such as new Russian Jewish immigrants (gathering wild mushrooms), Jewish immigrants from the Arab country's and local Arabs (wild plants). It also become a local gastronomical trend to serve traditional dishes from wild plants in fancy restaurants.

The nine edible plants of Israel presented above truly represent consumption traditions, some of them are of thousands of years according to remains found in archeological and pre-historical sites. Most of the plants presented above also bare medicinal qualities that had been known to the local inhabitants and were therefore used as drugs as well, in Israel and neighboring countries such as Lebanon, Syria (northern species) and Jordan, Egypt, Iraq and even Iran (southern and desert species).

Acknowledgments All images are from the author's private collections, except the images of the Spiny Asparagus and the Yellow Arum which are published with the courtesy of Dr. Ori Fragman-Sapir from his private collection. The picture of the Pricky Asparagus was taken by Ramon Casha and is published according to his permission on Flicker.

References

Abbo S, Zezak I, Schwatrz E, Lev-Yadun S, Gopher A (2008) Experimental harvesting of wild peas in Israel: implications for the origins of Near East farming. J Archaeol Sci 35:922–929

Abu-Rabia A (2005) Herbs as a food and medicine source in Palestine. Asian Pac J Cancer Prev 6:404–407

Al-Muqaddasi De G (ed) (1906) Ahsan al-Taqasim fi Marifat al-Aqalim. E.J. Brill, Leiden (Arabic)

Alon A (ed) (1982–1990) Plants and animals of the land of Israel: practical and illustrated encyclopedia, 12 vols. The Publishing House Society for Protection of Nature, Tel-Aviv (Hebrew)

Al-Rawi A, Chaakravarty HL (1964) Medicinal plants of Iraq, Bulletin no. 146. Ministry of Agriculture Technology, Baghdad

Amar Z (2000) Agricultural produce in the land of Israel in the Middle Ages. Yad Ben Zvi, Jerusalem (Hebrew)

Babai D, Molnár Z (2013) Multidimensionality and scale in a landscape ethnoecological partitioning of a mountainous landscape (Gyimes, Eastern Carpathians, Romania). J Ethnobiol Ethnomed 9:11

Bailey C, Danin A (1981) Bedouin plant utilization in Sinai and the Negev. Econ Bot 35:1–95

Benajahu M (1952) Rabbi Hayim Vital in Jerusalem. Sinai 30:65–75 (Hebrew)

Ben-Ya'akov A (1992) The traditional medicine of the Babylonian Jews. Yerid Hasefarim, Jerusalem (Hebrew)

Buchman J (1993) American aid to the tumble thistle. Teva Va-Eretz 248:46–47 (Hebrew)

Chizik B (1952) Otsar ha-Tsmahim [A treasury of plants]. Herzlia (Hebrew)

Crowfoot GM, Baldensperger L (1932) From cedar to hyssop. Sheldon Press, London
Dafni A (1984) The mandrakes give fragrance. Gestlit, Haifa (Hebrew)
Dafni A (1985) Edible plants of the land of Israel. The Society for the Protection of Nature in Israel, Tel-Aviv (Hebrew)
Diamond J (1997) Guns germs, and steel. W.W. Norton, New York
Feliks Y (1976) Plant world of the Bible. Masada, Ramat-Gan (Hebrew)
Feliks Y (1980–1987) The Jerusalem Talmud [Talmud Yerushalmi] Tractate Shev'iit [critically edited]. Jerusalem (Hebrew). Tsor-Or
Feliks Y (1987) Mishna Tractate Shevi'it. Rubin Mass, Jerusalem (Hebrew)
Goren-Inbar N, Sharon G, Melamed Y, Kislev ME (2002) Nuts, nut cracking, and pitted stones at Gesher Benot Ya'aqov, Israel. PNAS 99:2455–2460
Grieve M (1994) A modern herbal. Tiger Books, London
Gunther RT (trans) (1959) The Greek herbal of Dioscorides, vol II. Harper, New York, p 132
Herschfeld I (1991) Delicacies of thorns and thistles. Teva' va-Aretz 33:34–37
Hooper D (1937) Useful plants and drugs of Iran and Iraq, Publication 387. Field Museum of Natural History, Chicago
Huchberg M (2000a) Edible wild plants for keeping good health and shemita. Tezuna Plus 45:34–37 (Hebrew)
Huchberg M (2000b) Wild plants consumption is the best mean for the prevention and healing of the diseases of civilization. Tezuna Plus 44:26–29 (Hebrew)
Ibn Masawayh A (1932–1933) Kitab al-Azmina (Sbath P, ed & trans). Bulletin de l' Institute di Egypte 15:235–257 (Arabic)
Kislev M (1997) Cordia myxa and its use in birdcatching. In: Safrai Z, Friedman A, Schwartz J (eds) Hikrei Eretz, studies in the history of the land of Israel. Bar-Ilan University, Ramat-Gan (Hebrew)
Krispil N (1983–1989) Bag of plants. Yerid Hasefarim, Jerusalem (Hebrew)
Ledizinsky G (1985–1986) Legumes domestication in the Middle East. Mada 29(3):115–119 (Hebrew)
Lev E (2002) Medicinal substances of the Medieval Levant. Eretz Publication, Tel-Aviv (Hebrew)
Lev E (2006) Ethno-diversity within current ethno-pharmacology as part of Israeli traditional medicine – a review. J Ethnobiol Ethnomed 2(4):1–12
Lev E (2008) Ethnopharmacology in Israel and neighboring countries. In: Pretorius R (ed) Trends and developments in ethnopharmacology. Research Signpost, Kerala, pp 35–71
Lev Yadun S, Abbo S (1999) Traditional use of A'kub (Gundelia tournefortii, Asteraceae), in Israel and the Palestinian Authority area. Econ Bot 53(2):217–219
Low I (1924–1934) Die Flora der Juden, I–IV. Löwit, Vienna/Leipzig I
Mayer-Chissick U (2010) Wild edible plants. Mapa Publishing House, Tel-Aviv (Hebrew)
Mendelssohn H, Yom-Tov Y (1999) Fauna Palaestina: Mammalia of Israel. The Israel Academy of Sciences and Humanities, Jerusalem
Palevitch D, Yaniv Z, Dafni A, Friedman J (1985) Ethnobotanical survey of the flora of Israel as a source for drugs. Ministry of Science, Jerusalem (Hebrew)
Shmida A (1982) The endemic plants of Israel. Rotem 3:3–47 (Hebrew)
Taqi al-Din al-Badri (1980) Nuzhat al-Anam fi Mahasin al-Sham. Beirut (Arabic)
Uphof JC (1968) Dictionary of economic plants. Stechert-Hafner, New York
Zohary M (1966–1986) Flora Palaestina. The Israel Academy of Science and Humanities, Jerusalem
Zohary M (1982) Plants of the Bible. Cambridge University Press, London

Ethnobotany Among Bedouin Tribes in the Middle East

Aref Abu-Rabia

Abstract This paper describes beliefs and treatments for several illnesses and diseases in the Middle East, by traditional herbalists among the Bedouin tribes.

One of the most famous medicines in the Middle East is the arba'yn, which consists of a mixture of 40 different types of plants and considered a cure for all aches and pains. This medicine is sold by the 'attar (local herbalist/pharmacologist and vendor of medicinal spices). We found that the Bedouin use various parts of the plants, including leaves, flowers, barks, stems, stalks, roots, rhizomes, bulbs, tubers, fruit, corns, shells, seeds, stones/pits (in fruits), soft seed pods, grain buds, shoots, twigs, oils, resins and gums. These parts are used fresh and soft; cooked or dried. It should be noted, that some plants are used similarly throughout the Middle East, while some plants have different uses in different countries in the region.

Analysis of the findings shows that the Middle East is the geographic origin of both wild and cultivated medicinal plants. The author found that the most significant plants that used for medicine and/or food are found in the following families: Apiaceae, Asclepiadaceae, Capparidaceae, Compositae, Cruciferae, Euphorbiaceae, Gramineae, Iridaceae, Labiate, Liliaceae, Papilionaceae, Salicaceae, Solanaceae, Umbelliferae.

Keywords Artemesia herba-alba • Anastatica hierochuntica • Astomaea seselifolium • Bedouin • Capparis spinosa • Crocus sativus • Cymbopogon commutatus • Ethnobotany • Healers • Lactuca sativa • Lepidium sativum • Leptadenia pyrotechnica • Lilium candidum • Majorana syriaca • Middle East • Ocimum basilicum • Populus alba • Retama raetam • Ricinus communis • Salvia triloba • Thapsia garganica • Withania somnifera

A. Abu-Rabia (✉)
Department of Middle East Studies, Ben-Gurion University of the Negev, Beer-Sheva, Israel
e-mail: arefabu@gmail.com

Z. Yaniv, N. Dudai (eds.), *Medicinal and Aromatic Plants of the Middle-East*,
Medicinal and Aromatic Plants of the World 2, DOI 10.1007/978-94-017-9276-9_3

Introduction

This paper describes beliefs and treatments for several illnesses and diseases in the Middle East by traditional herbalists among the Bedouin tribes in the Middle East. In this study, we found that Bedouin use various parts of the plants, including leaves, flowers, barks, stems, stalks, roots, rhizomes, bulbs, tubers, fruit, corns, shells, seeds, stones/pits (in fruits), soft seed pods, grain buds, shoots, twigs, oils, resins and gums. These parts are used as fresh and soft, or cooked or dried. The dosages for patients with the same diseases or disorders may vary, according to the ages and the structures of the patients' bodies. The rich variety of approaches employed by different healers (men as well as women) to treat diseases and illnesses is indicative of the depth and breadth of indigenous medicine practiced among the Bedouin.

Analysis of the findings shows that the Middle East is the geographic origin of both wild and cultivated medicinal plants. In this paper the author found that the most significant plants that used for medicine and/or food are found in the following families: Apiaceae, Asclepiadaceae, Capparidaceae, Compositae, Cruciferae, Euphorbiaceae, Gramineae, Iridaceae, Labiate, Liliaceae, Papilionaceae, Salicaceae, Solanaceae, Umbelliferae.

Traditional Healers

There are many types of healers in the Middle East. Some of these healers are: midwifes (*daya*), who deliver babies and provide many types of "ethno gynecological" care women patients. Herbalists work with a rich ethno pharmacopeia of herbal and mineral substances and are often skilled in ethno-botany; spirits healers who are known for specializing the treatment of infertility, impotence and other ailments. One of the most famous medicines in use among the tribes in the Middle East is the *arba'yn*, which consists of a mixture of 40 different types of plants and is considered to be a cure for all aches and pains. In traditional and folk medicine, the tribal folk appeal not only to the herbalist, but also to the dervish, or to the *khatib* (the amulet writer); the cauterizer; the *mujabbir* (for setting broken or fractured bones); the *'Attar* (local pharmacologist and vendor of medicinal spices); holy tombs (of ancestors or prophets); the sea, rivers, holy springs, and so on. In addition, healers use techniques that stimulate physiological processes, including bathing, sweat-bathing, massage, cupping, emetics, incision, and bloodletting (Abu-Rabia 1999, 2005; Inhorn 2005). Bedouin classical medicine became the exclusive domain of traditional medicine and folk healers in the nineteenth and twentieth centuries. Most of the herbs were used both as food and as medicine. Many of the plants used by the Bedouin have direct effects on the body as purgatives, emetics, astringents, or tonics, or cause/prevent vomiting or diarrhea. This traditional medicine is based on a practical knowledge of plants and treatments over centuries of practice. It should be noted that some plants are used similarly throughout the Middle East, while some plants have different uses in different countries in the region (Abu-Rabia 2012).

Plants in Use by the Healers

In their work, traditional healers use many local plants. Some of these plants are brought here with their traits and uses.

Capparis spinosa *L. var.* aravensis *Zoh. [Capparidaceae]*

Arabic: *lasaf, qubbar*
English: caper bush, Common caper bush.
Plant parts: Flowers, leaves and fruit.
Preparation: soak the bulbs in water and drink one cup a day for 1 week,
Active constituents: Capric acid, rutin, pectin (Karim and Qura'an 1986); stachydrine, and material of antihistamine properties (Palevitch and Yaniv 2000); leaves and fruit contain glucosinolates, flavonoids and choline; mucilage consisting of glucose, arabinose, glucuronic acid, sterols, organic acids, fatty acids, coumarins, saponins and tannins (Abbas et al. 1992; Ghazanfar and Al-Sabahi 1993).
Ethno-botanical uses: diuretic, urinary tract infections and retention; expectorant, stimulant, diuretic, condiment, medicine for infertility among male and female, treat problems of erectile dysfunction; it is carminative and aphrodisiac.

Retama raetam *(Forssk.) Webb. [Papilionaceae]*

Arabic: *ratam, ratama*
English: white broom, ratame.
Plant parts: The whole herb, flowers
Active constituents: Essential oil (Karim and Qura'an 1986); alkaloids: retamine and sparteine (Palevitch and Yaniv 2000).
Ethno-botanical use: It is used as soothing inflamed eyes and sour throat (Karim and Qura'an 1986); Palevitch and Yaniv 1991). Boiled leaves stems and flowers in water, letting the body absorb the vapors, several times before sexual intercourse, is used to treat women infertility; washing the sexual organs as a treatment for syphilis. It is also used as a treatment for as abortive method.

Anastatica hierochuntica *L. [Cruciferae]*

Arabic: *kaf al-Rahaman, kaff Maryam.*
English: St. Mary's flower, rose of Jericho.
Plant parts: leaves, fruit.
Preparation: crush the dried fruit, soak them in water, and drink one cup each day for 1 week.

Active constituents: fruit contain glucoiberin; stems, leaves and root contain glucoiberin and glucocoheirolin; stems and roots contain flavone glycoside and four flavonol glycoside; and sterols. The fruit is a source of carbohydrate that comprises glucose, galactose, fructose, sucrose, raffinose and stachyose; and amino acids (Ghazanfar 1994).
Ethno-botanical uses: pounded seeds mixed with butter applied to anus to treat piles. Soaked dried flowers/seeds in water and drink, used to treat female infertility, uterus disorders, and prevent miscarriage; treat male impotency; and to stop bleeding after delivery. Infusion of dry plant reduces the pains and facilitates childbirth, and prevents miscarriage. It is believed that Mary clenched this plant in her hand when giving birth to Jesus.

Astomaea seselifolium *DC. [Umbelliferae]*

Arabic: *balbus, sumra*
English: Plant parts: bulbs.
Ethno-botanical use: tonic after childbirth; treat women infertility and treat potency; aphrodisiac.

Eating fresh, roasted or cooked bulbs, used to strengthen the body after childbirth. Boiled bulbs in water and drink, used to treat women infertility and increase sexual potency.

Crocus sativus *L. [Iridaceae]*

Arabic: *za'faran*
English: saffron
Plant parts: Style
Active constituents: Essential oil, crocin, picroerocin, lycopine, colouring matter (Karim and Qura'an 1986); and vitamin A (Lust 1980).
Ethno-botanical use: antispasmodic, aphrodisiac, carminative, diuretic, sedative, tonic, emmenagogue, and stimulant.

Among the Bedouin of the Negev, a woman after giving birth, that her genital have affected by 'cold inflammation' (*mariuha*), teaspoon of *za'faran* is mixed with a cup of honey, and a spoonful eaten early every morning for a week, immediately after menstruation; it is also a remedy for constipation; treat menstrual disorders, metrorrhagia; to stimulate menstrual flow, sedative at childbirth, and increase sexual potency.

Populus alba *L. [Salicaceae]*

Arabic: *safsaf abyad*
English: Abbey, Poplar
Plant parts: leaves, barks.
Preparation: soak in water and drink
Active constituents: Essential oil, gum, glycosides (Karim and Qura'an 1986).
Ethno-botanical use: diuretic, stimulant, anti-rheumatic (Karim and Qura'an 1986). Leaves soaked in water to drink one cup a day for 2 weeks to treat kidney infections and stone; it is also treat male infertility, diarrhea, and liver infections.

Lactuca sativa *L. [Compositae]*

Arabic: *khass*
English: lettuce
Plant parts: leaves and seeds.
Active constituents: Vitamins A and C, lactucon, lactucine, acids, latex, minerals (Karim and Qura'an 1986); Vitamin C, and chloride compound (Lust 1980).
Ethno-botanical use: seeds used as sedative. Leaves treat impotence, and calm excited women. Eat leaves, stems and stalks as a raw salad to treat kidney infections and stone. In the Negev and Sinai tribes seed oil treats female infertility, and known as aphrodisiac.

Lepidium sativum *L. [Cruciferae]*

Arabic: *habb al-rashad, hurf*
English: cress, garden cress.
Plant parts: seeds, leaves, and the whole herb.
Preparation: boil in water and drink
Active constituents: Lepidin, myrosin, minerals (Karim and Qura'an 1986). The seeds of the plant contain glucosinolates, glucotropaelin, gluconapin, glucoconasturtin and glucobrassicanapin (Ghazanfar 1994).
Ethno-botanical use: Appetizing, diuretic, tonic, and anti-scorbutic. It treats serious colds, cough and liver's pains; asthma and urinary retention and urinary tract infection, kidney stones and infection. It is a tonic and aphrodisiac, drunk after childbirth to clean the uterus; treating infertility among men and women, and syphilis. Boil seeds in water and drink one cup a day for 3 weeks.

Leptadenia pyrotechnica *(Forssk.) Decne. [Asclepiadaceae]*

Arabic: *Markh, Aghwaam*
English: broom bush, desert broom.
Plant parts: young flowers, fruit are also edible, stems.
Preparation: edible, to be eaten
Active constituents: the stems contain triterpenoids, taraxerol and fernenol, and B-sitosterol (Ghazanfar 1994).
Ethno-botanical use: to treat diuretic, kidney stone, infections; a fine silky tinder much used for catching the sparks of flint and steel in fire making. The dried hair tufts from the seeds were employed for this, and this use was presumably the basis for Forsskal's specific epithet, branches diuretic; used for retention of urine and to help expel uroliths; young flowers and fruit are edible; it is diuretic; infusion is used to treat kidney stone and infections. It is used to treat infertility.

Lilium candidum *L. [Liliaceae]*

Arabic: *zanbaq*
English: white lily
Plant parts: flowers and bulbs
Preparation: soak flowers and bulbs in water and drink a cup a day for 1 week.
Active constituents: Scillin, minerals, mucilage, pectin (Karim and Qura'an 1986).
Ethno-botanical use: antispasmodic; bulb-expectorant, demulcent, and diuretic. Among the tribes of the Negev it is used as treatment of infertility; kidney infection and pains.

Thapsia garganica *L. [Apiaceae]*

Arabic: *diryas al-abdan, yantun*
English: drias plant, smooth thapsia.
Plant parts: roots
Active constituents: thapsigargins, resin (Smitt et al. 1996).

Boil the dried roots in water and drink one cup to treat rabies. It is reported that roots used to treat infertility and vagina infections, rheumatism, bronchitis, and chest pains.

Withania somnifera *L. Dunal [Solanaceae]*

Arabic: *sam al-far*
English: withania

Plant parts: The whole herb, root, fruit.
Active constituents: somnirol, black resin, atropine (Karim and Qura'an 1986); lactones: withaferin A, withanolide D and E (Palevitch and Yaniv 2000); alkaloids, steroidal lactones, withanolides, and withferin A (Abbas et al. 1992).
Ethno-botanical uses: narcotic, analgesic, antibacterial, useful for impotency. Caution, it is a poisonous herb. Among the Jordanian tribes it is used for impotency (Karim and Qura'an 1986).

In Sinai, the plant used as calmative; leaves and fruits febrifuge, diuretic. It is also reported that leaves and fruits treat women infertility; skin allergy, and epilepsy.

Cymbopogon commutatus *(Steud.) Stapf. [Gramineae]*

Arabic: Sakhbar, Idhkhir
English: camel's hay, Geranium grass.
Plant parts: the whole plant.
Active constituents: geraniol, geraniol acetate, citral (Kumar et al. 2009).
Ethno-botanical use: perfume, use insect repellent, aromatic, principle; astringent, diuretic, emmenagogue. It is contains aromatic principle, small bunches sold in an herbalist shop in certainly has a history of medicinal use, probably as an infusion or tea, in Arabia (Mandaville 1990).

In Sinai tribes, infusion of this plant is diuretic, emmenagogue, and treats infertility. It is treat cold, influenza, diabetes, and stomachache.

Ocimum basilicum *L. [Labiate]*

Arabic: *Reihan, rihan*
English: sweet basil
Plant parts: leaves and seeds.
Preparation: boil in water and drink.
Active constituents: Essential oil, tannin (Karim and Qura'an 1986); oil is the active ingredient which consists: thymol, linalol, cineol, eugenol, terpenes, sesquiterpenes, and methylchavicol (Ghazanfar and Al-Sabahi 1993; Vernin et al. 1984).
Ethno-botanical use: In Jordan, it is used as stimulant, diuretic, carminative; seeds contain mucilage used to treat gonorrhea (Karim and Qura'an 1986). It is aphrodisiac and aromatic foliage (Mandaville 1990).

In Arabian Peninsula, basil is one of the most commonly used herbs, as a deodorant and aphrodisiac, worn by men in their visiting their women. The women mix it with oil into a paste and use it as a cosmetic (Ghazanfar 1994; Ghazanfar and Al-Sabahi 1993).

It is also reported that infusion of its seeds treats gonorrhea, infertility and sexual organs' infection; treat dysentery, diarrhea, and piles as a result of hard constipation.

Ricinus communis ***[Euphorbiaceae]***

Arabic: *khirwi'*
English: Palma Christi's castor/Castor oil plant
Plant parts: seeds, roots, and leaves; oil extracted from seeds.
Preparation: to be diluted in water and drank oil extracted from the seeds, to be diluted in water and to drink one table spoon, twice a week for 2 weeks.
Active constituents: Ricinine, resin, gum, castor oil (Karim and Qura'an 1986). Oil from castor bean contains enzymes (Lust 1980), and ricin; castor bean seeds form the raw material for the enzyme lipase (Palevitch and Yaniv 2000); seeds contain fixed oil (which consist glycerides of ricinoleic, insoricininoleic, stearic and dihydroxystearic acids); lipases, alkaloid ricinine. Ricinine is a cardiac stimulant (Ghazanfar and Al-Sabahi 1993).
Ethno-botanical use: Among the Bedouin of Negev and Sinai, castor oil used as purgative; treat kidney stones and kidney infection; treat syphilis and infertility. A piece of cotton wool is dipped in a mixture of castor oil, which has been heated over a fire. This is inserted into the womb (as suppository- *tahmila*) of a reclining woman and kept there for 1 h and removed; used as emmenagogue; the seeds through the mouth are used as a contraceptive.

Majorana syriaca ***(L.) Rafin. [Family: Labiatea]***

Arabic: Za'atar
English: Syrian Hyssop
Plant parts: leaves and flowers.
Active constituents: leaves contain essential oil: thymol, and carvacrol that belong to the phenols' group (Palevitch and Yaniv 2000).
Ethno-botanical use: colds; vomiting; allergic skin diseases, dermatitis, headaches, dental problems, tonsillitis and cough. The Bedouin of the Negev and Sinai use it as a treatment for colds by drinking tea brewed from this plant with a teaspoon of sugar. Vomiting is treated by soaking the leaves in water and dropping the water into the person's nostrils three times a day. Allergic skin diseases and dermatitis are treated as follows: leaves are boiled in water and given to the patient to drink; a child's body is also washed with this solution. Headaches, dental problems, tonsillitis, and cough are treated by a brew made from the leaves and are drunk.

Salvia triloba *L. [Family: Labiatae]*

Arabic: Meryamia, marmariya.
English: Sage.
Plant used: leaves, seeds
Active constituents: Essential oil, tannin, camphor, cineol, borneol, pinene, resin (Karim and Qura'an 1986).
Ethno-botanical use: For stomach disturbances, antiseptic carminative, sedative and wound healing. Treat tumors and cancer; stomach disturbances (Karim and Qura'an 1986).

It is used for irregular menstruation and dysmenorrhea; sage is given to help women recover from miscarriage and abdomen complaints; diarrhea, vomiting and nausea.

Artemesia herba-alba *Asso. [Family: Compositae]*

Arabic: Shih, sheeh
English: Worm wood
Plant parts: leaves, flowers
Preparation: boil in water and drink, or eat
Active constituents: essential oil, resin, pinene, cadinen, tannin (Karim and Qura'an 1986); Santonin, sterols and thujones (Atyat 1995; Tal 1981; Palevitch and Yaniv 1991); it also contains essential oils, sesquiterpene lactones and thymol; leaves and stems contain three non-glycosidic flavonoids (Ghazanfar 1994).
Ethno-botanical uses: to treat kidney pains and stones; emmenagogue and tonic, diuretic, healing wounds.

It is used as a medicine to treat diabetes, cough and impotence. To boil leaves and drink early on empty stomach. Boiling the leaves with olive is applied to wounds to treat and prevent infections.

References

Abbas JA, El-Oqlah AA, Mahasneh AM (1992) Herbal plants in the traditional medicine of Bahrain. Econ Bot 46(2):158–163

Abu-Rabia A (1999) The Bedouin's traditional medicine. Kinneret, Zmora-Bitan, Dvir Publishing, Tel-Aviv, pp 17–32 (Hebrew)

Abu-Rabia A (2005) The evil eye and cultural beliefs among the Bedouin tribes of the Negev, Middle East. Folklore 116:241–254

Abu-Rabia A (2012) Ethno-botanic treatments for paralysis (*falij*) in the Middle East. Chin Med 3(4):157–166

Atyat A (1995) Production and processing of medicinal and aromatic plants in the Arab world. Arab Institute for Research and Publishing, Beirut, pp 138–141

Ghazanfar S (1994) Handbook of Arabian medicinal plants. CRC Press, London, pp 35, 40, 60, 124

Ghazanfar S, Al-Sabahi A (1993) Medicinal plants of northern and central Oman (Arabia). Econ Bot 47(1):89–98

Inhorn M (2005) Healing and medicine: popular healing practices in Middle Eastern cultures. In: Jones L (ed) Encyclopedia of religion, 2nd edn. Macmillan, New York, pp 3834–3839

Karim FS, Qura'an S (1986) Medicinal plants of Jordan. Yarmouk University, Irbid, pp 39–46, 66–68

Kumar J, Verma V et al (2009) Genetics diversity analysis in Cymbopogon species using DNA markers. Plant Omics J 2(1):20–29

Lust J (1980) The herb book. Bantam Books, Toronto/New York, pp 501–506, 508

Mandaville JP (1990) Flora of Eastern Saudi Arabia. Kegan Paul International, London/New York, pp 268–340

Palevitch D, Yaniv Z (1991) Medicinal plants of the holy land. Tammuz Publishers Ltd/Modan Ltd, Tel-Aviv, pp 200–203 (in Hebrew)

Palevitch D, Yaniv Z (2000) Medicinal plants of the holy land. Modan Publishing House, Tel-Aviv, pp 68, 76, 200–203, 264, 283, 297, 307

Smitt U, Jager A, Nyman U (1996) Thapsia garganica L.: In Vitro culture, somatic embryogenesis, and the production of thapsigargins. Biotechnol Agric For 37:402–409

Tal P (1981) Medicinal plants. Rshafim, Tel-Aviv, pp 152–153 (Hebrew)

Vernin G, Metzger J, Vernin G, Fraisse D, Soun K, Schraff C (1984) Analysis of basil oils by GC-MS databank. Perfum Flavor 9(5):71–86

Endemic Aromatic Medicinal Plants in the Holy Land Vicinity

Nativ Dudai and Zohara Yaniv

Abstract The Holy Land has a long history of settlement by many different nations, cultures and religions, as well as a long tradition of using plants for culinary, medicinal and liturgical purposes. This chapter describes a selected list of promising local Medicinal Aromatic Plants (MAP) from this region. The focus is on the most important native MAP and on their potential as new cultivated modern crops. The large climatic variations of this region, present in a relatively small area, create a wide range of natural habitats and high biodiversity of wild plants. Due to the geographical location of this area, which is in the meeting of four phyto-geographical regions, there is a rich diversity of herbaceous plants. Part of the common MAP in the Holy Land vicinity are endemic to this area, such as *Micromeria fruticosa, Origanum dayi, O. ramonense, O. syriacumvar syriacum, Chiliadenus iphionoides* and *Salvia domenica.* The rest are local ecotypes of a wider dispersion area, such *as Salvia fruticosa, Artemisia judaica, Achillea fragrantissima, Asteriscus graveolens, Coridothymus capitatus, Foeniculum vulgare* and *Mentha longifolia.* Cultivation of medicinal plants that were traditionally collected from the wild is necessary both for protection of plant species in their native habitats, and as a response to the increased demand for uniform high-quality sources of medicinal herbs.

Keywords Aromatic plants • The holy land • Domestication • Breeding • Folk medicine • Biological activity • Origanum • *Salvia fruticosa* • *Salvia dominica* • *Micromeria fruticosa* • *Chiliadenus iphionoide*

N. Dudai (✉)
The Aromatic, Medicinal and Spice Crops Unit, Newe Ya'ar Research Center, Agricultural Research Organization, Ramat Yishay, Israel
e-mail: nativdud@gmail.com

Z. Yaniv
Department of Plant Introduction and Seed Research, Institute of Plant Sciences, Agricultural Research Organization, Bet-Dagan, Israel
e-mail: zoharayb@netvision.net.il

Z. Yaniv, N. Dudai (eds.), *Medicinal and Aromatic Plants of the Middle-East*,
Medicinal and Aromatic Plants of the World 2, DOI 10.1007/978-94-017-9276-9_4

Introduction

Medicinal plants, growing in the vicinity of the holy land, were traditionally used throughout history and many of them are mentioned in the scriptural sources (Duke 2007). The area traditionally considered to be the Holy Land of the Bible, is defined as the area from the mountains of Lebanon and Syria in the north, to the eastern banks of the Jordan River, to the northern reaches of the Sinai Peninsula in the south. Ethno-botanical surveys, conducted in recent years indicate that many of the traditional plants are still used in folk medicine today (Aburjai et al. 2007; Ali-Shtayeh et al. 2000; Dafni et al. 1984; Palevitch and Yaniv 2000; Lev and Amar 2000, 2002). Most of the plants are not common in the Western world today and are not cultivated in large scale, if at all. However, there is a great potential for future development of new crops and new products, mainly by domestication and breeding of selected unique local traditional Medicinal and Aromatic Plants (MAP). The great climatic variations in this region, in a relatively small area, create a wide range of natural habitats and high biodiversity of wild plants. Due to the geographical location of the Holy Land in the heart of the Levant, which is in the meeting area of four phyto-geographical regions, (Mediterranean, Irano-turanian, Saudi Arabian and Sudanian) there is a rich diversity of herbaceous plants. Mt. Hermon is in the north, where it snows in the winter. On this mountain, species such as *Nepeta italica L., Nepeta cilicia Boiss, Nepeta glomerata Bentham, and Trachonitica* can be found. Only 300–400 km away, in the desert region, we can find desert plants such as *Artemisia judaica, A. sieberi, Origanum dayi, O. ramonense, Achillea fragrantissima, Chiliadenus iphionoides* and *A. santolina* (Amzallag et al. 2005; Dudai et al. 2003; Putievsky et al. 1992, 1997; Tamir et al. 2011; Yaniv et al. 2011; Friedjung et al. 2013). In this chapter we describe a selected list of some interesting local medicinal species of this vicinity. We will focus on endemic aromatic plants and their potential as new cultivated crops.

Domestication and Breeding – Adapting the Traditional Former Glory to Modern Times

The increasing demand by industry for uniform and high-quality raw material, along with modern developments, results in the situation where fewer and fewer people still collect plants from the wild. In addition, some of the wild plant species are protected and as a result, not available. These facts have necessitated the cultivation of various species. This shortage of raw material has encouraged seed companies, researchers and farmers to select cultivars which could substitute for the raw material that once came from the wild. Therefore, the approach of the MAP research and development today is to develop modern agricultural methods for traditional crops. Most of them used to be collected by people in the wild, like oregano in Turkey and Greece or za'atar (*Origanum syriacum*) in Israel (Dudai 2012). The goal is the domestication of plants and the development of optimal growth methods for modern industry in order to replace traditional methods.

The challenge of domestication of wild plants includes the need to acclimatize plants outside of their native region. For instance, although land distances in Israel are short, the variability and changeability of climate makes this type of plant domestication all the more difficult. Desert plants from the Negev, native to an arid climate, were acclimatized to a Mediterranean climate in the north of Israel. Moreover, when considering domestication of foreign wild plants such as *Origanum vulgare* (Putievsky et al. 1997) and *Salvia officinalis* (Putievsky et al. 1986b), the process should be accompanied by introduction, acclimation, and selection of suitable varieties to the local conditions.

The danger of extinction and the need for domestication has promoted the collection and conservation of plants in all their natural biodiversity. This important activity has been done in the last few years by Gene Banks in various countries (Heywood 2002). These collections can serve as a base for selection by screening of various traits, such as desirable natural products, bio-activity or adaptation of cultural practices (Dudai 2012).

Wild Aromatic Medicinal Plants as Potential New Crops

Part of the common aromatic plants in the Holy Land vicinity are endemic to this area, such as *Micromeria fruticosa, Origanum dayi, O. ramonense, O. syriacum ve syriacum, Chiliadenus iphionoides* and *Salvia domenica.* The rest are local ecotypes of a wider dispersion area. A selected list of the main common traditional plants and their references is given in Table 1.

The following species were chosen as the most promising endemic plants to be developed as new crops.

Origanum spp.

The genus *Origanum* (Lamiaceae) is divided into ten sections: *Amaracus* Bentham, *Anatolicon* Bentham, *Brevifilamentum* Ietswaart, *Longitubus* Ietswaart, *Chilocalyx* Ietswaart, *Majorana* Bentham, *Campanulaticalyx* Ietswaart, *Elongatispica* Ietswaart, *Origanum* Ietswaart, *Prolaticorolla* Ietswaart, and contains about 20 species (Ietswaart 1980). *Origanum* species are perennials herbs, the majority of them are distributed over the Mediterranean (Kokkini 1997). The culinary and medicinal herbs contain mainly carvacrol and/or thymol ("Oregano"), or terpinene-4-ol, linalool, and sabinene hydrate ("Marjoram") (Kokkini 1997). In the Holy Land vicinity there are 7 known endemic Origanum species, some of them belong of the unique section: *Campanulaticalyx.*

Table 1 Selected list of the most common local aromatic medicinal species in the wild

Species	Family	References
Achillea fragrantissima (Forssk) Sch. Bip	Compositae	Fleisher and Fleisher (1993) and Ravid et al. (1995)
Artemisia arborecescence L.		Dudai and Amar (2005) and Yaniv et al. (2011)
Artemisia judaica L.	Compositae	Putievsky et al. (1992) and Yaniv et al. (2011)
Artemisia sieberi Besser	Compositae	Feuerstein et al. (1986), Fleisher et al. (2002), and Yaniv et al. (2011)
Asteriscus graveolens (Forssk.) Less	Compositae	
Chiliadenus iphionoides (Boiss. & Blanche) Brullo	Compositae	Tamir et al. (2011)
Coridothymus capitatus L.	Lamiaceae	Fleisher and Fleisher (2002)
Foeniculum vulgare Mill. var. *vulgare*	Umbelliferae	Barazani et al. (2002)
Mentha longifolia L.	Lamiaceae	Segev et al. (2012)
Micromeria fruticosa (L.) Druce	Lamiaceae	Dudai et al. (2001)
Origanum dayi Post.	Lamiaceae	Dudai et al. (2003) and Amzallag et al. (2005)
Origanum remonense Danin	Lamiaceae	Danin et al. (1997)
Origanum syriacum var *syriacum*	Lamiaceae	Dudai (2012)
Origanum isthmicum Danin	Lamiaceae	Danin and Künne (1996)
Origanum jordanicum Danin & Künne	Lamiaceae	Danin and Künne (1996)
Origanum petraum Danin	Lamiaceae	Danin (1990)
Origanum punonense Danin	Lamiaceae	Danin (1990)
Salvia domenica L.	Lamiaceae	Werker et al. (1985b) and Ravid and Putievsky (1985a)
Salvia fruticosa Mill	Lamiaceae	Putievsky and Ravid (1984) and Putievsky et al. (1986b)
Salvia sclarea L.	Lamiaceae	Elnir et al. (1991a, b)
Satureja thymbra L.	Lamiaceae	Ravid and Putievsky (1985b)
Thymbra spicata L.	Lamiaceae	Ravid and Putievsky (1985b)

Syrian marjoram

Origanum syriacum is one of the main important native plants in the Holy Land vicinity. Syrian marjoram is an aromatic perennial herb in the mint family, Lamiaceae. The local common name of this plant is "**Za'atar**" or "**Ezov matzuy**". This herb is a very important one for the local population, with a long history of traditional uses. This herb was identified as the biblical hyssop which is mentioned in the bible as a medicinal and antiseptic plant (Fleisher and Fleisher 1988), and fits today as a part of the trend of "natural products" and "functional food" (Dudai 2008).

Description

The Syrian marjoram undergoes dramatic morphological variation during the year, probably as a mechanism of adaptation for the extreme seasonal climatic changes in its dispersion area: During the winter and spring upright thick stems bear large (1–3 cm long) and soft leaves. In the summer these foliage abscise, while new small leathery leaves are expressed on thin prostrate stems. The flowering starts in the beginning of May. Full bloom is observed in the beginning of June. About 50 days later (end of July) most of the inflorescences dry up and abscise. New flowers are produced sporadically during the summer and early autumn. The last flowers appear in September (Dudai et al. 1989, 1992).

Classification

In the past Syrian marjoram was defined as *Majorana syriaca* (Labiatae). In the revision of Ietswaart (1980) it is classified as a species of the genus *Origanum*, and belongs to the section *Majorana*, named *Origanum syriacum.* In the Middle East there are three varieties of this species: *O. syriacum* var *syriacum O. syriacum* var *bevanii* (mainly in Turkey and Syria) (Lukas et al. 2009) and *O. syriacum* var *sinaicum* (in Egypt) (Başer et al. 2003).

Origin and Distribution

Majorana syriacum is endemic to the Middle East in Mediterranean maquis and forest, Batha, Phrygana, hard rock outcrops, mainly in Lebanon, Syria, Israel, Jordan and the Sinai Desert (Feinbrun-Dothan 1978; Ietswaart 1980).

Folk Medicine

The word *ezov* refers in the Bible to a plant tied into branches and used as a brush to sprinkle blood on the doorposts and lintels when the house was cleansed against leprosy (Leviticus 14:4), as well as for purposes of worship: “And a clean person shall take hyssop and dip it in the water and sprinkle it upon the tent, and upon all the vessels, and upon the persons that were there, and upon who touched a bone, or one slain, or one dead, or a grave” (Numbers 19:6) (Zohary 1982). Due to its association with cleaning, the hyssop plant was thought to possess powers of spiritual purification: “purge me with hyssop and I shall be pure: wash me, and I shall be whiter than snow” (Psalms 51, 7). At the same time it was supposed to exemplify the stunted discredited plants that grow out of walls: “From the cedar tree that is in Lebanon even unto the hyssop that springeth out of the wall” (I Kings 4:33), although it is actually a handsome plant, tall, and does not sprout on walls (Zohary 1982).

The identity of the Biblical *ezov* with *O. syriacum* is confirmed by a Samaritan custom whereby Origanum is traditionally used by Samaritans to sprinkle the blood of the Passover sacrifice. The hair on the stems is said to prevent coagulation of the blood (Zohary 1982).

The Syrian marjoram is known as a remedy for intestinal worms since Jesus' time (Palevitch and Yaniv 2000). Today, among Israeli Arabs, a tea is prepared from the leaves to relieve colds and stomach aches. Syrup prepared by scalding Syrian marjoram leaves is known as a remedy for several diseases, in the folk medicine of the Middle-East.

The Yemenite Jews use an infusion of hyssop to relieve labor pains, and use it in combination with sage and various aromatic leaves to reduce the pain of heart diseases, headaches and earache.

The plant is believed to promote verbal dexterity and can act as a charm against witchcraft.

An ethnobotanocal survey conducted in Israel revealed the following ethnic uses of Syrian marjoram (Dafni et al. 1984).

1. To relieve toothaches: Green leaves are crushed and placed on the aching tooth.
2. To treat gum infections: Dried leaves are crushed with salt and rubbed on the inflamed area.
3. To strengthen heart function: Tea is prepared with honey.
4. To treat infections of the digestive tract and urinary system: decoction is prepared and 2–4 spoons are taken daily.
5. To treat colds: tea is prepared.
6. To clear worms: An extract is made from leaves in olive oil, left for 2 weeks in the sun, the oil is absorbed in sugar lumps and 1–3 spoon are taken twice daily.

Biological Activities

The high quantities of carvacrol and thymol in oregano spp. oil, are responsible for its potent biological properties such as antioxidant (Yanishlieva et al. 1999), antibacterial (Ultee et al. 2002; Si et al. 2006; Esen et al. 2007), fungicide (Muller-Riebau et al. 1995), insecticide (Cetin et al. 2007), herbicide (Dudai et al. 1999b) and nematicide activities (Oka et al. 2000).

Many diverse activities of carvacrol, such as antimicrobial, antitumor, antimutagenic, antigenotoxic, analgesic, antispasmodic, anti-inflammatory, angiogenic, antiparasitic, antiplatelet, ache inhibitory, antielastase, insecticidal, antihepatotoxic and hepatoprotective activities have been described (Baser 2008). In addition, uses such as a feed additive, in honeybee breeding and in gastrointestinal ailments have been shown.

Major Chemical Constituents and Bioactive Compounds

O. syriacum essential oil is primarily comprised of the phenolic monoterpenes carvacrol and thymol (Lukas et al. 2009). Usually there are two identified chemotypes: one contains mainly carvacrol (60–80 %) and a small amount of thymol;

whereas the other contains mainly thymol and little carvacrol (Dudai et al. 1989). There is a huge seasonal variation in the essential oil content and composition: The essential oil content of plants harvested in February is only one-fifth of that in plants harvested in August. Their composition also varies significantly within the season; whereas during most of the Summer the main component of a thymol chemotype is thymol, in the Winter the thymol content falls to only 36 %, and p-cymene becomes the main component of the essential oil. The carvacrol chemotype shows a similar trend (Dudai et al. 1992). In addition, it contains other important phenolic compounds such as thumoquinon (0.04–23.7 % DW) (Lukas et al. 2009), oleanolic acid (0.15–1.2 % DW), ursolic acid (0.5–2.5 % DW) and rosmarinic acid (0.2–2.0 % DW) (Dudai 2012).

Development as a New Cultivated Crop

Lately, due to the danger of extinction as a result of the excess collection in the wild, *O. syriacum* has become a "protected species" by law, so its domestication for agricultural production becomes necessary. A project of domestication and breeding in Israel has been conducted over the last 30 years (Dudai 2012). First, vegetative propagation materials of *O. syriacum* were systematically collected to represent their natural diversity, and are grown in experimental fields under intensive conditions. The genetic variability of the essential oil content and compositions, as well as other ingredients such as ursolic acid and rosmarinic acid in various clones of the representative populations, grown in uniform conditions, was very high. The selection process yielded new high quality and uniform commercial varieties (Fig. 1). Two clones were registered as cultivars: "Carmel" (a carvacrol type) and "Tavor" (a thymol type) (Dudai 2012). Study of the factors affecting the yield components (Dudai et al. 1992) and a series of field agro-technical experiments were necessary to optimize the cultivation of the new crop in the various areas. Agronomic aspects studied included search of selective herbicides, the effect of fertilization, irrigation, and harvest regime and plant height on yield components during at least 2 years of growth. Most of the experiments were carried out with different clones in order to find the best treatments for each of the expected ones to be used for commercial purposes (Putievsky et al. 1997).

Other Endemic *Origanum* Species as Potential New Crops

In addition to *O. syriacum*, six species, *Campanulaticalyx* (Ietswaart 1980) are distributed in the limited desert area in Southern Israel, Jordan and Sinai (Egypt) *O. dayi* (Dudai et al. 2003; Amzallag et al. 2005; Friedjung et al. 2013), *O. remonense* (Danin et al. 1997), *O. jordanicum* (Danin and Künne 1996), *O. petraeum*, *O. punonense* (Danin 1990) and *O. isthmicum* (Danin and Künne 1996). Table 2 represents the main components of the essential oils extracted from dry material of the various species. *O. petraeum O. punonense, O. dayi* (Fig. 2) and *O. ramonenese* (Fig. 2),

Fig. 1 *Origanum syriacum* population in the wild (**a**) and a domesticated variety in an commercial open field in Israel (**b**) (Dudai, unpublished data)

containing substantial amounts of α-terpineol and terpinen-4-ol, belong to the "marjoram" culinary type, while *O. isthmicum and O. jordanicum*, belonging to "oregano" type, contain mainly carvacrol derivatives.

The common "oregano" (*Origaum vulgare* or other species used for this spice) is not native to the Holy Land vicinity, but was introduced to Israel due to its popularity in the world, together with some other Mediterranean herbs. The selection process yielded a few clonal cultivars which have been used for the production of dry or fresh herbs for export (Basker and Putievsky 1978; Putievsky et al. 1988, 1997). Recently, interspecific cross breeding yielded some interesting hybrids of *O. vulgare* × *O. syriacum* (Fig. 2), which are used for further breeding (Dudai, unpublished data).

Salvia spp.

The genus *Salvia* (Lamiaceae) includes more than 1,000 species (Alizar 1993; Rivera et al. 1994; Tucker et al. 1995). *Salvia* species are either herbaceous or shrubby perennials and many of them are used as herbs or ornamentals and garden plants (Clebsch and Barner 2003).

Table 2 Volatile composition of *Origanum* spp. of *Campanulaticalyx* section collected in Israel and Jordan[a]

Components	*O. ramonense*	*O. dayi*	*O. petraeum*	*O. punonense*	*O. isthmicum*	*O. jordanicum*
α-Pinene	1	tr	1	3	3	–
Camphene	–	1	1	1	4	2
β-Pinene				2		
3-Octanone					1	8
para-Cymene	4	10	8	8	13	7
Limonene	–	3				
1,8-Cineole	27	tr	10	22		
γ-Terpinene	2	1	2	9	tr	1
Linalool	–	tr	tr	1	tr	4
Borneol	2	9	3	3	9	4
Terpinen-4-ol	6	14	11	17	1	
α- Terpineol	41	43	38	18	1	
2α-Hydroxy 1,8-cineole	2	2	1			
Carvacrol methyl ether					tr	16
Thymoquinone					11	1
Linalyl acetate			7			
Bornyl acetate	tr	2	1	1		tr
Thymol				tr	2	
Carvacrol				–	44	17
iso-Ascaridol				2		
Carvacryl acetate						29
cis-p-Menth-2-en 1,8-diol	2					
trans-Caryophyllene	tr		1	1	3	2
Caryophyllene oxide					2	1
Cadinol			1			
Caryophylla-4(14),8(15)-dien-5-beta-ol					1	
Intermediol	–	1	2	2		

tr traces (<1 %), *n.d.* not detected (<0.05 %)
[a]Collected by Prof. Avinoam Danin and analyzed by GCMS on Rtx-5MS column
Larkov (2009)

*Greek Sage (*Salvia fruticosa, *syn.* S. triloba*)*

Salvia fruticosa is a common native herb and very popular which is growing in wide areas in the Holy land vicinity and the Mediterranean regions. Salvia is an aromatic perennial herb in the mint family, Lamiaceae (Rivera et al. 1994). It is known as

Fig. 2 Three endemic Origanum species to the Holy Land vicinity and a hybrid (*O. vulgare* × *O. Syriacum*) obtained by cross breeding in Israel. The pictures were taken in the Aromatic Plants Living Germplasm collection at the Newe Ya'ar Research Center, Israel (Dudai, unpublished data)

having a few local common name, depending on the region and the language, such as: Mariamia, Marmaria, Joa'sas, Shijeri (Arabic) and Marva – Meshuleshet (Hebrew).

There are many legends revolving this plant in the folklore of the region. There is a story about Mary, mother of Jesus, who, while fleeing to Egypt to escape Herod, with her son in her arms, wiped her face with sage leaves to refresh herself and her son in the oppressive heat. From this legend the Arabic name, "mariamiah" – "the blessed one" is derived. The Latin name: "Salvia" is a contraction of the Latin word Salvameaning "to rescue".

The three-lobed sage is now a protected plant, due to excessive gathering from the wild by healers and users. (Palevitch and Yaniv 2000).

Description

Salvia fruticosa is an evergreen shrub growing up to 1 m height. The unique leaf with three main parts (Fig. 3) is probably the reason of the former name *tri* (three) *loba* (lobes or leaves). The stems are upright and hairy, the flower color is lilac. Flowering starts in February–March. Anthesis and seed maturation continue until the end of June (Putievsky and Ravid 1984; Putievsky et al. 1986a).

Fig. 3 Sage "Newe Ya'ar": an Inter-specific hybrid (Dudai et al. 1999a) of local Israeli *S. fruticosa* with introduced Dalmatian sage

Folk Medicine

In ancient Greece a tea prepared from leaves of the plant was believed to purify the blood and impart a feeling of well-being.

The Yemenite Jews use the plant to relieve earache. Tunisian Jews use it to ease stomach ache, and the Jews of Morocco use it as a diuretic medication.

In the traditions of the Arabs in this region sage is used mainly to treat colds and disorders of the digestive system, including stomach ache, diarrhea, indigestion and flatulence (Dafni et al. 1984).

In Lebanon a water infusion of the leaves is used as a cure for liver diseases. Herbal healers in the Middle East recommend using sage to regulate menstruation, to enhance fertility, to strengthen the muscles of the womb before and after childbirth, and to treat problems associated with menopause. Sage helps opening the respiratory canals, and eases catarrh and sore throats. A gargle prepared from an infusion of sage leaves is used to treat coughs and painful gums.

Sage is attributed with the virtue of prolonging life, in both Arabic and European folklore.

In an ethnobotanical survey, conducted among the Arabs in Israel, the following uses were found to be the most popular (Dafni et al. 1984)

1. To treat mouth sores: fresh leaves are chewed.
2. To cure external sores: fresh leaves are placed on the sore.
3. To accelerate wound healing: dried crushed leaves are spread on the wound.
4. To treat colds: a vapor bath is prepared with sage leaves. The vapor is inhaled.
5. To relive aching joints and rectal pains: Sit over a vapor bath containing sage leaves.

Biological Activities and Modern Medicine

Thujones are some of the essential-oil components of *S. fruticosa*. They are known to depress the central nervous system, counter the activity of various poisons, and induce hallucinations. The use of large dosages of thujones is liable to induce side effects such as convulsions and loss of consciousness. However, using the plant according to tradition is usually safe.

The essential oil also contains phenols, which are active against microbes (Palevitch and Yaniv 2000 and human fungi (Adam et al. 1998).

The effect of extracts obtained from selected Salvia species was tested against oxidative and alkylation damage to DNA in human HCT15 and CO115 cells. Data showed that sage tea protected colon cells against oxidative and alkylating DNA damage and may also interfere with efficacy of alkylating agents used in cancer therapy (Ramos et al. 2012).

Antiproliferative activity of crude ethanol extracts from nine Salvia species grown in Jordan was studied against a panel of breast cancer cell lines.

Cytotoxic activity was evaluated in human tumor models of breast cancer; MCF-7, T47D, ZR-75-1, and BT 474 by the sulforhodamine B assay. In addition, the extracts were evaluated using a non-transformed cell line (Vero) and normal fibroblast cells in order to demonstrate their selectivity and safety (Abu-Dahab et al. 2012).

The results show that three of the plant extracts under investigation exhibited antiproliferative activity against breast cancer cells and were shown to be safe and selective. They are: *S. fruticosa, S. hormium* and *S. syriaca*. These could be considered as a potential source for novel anticancer therapy (Abu-Dahab et al. 2012).

Major Chemical Constituents and Bioactive Compounds

While the main components of the essential oil of *S. fruticosa* in Europe are 1, 8 cineol, thujone and camphor (Máthé et al. 2010), thujones were not reported as a significant component in plants sampled from wild populations in Israel (Putievsky and Ravid 1984). Antioxidant activity and phenolic antioxidants such as rosmarinic acid, caffeic acid, ursolic acid, oleanolic acid and carnosic acid in *S. fruticosa* were also reported (Máthé et al. 2010; Pizzale et al. 2002; Skoula et al. 2000).

Development as a New Cultivated Crop

Domestication and selection of *S. fruticosa* were studied in Israel in the 1980s (Putievsky and Ravid 1984), but any wide scale cultivation of this crop today does not exist. However, the main benefit of this work was obtained by interspecific hybridization with the European common Sage species (Putievsky et al. 1990). Crossing of *S. officinalis*, introduced to Israel from Dalmatia, (Putievsky et al. 1986b; Zutic et al. 2003) with local *S. fruticosa* (Putievsky et al. 1986a) yielded a new commercial cultivar of sage (Fig. 3) (Dudai et al. 1999a). This cultivar has aroma and appearance similar to those of *S. officinalis* (Dudai 2012) but grows in the Israeli Winter and Spring as well as the local *S. fruticosa*, and today it is the main cultivar grown for the fresh herbs market (Dudai et al. 1999a; Dudai 2012).

Salvia dominica

S. dominica is an endemic species to the Holy Land vicinity. Its local common names are Chuecha in Arabic and Marva Reichanit in Hebrew (Dafni et al. 1984). The aroma of its leaves, stems and inflorescences is strong and very similar to that of *S. sclarea* (Werker et al. 1985b).

Description

Salvia dominica is an evergreen shrub growing up to 0.8 m height. The stems are upright and hairy, the flower color is white. The flowering starts in February–March. Anthesis and seed maturation continue until the end of May. The trichoms are very similar to those of *S. sclarea* (Werker et al. 1985b).

Major Chemical Constituents and Bioactive Compounds

The essential oil content in leaves and flowers of *S dominica* is 0.2–0.3 %. The main components of the essential oil are (–) linalyl acetate, lonalool, trpineol, neryl acetate, geranyl acetate geraniol and nerol (Ravid and Putievsky 1985a; Ravid et al. 1994).

Folk Medicine

S. dominica is used to relieve pains, by means of poultices made of crushed leaves, placed on the painful limb. Sweet gall nuts of a peach-like taste develop on the stems of this plant. The local inhabitants are very fond of them. They are called: "little peach" (Palevitch and Yaniv 2000).

Biological Activity

Khalil et al. (2005) reported on anti-microbial activity of crude extract of *S. dominica* against *stapylococus aureu*, but not against other tested pathogenic bacteria. *S. dominica* extracts also showed cytotoxic and antiproliferative activity on choriocarcinoma, endometrium adenocarcinoma, B lymphoblast (Fiore et al. 2006) and breast adenocarcinoma (Abu-Dahab and Afifi 2007) tumor cell lines. Dal Piaz et al. (2009) identified 24 sesterterpene lactones isolated from *S. dominica*, 18 of them interact with tubulin – tyrosine ligase, an enzyme involved in the tyrosination cycle of the C-terminal of tubulin, and inhibit its activity in cancer cells.

A weak nematocidal activity was found in the essential oil components against root-kinot nematodes (Oka et al. 2000).

Clary Sage

A new and unique chemotype of clary sage, *Salvia sclarea* L., was identified recently in wild populations of this species in the northern part of Israel (Elnir et al. 1991a).

This chemotype contains relatively high citral, geranyl- acetate and geraniol, while the major essential-oil components of the common clary sage in the world are linalool and linalyl-acetete. Comparison of the new chemotype with a Russian type showed a great difference in composition and organoleptic character between the two oils. The relative quantity of most components of hybrid oils was intermediate between those of the parent plants (Elnir et al. 1991b).

White Micromeria

(*Micromeria fruticosa* (L.). Druce. Lamiaceae is a perennial herb found in Northern and central Israel, Lebanon, Syria, Jordan and Turkey. Thanks to its nice minty aroma, this is a popular herb, used for tea consumption, giving a feeling of coolness during the typical hot summer in the area. Its local common names are Zuta Levana in Hebrew and Zofa or Ishbat Al Shay in Arabic.

Description

White Micromeria is a dwarf- shrub growing up to 0.2–0.8 m height. The plant spreads vegetatively mainly at the end of winter and spring. The flowering starts in July and continues until November. The flower color is white (Dudai et al. 2001).

Major Chemical Constituents and Bioactive Compounds

The major component of the white micromeria is enantiomerically pure (+)-pulegone (Ravid et al. 1994). In addition to (+)-pulegone, major components of the essential oil are other monoterpenes such as isomenthol, isomenthone, limonene, menthol, α-pinene, β-pinene, piperitone, piperitenone oxide and the sesquiterpenes b-caryophyllene and germacrene. (Werker et al. 1985a; Fleisher and Fleisher 1991; Dudai et al. 2001; Kirimer et al. 1993). Dudai et al. (2001) found a significant seasonal variation in the essential oil composition: During the summer months, when growth rates are maximal, (+)-pulegone constituted up to 80 % of the essential oil, while in early winter, a period of growth-rest in Mediterranean climates, (+)-pulegone levels dropped dramatically to a few percent, while isomenthol constituted up to 80 % of the essential oil. Additionally, in this study there were marked differences in the extracts obtained from individual leaf along the stems: In young upper leaves, the main component was (+)-pulegone, constituting up to 70 % of the total essential oil extracted. During maturation, levels of this component dropped steadily, while reciprocally, levels of isomenthol increased steadily with leaf position, from 0 % in young leaves to more than 60 % in older leaves (Dudai et al. 2001).

Folk Medicine

Tea made of leaves is used to ease breathing of people suffering from colds and also to relieve stomach aches. In Nazareth they prepare an infusion of the leaves to relieve heart pains (Palevitch and Yaniv 2000). The aerial parts of *M. fruticosa* are widely used in the eastern Mediterranean region as medicinal teas, to treat many ailments including abdominal pains, diarrhea, eye infections, heart disorders, high blood pressure, weariness, exhaustion, colds and open wounds (Yaniv et al. 1982; Dafni et al. 1984).

Biological Activity

Evaluation of the allelochemical properties of white micromeria essential oils and its main component (+)-pulegone has demonstrated strong inhibition of seed germination and plant development (Dudai et al. 1999b, 2000, 2009). Pulegone and pulegone-derived lactones have anti-feedant, antibacterial, antifungal and insecticide activity (Szczepanik et al. 2005; Zielińska and Matkowski 2014).

Pulegone is also known as toxic, mainly hepato- and neurotoxicity (Moorthy et al. 1989). **Nevertheless,** De Sousa et al. (2011) reported that (+)–pulegone is a psychoactive compound and has the profile of an analgesic drug, it exhibited antinociceptive activity against chemically induced pain in mice, as well as CNS-depressing and anticonvulsant effects (de Sousa et al. 2011).

Sharp Varthemia

Chiliadenus (syn. *Varthemia*) *iphionoides* (Boiss. et Blanche) Brullo is a member of the Composites. *C. iphionoides* is an aromatic plant whose aerial parts were used in the traditional medicine of the eastern Mediterranean region, and is endemic to the Holy Land vicinity. Its local common names are Salimania in Arabic and Ktela Harifa in Hebrew.

Description

C. iphionoides is an evergreen dwarf shrub with small leaves covered with hairs and glands containing essential oil which gives the plant its unique aroma (Feinbrun-Dothan 1978). All the flowers are tubular. The flowering season is from September to December.

Classification

Chiliadenus iphionoides (Boiss. et Blanche) Brullo (syn. *Varthemia iphionoides* Boiss. et Blanche and Jasonia iphionoides (Boiss. & Blanche) Botsch), a member of the Asteraceae family, tribe Inuleae.

Origin and Distribution

It is found throughout most of Israel, from Mount Hermon to the southern Negev, as well as in Syria, Lebanon, Jordan, and Sinai (Greuter 2008), (Danin 1983).

Folk Medicine

Sharp varthemia is traditionally used for anti-diabetic treatment (Al-Mustafa and Al-Thunibat 2008). In an ethno-botanical survey, conducted among the Arabs in Israel, the following uses were found to be the most popular (Dafni et al. 1984)

1. To relieve headaches and stomach aches, and to clear flatulence: an infusion is prepared. Also fresh leaves are chewed.
2. To treat eye complaints: soaked leaves are placed on the eye.
3. To treat sick babies: The baby is washed in boiled leaves, and then covered to sweat.

4. To treat urine retention: Drink a water infusion of crushed leaves.
5. To relieve internal and external bodily pains: A steam bath is prepared from leaves.
6. To dress wounds: Fresh or dried leaves are placed on the wound.

The Yemenite Jews use sharp vartemia leaves mixed with honey to strengthen their heart and brain, as well as their mental power. They use the leaves cooked in water to relieve melancholia and nervousness, and to treat coughs and mild colds (Palevitch and Yaniv 2000). In traditional eastern Mediterranean medicine, *C. iphionoides* is used to treat influenza, fever, stomach ache, depression and nervousness (Palevitch and Yaniv 1991).

Biological Activity

Various studies demonstrated cytotoxic, anti-oxidant, anti-bacterial activities (Al-Dabbas et al. 2006; Al-Mustafa and Al-Thunibat 2008), anti-diabetic (Gorelick et al. 2011) and hypoglycemic activity (Afifi et al. 1997). Al-Dabbas et al. (2005) isolated eudesmane sesquiterpene with antibacterial activity.

T-cadinol, found in one chemotype, has muscle relaxing activity, possibly due to its calcium antagonistic properties (Zygmunt et al. 1993), and may be useful in immunotherapy (Takei et al. 2006).

Intermedeol has therapeutic potential for treatment human leukemia (Jeong et al. 2002), as well as repelling activity against mosquitoes (Cantrell et al. 2005), ticks (Soares et al. 2010), and ants (Chen et al. 2008).

Major Chemical Constituents and Bioactive Compounds

Avato et al. (2004) described the composition of *C. iphionoides*'s essential oil in a plant sampletaken from the Jordan Valley in Jordan, with borneol as the main compound of this essential oil. Other major compounds were 1, 8-Cineole, α-Terpinol, Camphor, Bornyl formate, Terpinen-4-ol and Bornyl acetate (Tamir et al. 2011). Even though several similar compounds were found, mostly monoterpens, these populations were rich in Intermedeol (not found in the Jordanian oil), Chrysanthemic acid and τ- Cadinol. In contrast to the Jordanian population, seven primary compounds were found in the essential oil of Israeli populations: Borneol, Camphor, 1,8-Cineole, τ-Cadinol, Chrysanthemic acid, fokienol and Intermedeol (Tamir et al. 2011). In 10 representative populations, chosen according to geographical location, climate and topography, 161 volatile compounds were found in the extracts, of which 75 were minor components and not identified. Considerable variation in essential oil composition was found among populations, enabling the identification of three main chemotypes: (A) camphor - α-pinene-fokienol chemotype,

(B) τ-cadinol-1,8 cineole-transchrysanthemol chemotype, and (C) intermedeol chemotype. The main compound of the southern populations was Camphor; therefore they belong to the "Camphorial" chemotype. Only the populations of Judea, Samaria and the northern Negev contain Intermedeol, therefore they belong to the "Intermedeolial" chemotype (Tamir et al. 2011).

Conclusion

Many special and unique local medicinal plants grow in the Holy Land, all with a long history of use, including scriptural documentation of many of them. However, most of them, although still used by the local population, are almost unknown in the western world. During the last decades a few ethno-botanical surveys have been conducted, in order to preserve the knowledge accumulated throughout the past. There is a great need to conserve the natural biodiversity of these plants, which has been damaged by recent intensive urban development. We think that these plants could be preserved by becoming agricultural commodities, since they have great potential as new crops and could become a source for novel bioactive compounds and drugs. In addition, some new culinary herbs could be developed by classical or by interspecific crossbreeding, as was demonstrated with oregano and sage. New products based on ancient plants and practices could bring the herbal traditions of the Holy Land to the rest of the world.

References

Abu-Dahab R, Afifi F (2007) Antiproliferative activity of selected medicinal plants of Jordan against a breast adenocarcinoma cell line (MCF7). Sci Pharm 75:121–136

Abu-Dahab R, Afifi F, Kasabri V, Majdalawi L, Naffga R (2012) Comparison of the antiproliferative activity of crude ethanol extracts of nine salvia species grown in Jordan against breast cancer cell line models. Pharmacogn Mag 32:319–324

Aburjai T, Hudaib M, Tayyem R, Yousef M, Qushawi M (2007) Ethnopharmacological survey of medicinal herbs in Jordan, the Ajloun Heights region. J Ethnopharmacol 110:294–304

Adam K, Sivropoulou A, Kokkini S, Lanaras T, Arsenakis M (1998) Antifungal Activities of *Origanum vulgare* subsp. *hirtum, Mentha spicata, Lavandula angustifolia*, and *Salvia fruticosa* essential oils against human pathogenic fungi. J Agric Food Chem 46:1739–1745

Afifi F, Saket M, Jaghabir M, Al-Eisawi D (1997) Effect of *Varthemia iphionoides* on blood glucose level of normal rats and rats with streptozonic-induced diabetes mellitus. Curr Ther Res 58:888–892

Al-Dabbas MM, Hashinaga F, Abdelgaleil SA, Suganuma T, Akiyama K, Hayashi H (2005) Antibacterial activity of an eudesmane sesquiterpene isolated from common Varthemia, *Varthemia iphionoides*. J Ethnopharmacol 97:237–240

Al-Dabbas MM, Suganuma T, Kitahara K, Hou DX, Fujii M (2006) Cytotoxic, antioxidant and antibacterial activities of *Varthemia iphionoides* Boiss. extracts. J Ethnopharmacol 108(2):287–293

Ali-Shtayeh MS, Yaniv Z, Mahajna J (2000) Ethnobotanical survey in the Palestinian area: a classification of the healing potential of medicinal plants. J Ethnopharmacol 73:221–232

Alizar G (1993) Catalogue Synonymique des Salvia du Monde. (Lamiaceae) VI. Biocosme Mésogéen. Nice 10(3–4):33–117

Al-Mustafa AH, Al-Thunibat OY (2008) Antioxidant activity of some Jordanian medicinal plants used traditionally for treatment of diabetes. Pak J Biol Sci 11:351–358

Amzallag N, Larkov O, Ben Hur M, Dudai N (2005) Soil microvariation in the wild as a source of variability: the case of secondary metabolism in *Origanum dayi* Post. J Chem Ecol 31(6):1135–1254

Avato P, Raffo F, Aldour NA, Vartanian ST (2004) Essential oil of *Varthemia iphionoides* from Jordan. Flavour Fragr J 19:559–561

Barazani O, Cohen Y, Fait A, Diminstein S, Dudai N, Ravid U, Putievsky E, Friedman J (2002) Chemotypic differentiation in indigenous populations of *Foeniculum vulgare* var. *vulgare* in Israel. Biochem Syst Ecol 30:721–731

Baser KHC (2008) Biological and pharmacological activities of carvacrol and carvacrol bearing essential oils. Curr Pharm Des 14:3106–3119

Başer KHC, Kürkçüoğlu M, Demirci B, Özek T (2003) The essential oil of *Origanum syriacum* L. var. *sinaicum* (Boiss.) letswaart. Flavour Fragr J 18(2):98–99

Basker D, Putievsky E (1978) Seasonal variation in the yield of herb and essential oil in some Labiatae species. J Hortic Sci 53:179–183

Cantrell CL, Klun JA, Bryson CT, Kobaisy M, Duke SO (2005) Isolation and identification of mosquito bite deterrent terpenoids from leaves of American (*Callicarpa americana*) and Japanese (*Callicarpa japonica*) beautyberry. J Agric Food Chem 53:5948–5953

Cetin H, Erler F, Yanikoglu A (2007) A comparative evaluation of *Origanum onites* essential oil and its four major components as larvicides against the pine processionary moth, *Thaumetopoea wilkinsoni* Tams. Pest Manag Sci 63:830–833

Chen J, Cantrell CL, Duke SO, Allen ML (2008) Repellency of callicarpenal and intermedeol against workers of imported fire ants (Hymenoptera: Formicidae). J Econ Entomol 101:265–271

Clebsch B, Barner CD (2003) The new book of Salvias. Timber Press, Portland, pp 125–127. ISBN 978-0-88192-560-9

Dafni A, Yaniv Z, Palevitch D (1984) Ethnobotanical survey of medicinal plants in northern Israel. J Ethnopharmacol 10(3):295–310

Dal Piaz F, Vassallo A, Lepore L, Tosco A, Bader A, De Tommasi N (2009) Sesterterpenes as tubulin tyrosine ligase inhibitors. First insight of structure-activity relationships and discovery of new lead. J Med Chem 52:3814–3828

Danin A (1983) Chilliadenus iphionoides. In: Flowering plants, the fauna and flora of Israel. Ministry of Defense – Publications, Tel Aviv, p 138 (Hebrew)

Danin A (1990) Two new species of *Origanum* (Labiatae) from Jordan. Willdenowia 25:601–602

Danin A, Künne I (1996) *Origanum jordanicum* (*Labiatae*), a new species from Jordan, and notes on other species of *O.* sect *Campanulaticalyx*. Willdenowia 25:601–611

Danin A, Ravid U, Umano K, Shibamoto T (1997) Essential oil composition of *Origanum ramonense* Danin leaves from Israel. J Essent Oil Res 9:411–417

De Sousa DP, Nobrega FF, de Lima MR, de Almeida RN (2011) Pharmacological activity of (R)-(+)- pulegone, a chemical constituent of essential oils. Z Naturforsch 66(7–8):353–359

Dudai N (2008) Optimization and improvement of phenolic monoterpenes production by oregano (*Origanum* sp.). Acta Hortic 778:15–28

Dudai N (2012) Domestication and breeding of wild medicinal and aromatic plants – 30 years of experience in Israel. Acta Hortic (ISHS) 955:175–183

Dudai N, Amar Z (2005) Tree wormwood (*Artemisia arborescens*) – a medicinal plant that arrived in the region of Israel from Western Europe during the Crusaders. Haya'ar 7:35–40 (Hebrew)

Dudai N, Putievsky E, Palevitch D, Halevy AH (1989) Environmental factors affecting flower initiation and development of *Majorana syriaca* L. (syn. *Origanum syriacum* var. *syriacum*). Isr J Bot 38:229–239

Dudai N, Putievsky E, Ravid U, Palevitch D, Halevy AH (1992) Light, temperature and flowering affecting monoterpenes content in *Origanum syriacum* L. Physiol Plant 84:453–459

Dudai N, Lewinsohn E, Larkov O, Katzir I, Ravid U, Putievsky E (1999a) Dynamics of yield components and essential oil production in a commercial hybrid sage (*Salvia officinalis × fruticosa* var. "Newe Ya'ar No 4"). J Agric Food Chem 47:4341 ± 4345
Dudai N, Poljakoff-Mayber A, Mayer AM, Putievsky E, Lerner HR (1999b) Essential oils as allelochemicals and theirs potential use as bioherbicides. J Chem Ecol 25:1079–1089
Dudai N, Larkov O, Putievsky E, Lerner HR, Ravid U, Lewinsohn E, Mayer AM (2000) Biotransformation of constituents of essential oils by germinating wheat seed. Phytochemistry 55:375±382
Dudai N, Larkov O, Ravid U, Putievsky E, Lewinsohn E (2001) Developmental control of monoterpenes content and composition in *Micromeria fruticosa* (L.) Druce. Ann Bot 88(3):349–354
Dudai N, Larkov O, Chaimovitsh D, Lewinsohn E, Freiman L, Ravid U (2003) Volatile compounds of *Origanum dayi* Post. Flavour Fragr J 18(4):334–337
Dudai N, Chaimovitsh D, Larkov O, Ficsher R, Blaicher Y, Mayer AM (2009) Allelochemicals released by leaf residues of *Micromeria fruticosa* in soils, their uptake and metabolism by inhibited wheat seed. Plant Soil 314:311–317
Duke JA (2007) Medicinal plants of the bible. CRC Press, Atlanta, 488 pp
Elnir O, Ravid U, Putievsky E, Dudai N, Ladzinsky G (1991a) Genetic affinity between two chemotypes of clary sage (*Salvia sclarea* L.). Euphytica 54:205–208
Elnir O, Ravid U, Putievsky E, Dudai N, Ladezinsky G (1991b) The chemical composition of two clary Sage chemotypes and their hybrids. Flavour Fragr J 6:153–155
Esen G, Azaz AD, Kurkcuoglu M, Baser KHC, Tinmaz A (2007) Essential oil and antimicrobial activity of wild and cultivated *Origanum vulgare* L. subsp. Hirtum letswaart from the Marmara region, Turkey. Flavour Fragr J 22:371–376
Feinbrun-Dothan N (1978) Flora Palestina. Part III, The Israel Academy of Science and Humanity, Jerusalem
Feuerstein I, Mueller D, Hobert K, Danin A, Segal R (1986) Constitution of essential oils from *Artemisia herba-alba* populations of Israel and Sinai. Phytochemistry 25:2343–2347
Fiore G, Nencini C, Cavallo F, Capasso A, Bader A, Giorgi G, Micheli L (2006) *In vitro* antiproliferative effect of six *salvia* species on human tumor cell lines. Phytother Res 20:701–703
Fleisher A, Fleisher Z (1988) Identification of biblical hyssop and origin of the traditional use of oregano-group herbs in the Mediterranean region. Econ Bot 42(2):232–241
Fleisher Z, Fleisher A (1991) The essential oil of *Micromeria fruticosa* (L.) Druce subsp. Barbata (Boiss et Kotschy.), P.H. Davis: aromatic plants at the Holy Land and the Sinai, Part VII. J Essent Oil Res 3:477–479
Fleisher Z, Fleisher A (1993) Volatiles of *Achillea fragrantissima* (Forssk.) Sch. Bip. Aromatic plants of the Holy Land and the Sinai. XL. J Essent Oil Res 5:211–214
Fleisher Z, Fleisher A (2002) Volatiles of *Coridothymus capitatus* chemotypes growing in Israel: aromatic plants of the Holy Land and Sinai part XV. J Essent Oil Res 14:105–106
Fleisher Z, Fleisher A, Nachbar RB (2002) Chemovariation of *Artemisia herba-alba* Asso. Aromatic plants of the Holy land and the Sinai. Part XVI. J Essent Oil Res 14:156–160
Friedjung A, Choudhary S, Dudai N, Rachmilevitch S (2013) Physiological conjunction of allelochemicals and desert plants. PLoS ONE 8(12):e81580. doi:10.1371/journal.pone.0081580
Gorelick J, Kitron A, Pen S, Rosenzweig T, Madar Z (2011) Anti-diabetic activity of *Chiliadenus iphionoides*. J Ethnopharmacol 137(3):1245–1249
Greuter W (2008) Med-checklist. A critical inventory of vascular plants of the circum-Mediterranean countries, 2nd edn, Dicotyledones (Compositae). Palermo, Genève/Berlin
Heywood VH (2002) The conservation of genetic and chemical diversity in medicinal and aromatic plants biodiversity. In: Şener B (ed) Biodiversity. Springer, New York, pp 13–22
Ietswaart JH (1980) A taxonomic revision of the genus Origanum (Labiatae). Leiden University Press, The Hague
Jeong SH, Koo SJ, Choi JH, Park JH, Ha J, Park HJ, Lee KT (2002) Intermedeol isolated from the leaves of *Ligularia fischeri* var. *spiciformis* induces the differentiation of human acute promyeocytic leukemia HL-60 cells. Planta Med 68:881–885
Khalil A, Hassawi DS, Kharma A (2005) Genetic relationship among salvia species and antimicrobial activity of their crud extracts against pathogenic bacteria. Asian J Plant Sci 4:544–549

Kirimer N, Tumen G, Ozek T, Baser KHC (1993) The essential oil of *Micromeria fruticosa* (L.) Druce subsp. Barbata (Boiss et Kotschy.) P.H. Davis of Turkish origin. J Essent Oil Res 5:79–80

Kokkini S (1997) Taxonomy, diversity and distribution of *Origanum* species. In: Proceedings of the IPGRI international workshop on Oregano, 8–12 May 1996, CIHEAM, Valenzano, Bari, Italy, p 2–12

Larkov O (2009) Aromatic plants as a source of natural chiral semiochemicals. Ph.D. thesis, Hebrew University of Jerusalem (submitted)

Lev E, Amar Z (2000) Ethnopharmacological survey of traditional drugs sold in Israel at the end of the 20th century. J Ethnopharmacol 72:191–205

Lev E, Amar Z (2002) Ethnopharmacological survey of traditional drugs sold in the Kingdom of Jordan. J Ethnopharmacol 82:131–145

Lukas B, Schmiderer C, Franz C, Novak J (2009) Composition of essential oil compounds from different Syrian populations of *Origanum syriacum* L. (Lamiaceae). J Agric Food Chem 57(4):1362–1365

Máthé I, Máthé Á, Hohmann J, Janicsák G (2010) Volatile and some non-volatile chemical constituents of Mediterranean Salvia species beyond their native area. Isr J Plant Sci 58(3–4):273–277

Moorthy B, Madyastha P, Madyastha KM (1989) Hepatotoxicity of pulegone in rats: its effects on microsomal enzymes *in vivo*. Toxicology 55:327–337

Muller-Riebau F, Berger B, Yeged O (1995) Chemical composition and fungitoxic properties to phytopathogenic fungi of essential oils of selected aromatic plants growing wild in Turkey. J Agric Food Chem 43:2262–2266

Oka Y, Nacar S, Putievsky E, Ravid U, Yaniv Z, Spiegel Y (2000) Nematicidal activity of essential oils and their components against the root-knot nematode. Phytopathology 90:710–715

Palevitch D, Yaniv Z (1991) Medicinal plants of Israel. Tamuz-Modan Publishing, Tel-Aviv

Palevitch D, Yaniv Z (2000) Medicinal plants of the Holy Land. Modan, Tel-Aviv, 358 pp

Pizzale L, Bortolomeazzi R, Vichi S, Uberegger E, Conte LS (2002) Antioxidant activity of sage (*Salvia officinalis* and *S. fruticosa*) and oregano (*Origanum onites* and *O. indercedens*) extracts related to their phenolic compound content. J Sci Food Agric 82:1645–1651

Putievsky E, Ravid U (1984) Selection and cultivation of *Salvia fruticosa* from wild population in Israel. In: Genetic resources of aromatic and medicinal plants, Organized by EUCARPIA, F.A.O., Oeiras, Portugal, p 87–94

Putievsky E, Ravid U, Dudai N (1986a) The essential oil and yield components from various plant parts of *Salvia fruticosa* Mill. (syn. S. *triloba*). J Nat Prod 49:1015–1017

Putievsky E, Ravid U, Dudai N (1986b) The influence of season and harvest frequency on essential oil and herbal yield from a pure clone of sage (*Salvia officinalis*) grown under cultivated conditions. J Nat Prod 49:326–329

Putievsky E, Ravid U, Dudai N (1988) Phenological and seasonal influences on essential oil of a cultivated clone of *Origanum vulgare* L. J Sci Food Agric 43:225–248

Putievsky E, Ravid U, Diwan-Rinzler N, Zohary D (1990) Genetic affinities and essential oil composition of *Salvia officinalis* L., *S. fruticosa* Mill., *S. tomentosa* Mill. and their hybrids. Flavour Fragr J 5(2):121–123

Putievsky E, Ravid U, Dudai N, Katzir I, Carmeli D, Eshel A (1992) Variation in essential oil of *Artemisia judaica* chemotypes related to phenological and environmental factors. Flavour Fragr J 7:253–257

Putievsky E, Dudai N, Ravid U (1997) Cultivation, selection and conservation of oregano species in Israel. In: Padulsi S (ed) Proceedings of the IPGRI international workshop on Oregano, (CIHEAM, Valenzano (Bari), Italy, p 103–110

Ramos AA, Pedro D, Collins AR, Pereira-Wilson C (2012) Protection by Salvia extracts against oxidative and alkylation damage to DNA in human HCT15 and CO115 cells. J Toxicol Environ Health A 75(13–15):765–775

Ravid U, Putievsky E (1985a) Essential oils of Israeli wild species of Labiatae. In: Baerheim Svendsen A, Scheffes JC (eds) Essential oils and aromatic plants. Martinus Nijhoff, Dordrecht, pp 155–161

Ravid U, Putievsky E (1985b) Composition of essential oils of *Thymbra spicata* and *Satureja thymbra* chemotypes. Planta Med 51:337–339

Ravid U, Putievsky E, Katzir I (1994) Chiral GC analysis of (R)(+) pulegone with high enantiomeric purity in essential oils of some lamiaceae aromatic plants. Flavour Fragr J 9:205–207

Ravid U, Putievsky E, Katzir I (1995) Determination of the enantiomeric composition of α-terpineol in essential oils. Flavour Fragr J 10(4):281–284

Rivera D, Obon C, Cano F (1994) The botany, history and traditional uses of three-lobed sage (Salvia fruticosa Miller) (Labiatae). Econ Bot 48:190–195

Segev D, Nitzan N, Chaimovitsh D, Eshel A, Dudai N (2012) Chemical and morphological diversity in wild populations of *Mentha longifolia* in Israel. Chem Biodivers 9(3):577–588

Si W, Gong J, Chanas C, Cui S, Yu H, Caballero C, Friendship RM (2006) In vitro assessment of antimicrobial activity of carvacrol, thymol and cinnamaldehyde towards *Salmonella* serotype Typhimurium DT104: effects of pig diets and emulsification in hydrocolloids. J Appl Microbiol 101:1282–1291

Skoula M, Abbes JE, Johnson CB (2000) Genetic variation of volatiles and rosmarinic acid in populations of *Salvia fruticosa* mill growing in Crete. Biochem Syst Ecol 28(6):551ERLI

Soares S, Borges L, Braga R, Ferreira L, Louly C, Tresvenzol L, de Paula J, Ferri P (2010) Repellent activity of plant-derived compounds against *Amblyomma cajennense* (Acari: Ixodidae) nymphs. Vet Parasitol 167:67–73

Szczepanik M, Dams I, Wawrzenczyk C (2005) Feeding deterrent activity of Terpenoid Lactones with the *p*-Menthane system against the Colorado Potato Beetle (Coleoptera: Chrysomelidae). Environ Entomol 34(6):1433–1440

Takei M, Umeyama A, Arihara S (2006) T-cadinol and calamenene induce dendritic cells from human monocytes and drive Th1 polarization. Eur J Pharmacol 537:190–199

Tamir H, Satovicb Z, Gorelick J, Danin A, Fischer R, Chaimovitsh D, Dudai N (2011) Intraspecific variation of *Chiliadenus iphionoides* essential oil in Israel. Chem Biodivers 8:1065–1082

Tucker AO, Maciarello MJ, Clebsch BB (1995) Volatile leaf oils of cultivars of *Salvia clevelandii* (Gray) Greene X S. *leucophylla Greene*. J Essent Oil Res 7:101–104

Ultee A, Bennik MHJ, Moezelaar R (2002) The phenolic hydroxyl group of carvacrol is essential for action against the food-borne pathogen *Bacillus cereus*. Appl Environ Microbiol 68:1561–1568

Werker E, Putievsky E, Ravid U (1985a) Structure of glandular hairs and identification of the main components of their secreted material in some species of Labiatae. Isr J Bot 34:35±45

Werker E, Ravid U, Putievsky E (1985b) Glandular hairs and their secretions in the vegetative and reproductive organs of *Salvia sclarea* and *S. dominica*. Isr J Bot 34:239–252

Yanishlieva NV, Marinova EM, Gordon MH, Raneva VG (1999) Antioxidant activity and mechanism of action of thymol and carvacrol in two lipid systems. Food Chem 64:59–66

Yaniv Z, Dafni A, Palevitch D (1982) Labiateae as medicinal plants in Israel. In: Margaris N, Koedam A, Vokou D (eds) Aromatic plants: basic and applied aspects. Martinus Nijhoff Publishers, The Hague, pp 265–269

Yaniv Z, Dudai N, Bacharach U (2011) *Artemisia* spp. In: Singh RJ (ed) Genetic resources, chromosome engineering and crop improvement: medicinal crops, vol 6. CRC Press, Boca-Raton, pp 338–359

Zielińska S, Matkowski A (2014) Phytochemistry and bioactivity of aromatic and medicinal plants from the genus *Agastache* (Lamiaceae). Phytochem Rev 13(2):391–416. doi:10.1007/s11101-014-9349-1

Zohary M (1982) Plants of the bible. Cambridge University Press, New York

Zutic I, Putievsky E, Ravid U, Dudai N (2003) Influence of harvest dynamics and cut height upon yield components of sage (*Salvia officinalis* L.). J Herb Spices Med Plant 10(4):49–61

Zygmunt PM, Larsson B, Sterner O, Vinge E, Hogestatt ED (1993) Calcium antagonistic properties of the sesquiterpene T-cadinol and related substances: structure-activity studies. Pharmacol Toxicol 73:3–9

Capparis spinosa – The Plant on the Wall

Mina Faran

Abstract *Capparis spinosa* a native Mediterranean plant is common in Israel. It grows flowers and gives fruits in the hot and dry summer.

In this review article its morphological properties, chemical constituents, biological activities, traditional uses, culinary uses, medicinal use and pharmacological studies are reported.

Based on the literature we conclude that *Capparis spinosa* has highly diverse economic and medicinal value in different systems of medicine, including Chinese medicine, the Ayurveda, the Unani and the Siddha medicine.

It is recommended to encourage studies in Israel, and to investigate ways to use the Caper for medicine, cosmetics and food, as a natural plant resource of our country.

Keywords *Capparis spinosa* • Caper • Capparidceae • Culinary uses • Traditional uses

Plant Description

Common name: Caper
English name: Caper, Caperberry, Caperbush
Hebrew name: Tzalaph
Family: Capparidaceae/Capparaceae
Genus: Capparis
Species: *Capparis spinosa* (Picture 1)

M. Faran (✉)
The Israeli Association of Medicinal Plants, Jerusalem, Israel
e-mail: faranm@smile.net.il

Z. Yaniv, N. Dudai (eds.), *Medicinal and Aromatic Plants of the Middle-East*, Medicinal and Aromatic Plants of the World 2, DOI 10.1007/978-94-017-9276-9_5

Picture 1 *Capparis spinosa* (Photography: Ishai Shamidov)

Other Species Growing in Israel

C. decidua
C. sinaica
C. sicula
C. aegyptia
C. ramonensis

Capparis spinosa (Capparaceae) is a perennial, winter deciduous shrub found throughout the tropical and subtropical regions of the world.

The plant is widely spread on the arid soil of the Mediterranean basin. It grows flowers and gives fruits entirely during summer (Zohary 1969).

The plant has numerous branches, thick and violet colored, bearing a pair of hooked spines at the base of each leaf stalk.

The leaves are alternate, shiny, round to ovate. The roots are large and penetrate deeply into the soil.

The flowers are hermaphrodite, with sweet fragrance, with four sepals, and four white petals, many long violet colored stamens, and a single stigma usually rising above the stamens (Sozzi 2001). Seeds are large, kidney shaped. There are many gray-brown seeds in every fruit (Alkire 1998).

Geographical Distribution

C. spinosa grows in East Africa, Madagascar, Israel, Turkey, Central Asia, Iran, the Pacific Islands and Australia. It is unclear if the plant is indigenous to these regions. The Caper could originated in the tropics and later extend to the Mediterranean basin and central Asia (Zohary 1960).

Dry heat and intense sunlight make the preferred environment for caper plants. They survive the hot summer with high temperatures and no rain during several months.

C. spinosa is abundant in the Mediterranean area of Israel. Other species grow in the desert of Judea and in the Negev. The plants grow on poor and dry soil, on stone walls, in cracks and crevices of rocks.

In the last four decades *C. spinosa* is cultivated in some European countries, and in Morocco and Turkey (Sozzi 2001). The plant adapts easily to different countries, soils and climate.

Major Chemical Constituents

The aerial parts of the caper include the leaves, flowers and buds. The ingredients that are found in the alcoholic extract of these parts are:

Flavonoids: rutin, quercerin-3-O-glucoside, quercetin-3-O-glucoside-7-O-rhamnoside, a new flavonoid quercetin3-O-(6‴-alpha-L-rhamnosyl-″-beta-D-glucosyl)-beta-D-glucoside, kaempferol glycosides (Sushila et al. 2010).

Lipids, glucocapparin (methyl glycosinolate) methyl isothiocyanate, isopropyl isothiocyanate, sec-butyl isothiocyanate, benzyl-isoithiocyanate beta-sitosterylglucoside-6′-octadecanoate, 3-methyl-2-butanyl-beta-glucoside.

Alkaloids: stachydrine, cadabicine (Sher and Al-Yemeni 2010b) The leaf oil contains: n-alkanes, terpenoids, isothiocyanates a phenyl propanoid, an aldehyde, a fatty acid.

The main components of this oil were thymol (36.4 %).

Isopropyl isothiocyanate (11 %) 2-hexenol (10.2 %) and butyl isothiocyanate (6.3 %).

The volatile oil of the ripe fruit and the roots are composed mainly of the methyl isothiocyanate, isopropyl isothiocyanate and sec-butyl isothiocyanates (Behnaz et al. 2012).

In buds and flowers there are both alpha and gamma-tocopherols (Behnaz et al. 2012).

The flower buds include protein, crude fiber, crude oil, total carotenoids and starch (Behnaz et al. 2012).

In general, highest crude protein and ash are found in small buds. Crude contents of small and middle size buds in June are higher than that of August. June was established as most suitable harvest time of the flower buds (Polat 2003).

Biological Activities

Flavonoids and other phenolic compounds in the leaves and flower buds have strong antioxidant activity (Polat 2003).

Anti-inflammatory and antioxidant activity were found in the roots of other species of caper, *C. tenera* and *C. deciduas* (Sher and Al-Yemeni 2010b).

Cappaprenol isolated from *C. spinosa* showed significant anti-inflammatory and antioxidant activity (Sher and Al-Yemeni 2010b).

Antimicrobial, antiviral, immunomodulatory and significant activity against Plasmodium falciparum was shown in extracts of *C. spinosa* (Arena et al. 2008).

Anti-diabetic activity: an aqueous extract of the aerial parts was examined in diabetic rats. Treatment with *C. spinosa* reduced plasma glucose, triglyceride levels and cholesterol (Sher and Al-Yemeni 2010b).

Antiallergic and anti-histaminic activity was shown in the flower buds (Arena et al. 2008).

Ethanol extract of the plant inhibit the fibroblast proliferation and type 1 collagen production in progressive systematic sclerosis (Sher and Al-Yemeni 2010a).

Antihepatotoxic activity was shown with p-methoxy benzoic acid that was isolated from *C. spinosa* (Gadgoli and Mishra 1999).

Traditional Use

C. spinosa is known as an important medicinal and culinary plant for thousands of years. It was used in all the different traditional systems of medicines, including the Unani, Chinese, Ayurvedic and Siddha medicine, and in ancient Greek and Rome.

Capers are mentioned as medicinal and culinary by Dioscorides, Pliny and Theophrastus.

Several parts of the plant, such as roots, leaves, sprouts and seed were proposed by Dioscorides as anti-inflammatory (Rivera et al. 2003).

Infusions of the caper root and young shoots were used in Greek as a treatment of rheumatism.

Fruits and root bark were used in ancient Greek to treat spleen disease, expel urine, and drive out the menstrual discharge.

Root bark and leaves were used with honey for cleaning ulcers (Rivera et al. 2003).

In China the root bark was used to treat cough and asthma, to kill worms in the intestine, soothe pain and release mucus. It was also used to treat paralysis, spleen disease, skin disease and toothache.

In Arab countries plants were used to cure wounds and problems in the spleen, liver, kidneys and intestines, to dispel gases, to treat skin diseases, to strengthen teeth and relieve backaches and joint pain.

The unopened flower buds were pickled in ancient time, just as it is done today. The buds have a pungent taste and smell and they are used as spice and food.

Infusions of the leaves and buds were used as treatment for cold and upper respiratory infections (Sher and Al-Yemeni 2010b). A paste prepared from the root bark was used externally to treat swollen joints, skin rashes and dry skin (Sher and Al-Yemeni 2010b).

The unopened flower buds were used externally to treat eye infections (Sher and Al-Yemeni 2010b).

Rabbi Moshe Ben- Maimon (Maimonides) that lived in the twelfth century used the roots to prepare a poultice to bandage a hardened spleen. The plant infused in vinegar and honey was used for an immediate treatment of an obstruction in the liver and for kidney stones (Ben Maimon 1971).

In a research that was done in Israel on plants known in folk medicine for their therapeutical properties, it was shown that roots, leaves, flowers and fruits were used by Arab and Bedouin healers (Palevitch et al. 1984). The different parts of the caper bush were used to cure cough, rheumatism, diabetes, lung disease, hardness of hearing, nerve illness, sooth pain, treat open wounds, clear the respiratory tract, and to treat infertility of men and women (Dafni et al. 1984).

The caper bush is mentioned several times in the ancient books of the people of Israel, the Mishnah and the Talmud. The name of the plant was Tzalaf already thousands of years ago.

The people of Israel used the different parts of the plant, and gave them different names. The flowers buds were called Kaphrisin. They were pickled in salt water and were used as food. Wine was prepared from the flower buds, and one of the uses of this wine was for the preparation of the incense for the Temple. The Clove was steeped in the Kaphrisin wine to invigorate it.

The tips of the branches and the short non-lignified stems with young leaves were called Timorot. They were pickled and used as food.

The fruits were called Evionot. The young fruits were pickled and eaten.

It is mentioned in the Talmud (Baiza, p. 25, 2) that Rabbi Shimon Ben Lakish says that the Tzalaf is the most powerful plant. The sages try to explain what is so special about him. They explain that his survival in difficult conditions and its vitality are extraordinary. The Tzalaf survives even after it is burnt.

Culinary Uses

The small flower buds are collected in June, and pickled in salty water. They are used as a condiment in many Mediterranean countries. Young fruits and young shoots with small leaves may also be pickled for use as condiment.

Capers have a sharp piquant flavor, which comes from methyl isothiocyanate, arising from crushed plant tissues.

The tender young shoots and the small leaves may also be cooked and eaten as a vegetable (Sher and Al-Yemeni 2010b).

More rarely mature and semi mature fruits are eaten as a cooked vegetable.

The seeds have various uses. They are rich in protein, oil and fiber, and have a potential as food (Jiang et al. 2007). It has a peppery flavor, and could be used as a condiment. In Egypt they were added to wine to keep it from deteriorating (Jiang et al. 2007). Pickled caper buds and berries became an important export product in Turkey in the last 15 years (Polat 2003). They are exported to Mediterranean countries, mainly to Spain. The best quality products are obtained from the Aegean region. Capers are very popular for flavoring in the Turkish cuisines (Polat 2003).

Medicinal Use and Pharmacology

Medicinal use is based mostly on traditional medicine and limited research.

The root bark is analgesic, anthelmintic, diuretic and expectorant. It is used for gastrointestinal infections, gout and rheumatism. Externally it is used to treat skin conditions and capillary weakness.

The young flower buds are laxative. They are used externally to treat eye infections.

The leaves are applied as a poultice in the treatment of gout. Methanol extracts of the flower buds have been evaluated for antioxidant activity. Inhibition of lipid oxidation was demonstrated in vitro. The tocopherol, flavonoid and isothiocyanate are responsible for this activity.

Antiviral and immunostimulant activity were shown in vivo with a methanolic extract of the buds (Arena et al. 2008). The results of the study have shown that the buds' extract interferes with Herpes simplex virus replication. The conclusion of this study is that *Capparis spinosa* flower buds may contribute in improving immune surveillance toward viral infection, and that it should be employed for treatment of Herpes simplex type 2 infections.

Seeds of caper contain ferulic acid and sinapic acid. Seeds boiled in vinegar can be used to relieve toothache.

Conclusion

Based on the literature review we can conclude that *Capparis spinosa* has economic and medical importance. The plant was used by the ancient medicines, the Chinese medicine, the Ayurveda, the Unani and the Siddha medicine. Various parts of the plants were used as culinary, medicinal and cosmetic.

Capparis spinosa gives some green color to our yellow fields in the summer. The plant is green and flowering in the dry fields, and unfortunately people don't pay much attention to it.

We have to encourage people to learn more about this plant, about its ingredients and about its medicinal value.

References

Alkire B (1998) New crop factsheet: capers, Center of New Crops and Plant Products, Purdue University, USA

Arena A, Bisifnano G, Pavone B et al (2008) Antiviral and buds. Phytother Res 22:313–317

Behnaz M, Asli E, Anoosh E (2012) Caper the Mystique of the recent century. Int J Agric Crop Sci 4:604–608

Ben Maimon M (1971) The medical aphorisms of Moses Maimonides (trans: Rosner E, Muntner S). Yeshiva University Press, New York

Dafni A, Yaniv Z, Palevitch D (1984) Ethnobotanical survey of medicinal plants in northern Israel. J Ethnopharmacol 10:295–310

Gadgoli C, Mishra SH (1999) Antihepatotoxic activity of p-methoxy benzoic acid from *Capparis spinosa*. J Ethnopharmacol 66:187–192

Jiang H-E, Li X, Ferguson DK et al (2007) The discovery of *Capparis spinosa* L. (Capparidaceae) in the Yanghai Tombs (2800 years B.P.), NW China, and its medicinal implications. J Ethnopharmacol 113:409–420

Palevitch D, Yaniv Z, Dafni A, Friedman Y (1984) Medicinal plants of Israel: an ethnobotanical survey. In: Craker LE, Simon JE (eds) Herbs, spices, and medicinal plants, recent advances in botany, horticulture, and pharmacology, vol 1. Food Products Press, New York, pp 281–345

Polat M (2003) *Capparis spinosa* L. (Capparidaceae): a review. Afyon Kocatepe Univ J Sci 7:35–48

Rivera D, Inocencio C, Obon C, Alcaraz F (2003) Review of food and medicinal uses of *Capparis* L. (Capparidaceae). Econ Bot 57:515–534

Sher H, Al-Yemeni MN (2010a) Forest resources utilization assessment for economic development of rural community, Northern parts of Pakistan. J Med Plant Res 4:1197–1208

Sher H, Al-Yemeni MN (2010b) Ethnobotanical and pharmaceutical evaluation of *Capparis spinosa* L. validity of local folk and Unani system of medicine. J Med Plants Res 4(17):1751–1756

Sozzi GO (2001) Caper bush: botany and horticulture. In: Horticultural reviews, vol 27. Wiley, Hoboken, pp 125–188

Sushila R, Dharmender R, Deepti R, Vikash K (2010) Phytochemical and pharmacological potential of Kair (*Capparis decidua*). Int J Phytomed 2:10–17

Zohary M (1960) The species of Capparis in the Mediterranean and the Near Eastern countries. Bull Res Counc Isr 8:49–64

Zohary M (1969) The species of Capparis in the Mediterranean and the Near Eastern countries. Bull Res Counc 8D:49–64

Apharsemon, Myrrh and Olibanum: Ancient Medical Plants

Shimshon Ben-Yehoshua and Lumír Ondřej Hanuš

Abstract Among the most reputed ancient medical plants were the: olibanum – frankincense, derived from *Boswellia* spp., myrrh, derived from *Commiphoras* spp., both from southern Arabia and the Horn of Africa, and apharsemon of Judea, derived from *Commiphora gileadensis* that had its origin also in these territories. The demand for these medical plants that were also important spices was met by scarce and limited sources of supply. The incense trade and trade routes were developed to carry this precious cargo over long distances through many countries to the important foreign markets of Egypt, Mesopotamia, Persia, Greece, and Rome. The export of the frankincense and myrrh made Arabia extremely wealthy, so much so that Theophrastus, Strabo, and Pliny all referred to it as *Felix* (fortunate) *Arabia*. At present, this export hardly exists, and the spice trade has declined to around 1,500 t, coming mainly from Somalia; both Yemen and Saudi Arabia import rather than export these frankincense and myrrh.

Apharsemon, known also as the Judaean balsam, grew only around the Dead Sea Basin in antiquity and achieved fame by its highly reputed aroma and medical properties but has been extinct in this area for many centuries. The resin of this crop was sold, by weight, at a price twice that of gold, the highest price ever paid for an agricultural commodity. This crop was an important source of income for the many rulers of ancient Judea; the farmers' guild that produced the apharsemon survived over 1,000 years. Currently there is interest in a revival based on related plants of similar origin. These three ancient plants now are under investigation in many countries for medicinal uses. Many publications and patents on these three plants appeared in recent years.

Keywords *Commiphora gileadensis* • *Boswellia* spp. • *Commiphora* spp. • *Judaean balsam* • Frankincense • Spice trade • Traditional medicine • Modern medicine • Current research

S. Ben-Yehoshua (✉)
Department of Postharvest Science, Volcani Center, Agricultural Research Organization, Bet Dagan 50250, Israel
e-mail: sbenyehoshua@gmail.com

L.O. Hanuš
Institute of Drug Research, School of Pharmacy, Faculty of Medicine, Hebrew University, Ein Kerem, Jerusalem 91120, Israel

Z. Yaniv, N. Dudai (eds.), *Medicinal and Aromatic Plants of the Middle-East*, Medicinal and Aromatic Plants of the World 2, DOI 10.1007/978-94-017-9276-9_6

Taxonomy Characteristics (Botanical Classification)

Taxonomical Characteristics: Apharsemon

The history of the plant, Apharsemon, is a complicated and ancient story, involving ancient sources and exciting medical plant with long historical documents as a source of drugs with many medical benefits recorded by eminent physician along over two millennia. This plant was described in much detail by the great biologists as Theophrastus – (372–287 BCE) (Birdwood 1862; Hort 1916) and Pliny (23–79 AD) (Bostock 1855) and many others whose reports would be discussed later (Ben-Yehoshua et al. 2012). The plant is known in the Bible as tzori or *tzori Gilead* (Hebrew) (Genesis 37, 25; Jeremiah 8, 22) which is derived from the Hebrew word *tzori*, meaning balm (Miller 1998; Amar 2002), and Gilead, a geographical area east of the Jordan River in the center of present day Jordan. The apharsemon was identified as *Commiphora opobalsamum* (Forssk) *Engl.* (Pictures 1 and 2), and belonged to the family *Burseraceae*. The word apharsemon is probably related to the similar word opobalsamum in Greek (Feliks 1968). It has many synonyms: *C. gileadensis* (L.) Chr., *C. gileadensis* (L.), *Balsamodendron opobalsamum* Kunth, and *Amyris*

Picture 1 *Commiphora opobalsamum* fruits (Taken by Muhammad Al Shanfari on May 20, 2011 at Shihait Sinkhole, Taqah, Dhofar Governorate, Sultanate of Oman)

Picture 2 *Commiphora opobalsamum* plant (Taken by Muhammad Al Shanfari on July 10, 2011 in Shaat, Shahab Asaib, Rakhyout, Dhofar Governorate, Sultanate of Oman)

gileadensis L. Due to the many names that this plant bears, in this chapter the Latin names used are *C. gileadensis*, *C. opobalsamum*, and *C. gileadensis opobalsamum*. The term "opobalsamum" refers to the fact that the resin of this plant is juice (opo). This species is known for its fragrant resin (Wood 1997). Linnaeus (1767) claimed it was the source of Tzori and Mecca balsam.

The identification of this plant with the Hebrew names *Apharsemon*, *kataf*, *nataf*, and *tzori Gilead* can be traced to several sages, including Shimon Ben-Gamliel, Rambam (Maimonides), Saadia Gaon, and the modern biblical botanist Yehuda Feliks. The identification of these Hebrew names with the botanical classification of Forsskal and Linnaeus was done by Zohary (1982).

The present authors used *Commiphora gileadensis* as the Latin name of the apharsemon plant that the Queen of Sheba brought to King Solomon (Chronicles II 9:9) and that was domesticated in the Dead Sea Basin. As discussed by Ben-Yehoshua and Rosen (2009), it appears that the apharsemon that grew in Judea was a new variant or cultivar, much improved over its original ancestor. We believe that this ancestor, *C. gileadensis*, from Yemen, had been naturalized in Judea and became the apharsemon after over 1,000 years of cultivation around the Dead Sea by a special guild of farmers who aimed at achieving the best yield of the specific products they had derived from this plant: incense, perfume, and specific medicinal drugs. The present authors conclude that these *Commiphora* plants introduced from the Arabian Desert were domesticated and continuously improved over about 1,000 years in the Dead Sea Basin, to become the true apharsemon. This plant was a unique cultivar, not found in other places, as already suggested by several Greek and Roman experts in this subject (Ben-Yehoshua and Rosen 2009). Although the ancient apharsemon may be an improved cultivar of the tree identified by Forsskal, we suggest that all these plants – those identified by Forsskal in Yemen as well as the ancient plants grown in Judea – be referred to henceforth as apharsemon. Relevant contradicting opinion in this respect was given by Groom (1981). He said that the "Apharsemon of classical times was a very different product than the Arabian tree, that had quite different qualities, and that the *Commiphora* grew only in southern Arabia, Somalia and parts of Ethiopia. However, Groom ignored Josephus information and the Biblical report about the Queen of Sheba visit to King Solomon. Although the timing of her visit to the Kingdom of Israel is controversial, the rationale for the visit of a queen of a kingdom that sells spices to a country that has just established a new temple, which needs large quantities of spices for routine rituals, cannot be disputed. Recently, Lemaire (2010) wrote a special essay about Queen of Sheba in Yemen who traveled to King Solomon and reported there several archeological findings that support this Biblical story.

The Forsskal Taxonomical Expedition

With the aim of identifying the *opobalsamum*, "the apharsemon" of the Bible (1911), which had been produced in Jericho and Ein Gedi around the Dead Sea in Judea, in 1763, Peter Forsskål, on behalf of the King of Denmark and Norway, collected and described an apharsemon tree on an expedition to Oude, Yemen. Following the biblical stories and also those of the many Greek and Roman writers, geographers, and historians, including Josephus Flavius, Forsskål had knowledge of philology, Arabic and natural sciences. He traveled to Yemen, where the Kingdom of Sheba was located, hoping to find this tree, which had become extinct in Judea. Many of the eminent botanical writers of antiquity, such as Theophrastus, had reported that the apharsemon, from which the opobalsamum was collected, grew only at two sites in Judea (Birdwood 1862) but over time, this plant had become extinct in Judea. Knowing the biblical story of the gift of spices by the Queen of Sheba to King Solomon, Forsskal's journey to Yemen in search for the Kingdom of Sheba was a logical choice. The known features that could help his search were fragrance, exudation of a liquid resin – the opobalsamum – and the medicinal tradition for which the apharsemon was famous. The local Yemen experts were not aware of all these apharsemon stories according to Niebuhr (1792). After a long and stressful journey, Forsskål eventually found one small tree at Oude whose leaves (Picture 2) emitted a special fragrance when crushed. Forsskål sent his "eureka" message to his respected mentor Linnaeus: "Now I know the genus of the '*opobalsamum*.' The tree grows in Yemen. It is not *Pistacia*, not *lentiscus*" (Picture 1) (Hepper and Friis 1994).

However, the death of Forsskål during this expedition, and the subsequent publication of his results by others, has made the herbarium specimen and the publications complicated to use. Although Linnaeus based *A. gileadensis* on a specimen sheet by Forsskål sent from Yemen with a letter dated June 9, 1763, no type specimen is in herb. Linn.

Carsten Niebuhr was the Forsskål expedition's cartographer/astronomer. Following the death of Forsskål, Niebuhr took upon himself to summarize the expedition's report. The preparation of Forsskål's research for publication constituted a distinct challenge for Niebuhr and the unknown assistant who aided him. Sifting through unorganized bundles containing almost 2,000 slips of paper, with notes on a particular species sometimes on multiple sheets the task of editing was made very difficult. In his book "Travels to Arabia and other countries in the East" Niebuhr (1792) said: "An Arabian tree, famous from the most remote antiquity, and nevertheless but little known, is that from which the balsam of Mecca is obtained". We found one of these trees in the open fields; and under its shade Mr. Forsskal wrote the first botanical description of the species. He [...] named it, as a new species, *Amyris*; a name which has since been adopted by other botanists. The tree has not a beautiful appearance and what is surprising, its qualities are not known to the inhabitants of Yemen, in which we met with it..." (Niebuhr 1792. Vol. 2, p. 355–6)

Hepper and Friis (1994), in the preface to their book, *The Plants of Pehr Forsskål's Flora Aegyptiaco-Arabica*, summarized Forsskål's botanical results and the assertion of Linnaeus (1764) that the *A. gileadensis* produces the balm or opobalsamum of the Bible. However, since the plant that produced opobalsamum in Judea was extinct then, Forsskål's assertion is open to question (Hepper and Friis 1994). In a subsequent letter to his mentor Linnaeus, Forsskål was able to give more specific information, and pointed out that opobalsamum belonged to the genus *Amyris* P. Browne. Other reports (Arnott 1839) bring out that several previous suggestions were presented by Alpinus and others, but Linnaeus did not accept them as wholeheartedly as he did with Forsskål.

Baack (2013) wrote a monograph about Forsskål in which he said:

> His [Forsskål] biological work stands out for the large number of species identified, its attention to detail, the expansiveness of his descriptions, his knowledge and use of Arabic and his early ideas on plant geography.
>
> "Niebuhr was not a natural scientist" – he continued. "Thus despite his dedication to the task [...] the published work has deficiencies. The terminology and descriptions are inconsistent, its organization is confusing and unwieldy, there is no concordance with the surviving herbarium, and the volume has no index to aid the reader".

The taxonomic and nomenclatural aspects regarding the plant discovered by Forsskål which is currently known as *Commiphora gileadensis* (L.) Chr.Engl remained rather controversial. Forsskål and Linnaeus named the genus *Amyris* and related it in the Octandria class.

However, there is a difference in the epithet of the species' name: Forsskål named it *opobalsamum*, whereas Linnaeus gave the name *gileadensis*. Linnaeus used the term *opobalsamum* for the plant with the pinnate leaves.

In 1782 the botanist J.G. Gleditsch obtained dried specimens from Achmet Effendi near Mecca and named the plant *Balsamea meccanensis*. Gleditsch contrasted the characteristics of this plant with those of *Amyris gileadensis*, Linn. He claimed that the two plants could not belong to the same genus. The leaves of his plant are bipinnate, the calyx and corolla each of five parts, and larger than in the Linnean plant; the stamens ten, though they may be eight or nine, exceeding the corolla, and the immature fruit with a basal pentagonal stipe. He also points out that Linnaeus's plant is not *Amyris* (see Balfour 1888). Gleditsch's plant has been a subject of controversy in the botanical community.

Willdenow (1799) considered that the problem causing the difficulty in identifying the plant lies in the manner in which the filaments are united to the flower: "It is difficult to ascertain the true position of the filaments; because the boundary between the calyx and the receptacle cannot, in all instances, be accurately distinguished."

According to Linnaeus, the filaments are joined to the edge of the receptacle, while Gleditsch insists that this edge is a part of the calyx itself. Willdenow concludes that the plant described by Gleditsch does not form a distinct genus, but is the same as that which Linnaeus has given under the name of *Amyris*.

Moreover, Willdenow showed how leaves change their form in the different stages of the tree's age. Thus, Forsskål's plant was probably an old tree, while Gleditsch's plant could have been a very young tree.

Woodville (1810) also expresses uncertainty regarding Gleditsch's plant: "The description of the Balsam of Mecca-tree, lately given by Gleditsch, differs from that of all other writers."

The authors of the Dictionnaire des sciences naturelles (1825) note that the successors of Linnaeus didn't accept the name *Balsamea meccanensis*.

Meanwhile, the generic name *Balsamodendron* ("balsam tree" Kunth 1824) has come into general use (Balfour 1888). The names given by Forsskål and Linnaeus became synonyms.

Bentley and Trimen (1880) claim, in the section of *Balsamodendron opobalsamum*:

> Prof. Baillon adopts the name *Balsamea* Gleditsch, for this genus, the date of which is 1782 (Berlin Gesellsch. Naturforsch. Freunde, vol. iii, p. 103, t.3, fig. 2). It does not, however, appear from the description and figure given, that Gleditsch really had this plant under his observation.

Balfour (1888) doesn't recognize in Gleditsch's description and figure the characteristics of a *Balsamodendron*. In a correspondence with Engler on this issue, Engler admits that the floral characters of the *Balsamea* are not met with any of the species of balsamodendron. The flowers do not belong to the *Burseraceae* at all. But Engler considers the branching and inflorescence agree well with the characters of the Balsamodendron, and the bipinnate leaves are no barrier to such an identification.

The nineteenth century has seen the abundance of botanical systems, each influenced by its own philosophical concept.

For example, between 1825 and 1845 various botanists proposed some 24 or more systems of classification (Bell 1967).

The shift from *Amyris* to *Commiphora* took place in 1883, when Engler published the section on the *Burseraceae* in the framework of De Candolle's Monographiae phanerogamarum, Vol. 4.

Engler makes changes within the family: he deviates from the classification of Bentham and Hooker (1862–1883) by excluding the Amyridaceae and assigning them to the Rutaceae. One of these excluded species is *Amyris* L.

By this change the family of *Burseraceae* becomes a very natural group that differs enough from the Rutaceae and Simarubaceae in terms of anatomical structure, but is still closer to these two families than to the Anacardiaceae, regarding the arrangements of ovules, the ventral raphe and the micropyle.

Most of the 13 genera recognized by Engler are very closely related and can be distinguished almost only by the fruit.

The number of floral parts varies often within the same genus very strongly, so that generic differences can't be based on the number of petals and stamens (Uhlworm and Behrens 1883).

One of the genera on Engler's list is *Commiphora* Jacq. The genus *Commiphora* was discovered by N.J. Jacquin in 1797.

In Engler's work the genus *Commiphora* includes *Balsamea meccanensis* Gleditsch and *Balsamodendron* Kunth.

As to Gleditsch, he adds the following observation: Gleditsch is right when he claims that *Amyris gileadensis* doesn't belong to the American Amyrids.

On the other hand, the Balsamea of Gleditsch doesn't belong to the genus Balsamodendron Kunth. Finally, Engler decided to prefer the name *Commiphora* over *Balsamodendron*, because Jacquin proposed the best description and figure.

The species *Commiphora opobalsamum* (Engl.) comprises the following synonyms: *Balsamea meccanensis* Gleditsch *Balsamodendron opobalsamum* Oliver

The variants: kunthii (pinnate leaves, rarely ternate)

Amyris opobalsamum Linn

Balsamodendron opobalsamum Kunth

Gileadensis (ternate leaves, rarely pinnate)

Amyris opobalsamum Forsk

Amyris gileadensis Linn

Balsamodendron gileadense Kunth

Balfour (1888) criticizes Engler for preferring the name *Commiphora* over *Balsamodendron*:

> Undoubtedly the plant described and figured by Jacquin [...] as *Commiphora madagascariensis* is a *Balsamodendron* [...] The significant name Balsamodendron is now commonly accepted, not only by botanists but by pharmacists and physicians, and is indeed current in general literature, and the substitution of another name would be almost impossible, and would certainly lead to much confusion. How poor, too, is the name *Commoiphora* beside the suggestive *Balsmodendron* [sic]!
>
> Bentham and Hooker [1862–1883], with set purpose, place Jacquin's name as a synonym of *Balsamodendron*, and their lead will be generally followed. Were *Commiphora* to be now accepted, it would entail the renaming of all the species, some thirty-six, and as they have been already renamed by Baillon and Engler under Balsamea, we should have an addition of some seventy specific names to the nomenclature.

Despite the above objection, the genus *Commiphora* was accepted.

When botanists describe the natural order of Amyridaceae, they refer to trees and shrubs, abounding in a balsamic juice, and having alternate or opposite leaves, which are ternate or unequally pinnate, sometimes with stipules, and occasionally with pellucid dots. According to Griffith (1847), they are all natives of tropical climates; one species only is found in the United States.

Nevertheless, the status of the Amyrids is still problematic.

The genus *Amyris* P. Browne has enjoyed a confused taxonomic history, having been classified in both the *Burseraceae* and in the family *Rutaceae*. It is now considered to belong to the latter family, but certain species have been transferred to the genus *Commiphora*, which belongs to the *Burseraceae*. Thus, Yucatan elemi derived from *Amyris plumieri* DC. is a product of the family Rutaceae, while balm of Mecca is derived from *Amyris opobalsamum* L. which is now considered to be a synonym of *Commiphora opobalsamum* Engl., fam. *Burseraceae*. (Botanical dermatology database)

The controversy persists along the twentieth and twenty-first centuries:

The Rutaceae family has been placed in many orders including the Sapindales, the Geraniales and the Rutales.

The first comprehensive classification within the Rutaceae was made by Engler in 1896. He later divided the family into seven sub-families. Four of these subfamilies have been subject to significant controversy with regard to their placement, and two of the subfamilies have since been combined. The seventh subfamily, Citroideae, has remained unchanged.

> The controversial history of the subfamilies has resulted in considerable movement of subfamilies in and out of the Rutaceae. (Scott et al. 2000)

In a research done by Clarkson et al. (2002), the authors studied the phylogenetic relationships in *Burseraceae*.

Burseraceae were first described by Kunth in 1824. In more recent classifications they have been placed in Sapindales. The family is closely allied with Anacardiaceae, Rutaceae, Simaroubaceae and Meliaceae and comprises trees or shrubs, with the inner bark having resin ducts and, unlike Rutaceae, distinctively absent foliar pellucid glands. The family is distributed throughout the tropics but is especially diverse in Malaysia, South America and Africa.

There have been various attempts to divide the family, each using different characters or character combinations such as flowers, embryos, anatomy, fruit, germination and seeds. The first comprehensive subdivision of the family was made by Engler in 1931. He split the taxa into three tribes based exclusively upon the structure of the fruits: *Protieae*, *Boswellieae* and *Canarieae*. This system was accepted and slightly modified by Lam in 1932, producing the classification used today. Lam renamed the tribe *Boswellieae* as *Bursereae* (because the latter is older than the former).

Since then new taxa have been discovered and assigned to these tribes, but some of them were so morphologically distinct that they justified the creation of entirely new genera.

Authors of more recent works on the family have concentrated on particular geographic areas or genera, often using newer characters, such as pollen morphology.

Clarkson and his colleagues, who used genetic methods in their study, assert that there is a close association between the African genus *Commiphora* and the South American genus *Bursera*.

Present reviewers consider the recent article by Gachathi (1997) who also examined the *Burseraceae* as a base to stress the problem of the present controversy:

> *Burseraceae* is a family of 17 genera with some 560 species which are widespread in the tropics especially in Africa, Malaysia and South America. These are trees or shrubs characterised by aromatic resins from the bark used even in Biblical times for frankincense, myrrh and perfumes. The main resin-producing species are found in the genera *Boswellia* and *Commiphora* and others which are common in the hot drylands. Despite their early recognition, classification and nomenclature of members of the two genera, and particularly those of *Commiphora* have remained unstable. They have been described by botanists as taxonomically difficult, frustrating or confusing. This is largely because of the nature of the plants themselves, appearing leafless and in a drought-dormant condition for much of the year. This situation has led to the practice of describing species from inadequate and often sterile material. As a result, some species have been described by different botanists under different names. Moreover, sterile plants from other genera have been described as species of either *Boswellia* or *Commiphora*. For example, six plants described by Engler (the chief worker on the genus *Commiphora*) as new species of *Commiphora* belonged in fact to other genera and in other families. Several plants within the two genera therefore have been known by two or more different names. This instability of plant names is a real disadvantage as all information about plants and their products is communicated by name. Recent taxonomic revisions of the family *Burseraceae* have resulted in the union of two or more species previously considered distinct, splitting what was considered previously to be one species into two or more species or outright rejection of wrong names brought about by misidentification. Most names of the members of the family *Burseraceae* are therefore presented with numerous synonyms, subspecies, varieties, long descriptions and additional notes.

Furthermore, different plants at different parts of the world are called at present *Commiphora gileadensis* which definitely do not belong to the same species.

In conclusion, the disorder in the taxonomy and nomenclature of the resinous plants presents problems in determining the true balm of Gilead. Therefore, there is a need to examine carefully and thoroughly the different species and separate those mistakenly put together.

This is the old classification of the Apharsemon of Stephenson (1831)

Class VIII. OCTANDRIA – *OrderI.* MONOGYNIA. *Nat. Ord.* TEREBINTACEAE, *Juss.*

GEN. CHAR. *Calyx* four-toothed. *Petals* four, oblong. *Stigma* quadrangular. *Berry* drupaceous.

SPEC. CHAR. *Leaves* ternate; leaflets entire; peduncles, one-flowered, lateral.

Syn.-Balsamum. *Theophr. I.* 9. c. 6; *Plin. I.* 12. c. 25; *Justin, I.* 36. c. 3; *Bellon.* 110.

Balsamum syriacum, rutae folia. *Bauh. Pin.* 400.

Balsamum verum. *Bauh. Hist.* 1.298; Raii. *Hist.* 1755.

Balsamum Alpini cum Carpobalsamo. *Ger. Em.* 1528.

Balsamum, ab sagyptiis Balessan. *Alpin.AEgypt, p.* 48. *t.* 60.

Balsamea meccanensis. *Gleid. Act. Soc. Berol.* 3. *p.* 127. *t. 3.f.* 2.

Balsamodendron Gileadense. *Decand. Prodr. t. 2.p.* 76.

Amyris Opobalsamum. *Forsk. Aegypt. p.* 79; *Niebuhr. v. i.* 307.
Theophrasti et Dioscoridis.
Amyris gileadensis. *Lin. Mantis.* 65; *Diss, de Opobals.* 1764; *Willd.v.2.*
p. 333. *Vahl. Symb.* i. 28 t. II; *Lam. III. t.* 303.f. 2;Woodv. *v.* 3. t. 192;
Stokes, 2.357.
Foreign. – *Balsamier de la Mecque*, *Fr.*; *Balsamino di Gilead*, It.; *Gileadischer Balsamstrauch*, Ger.
Present Taxonomy
http://www.itis.gov/servlet/SingleRpt/SingleRpt?search_topic=TSN&search_value=896045. Accessed July 14, 2014
***Commiphora gileadensis* (L.) C. Chr.**

Taxonomy and Nomenclature

Kingdom:	**Plantae**:
Taxonomic rank:	Species
Synonym(s):	*Commiphora opobalsamum* (L.) Engl.
Common name(s):	
Taxonomic Status:	
Current standing:	Accepted
Data quality indicators:	
Record credibility rating:	verified – standards met

Taxonomic Hierarchy

Kingdom	**Plantae** – plantes, Planta, Vegetal, plants
Subkingdom	Viridaeplantae – green plants
Infrakingdom	Streptophyta – land plants
Division	Tracheophyta – vascular plants, tracheophytes
Subdivision	Spermatophytina – spermatophytes, seed plants, phanérogames
Infradivision	Angiospermae – flowering plants, angiosperms, plantas com flor, angiosperma, plantes à fleurs, angiospermes, plantes à fruits
Class	Magnoliopsida
Superorder	Rosanae
Order	Sapindales
Family	*Burseraceae* – burseras
Genus	*Commiphora* Jacq. – myrrh
Species	*Commiphora gileadensis* (L.) C. Chr.

Reference
Germplasm Resources Information Network (GRIN), 2007–2011, database (version 2011)
USDA, ARS, National Genetic Resources Program. Germplasm Resources Information Network – (GRIN) [Online Database]. National Germplasm Resources Laboratory, Beltsville, Maryland
URL: http://www.ars-grin.gov/cgi-bin/npgs/html/paper.pl?language=en

Taxonomical Identification: Myrrh

The oleo-gum resin myrrh (so named from the ancient Acadian *murru*, Arabic mur, and Hebrew mar, meaning "bitter") is obtained from the *Commiphora myrrha* (Nees) Engler tree (Picture 3) and called *Apharsemonodendron myrrha* by Nees von Esenbeck in 1826 (Orwa et al. 2009). It also has other synonyms, such as: *Commiphora molmol* (Engl.) Engl. and *Commiphora myrrha* var. *molmol* Engl.

There is still discussion regarding from which *Commiphora* tree myrrh was harvested in biblical times. Feliks (1968) and Zohary (1982) identified the myrrh of the Bible with the *C. abyssinica* Berg and *C. schimperi* Berg, both growing in Africa, where the resin is still used today medicinally and socially among local traditional communities; others suggest *C. africana*. Engl. (African myrrh) and *C. myrrha* (Nees) Engl. (Moldenke and Moldenke 1952).

Taxonomical Identification: Olibanum

The genus *Boswellia* was named after John Boswell in 1846, and his associate H. J. Carter (1851) was responsible for the first scientific survey of these trees in the same year, refined by Birdwood in his article in 1870. According to van Beek (1958), there are five species, but only *B. carterii* (Picture 4) and *B. frereana* yield frankincense of commercial value.

Picture 3 Myrrh, *Commiphora myrrha* – drawings of botanical details (Köhler), tree, and natural aromatic resin

Picture 4 Frankincense, *Boswellia carterii* – drawings of botanical details (Köhler), tree, and natural aromatic resin

Boswellia carterii is also referred to as olibanum and dragon's blood. Some authorities regard this species as the biblical frankincense and the same species as the *B. sacra*, but there is some dispute over this. Thulin and Warfa (1987) determined that the *B. carterii* is a variable form of *B. sacra*.

Boswellia frereana (Birdw.) and *B. thurifera* (Roxb. Ex Flem. 1810) grow in northern Somalia (Thulin and Warfa 1987) and are the source of the Maydi frankincense, also called Coptic frankincense, as it is highly esteemed by the Coptic church, but the main part of its production is purchased by Muslim pilgrims in Saudi Arabia. These resins have a pleasant lemon scent and are also manufactured into a popular chewing gum. The smell of the *B. frereana* is different from *B. sacra*, but they are often marketed together for different uses (Bowen 1989).

Boswellia papyrifera Hochst grows in Ethiopia and Sudan but is not marketed in the western world. The resin is transparent and oilier than the other resins.

Boswellia serrata Triana & Planch. (Roxb.) Colebr. is the Indian frankincense, considered by some to be of inferior quality. The golden brown color resin is soft and hardens slowly; it is mainly burned as incense but also used in Ayurvedic medicine (Miller and Morris 1988). *B. serrata* is taller than the other *Boswellia* trees and has a straight trunk. The scent of the resin extracted from *B. serrata* is quite distinct from that obtained from the other *Boswellia* trees and is heavier than the African resins, more of an orange type of scent, while the *B. sacra* resins have a lighter, lemon scent. The difference in odor between the various *Boswellia* tree resins is due to their complex sesquiterpenes (Tucker 1986).

Crude Drug Uses

Apharsemon

The crude drug of apharsemon was a special fragrant resin that exuded from the branches after cutting the branch. The exudates were processed into various products: incense, perfume, and different medical drugs. This resin was the most expensive agricultural product, with a price twice its weight in gold during the Middle Ages and twice its weight in silver during the Roman period. Documents show that the apharsemon plants were guarded in order to prevent theft (Picture 5).

Pliny describes just how expensive this rare spice was in classical times: For a sextarius (equaling about 20 fluid ounces or half a liter) of apharsemon which is sold by the fiscal authorities at 300 dinars (denarii), is sold again for a thousand, so vast is the profit to be derived from increasing this liquid by sophistication. The price of xylobalsam is six dinars per pound (Book 12, Chapter 34). In other words, a sextarius of apharsemon sold at the source for the equivalent of nearly the yearly wages of one laborer in the early Roman period and later sold for over three times that amount. Even the wood cuttings of the plant (xylobalsam) were coveted and sold for the price of 6 days wages. Oil of Apharsemon was considered to be the most valuable oil used for medicinal purposes. Strabo refers to it as a remedy for headaches, cataracts, and dimness of sight (Jones 1924). Pliny lists 15 different ailments that could be cured with apharsemon oil (Book 12, Chapter 54).

Picture 5 A Janissary guarding the balsam tree (Cartwright 1760)

Apharsemon in Judea

The Bible refers to the transport and trade of apharsemon (*tzori*) in the time of the Patriarchs, about 1850–1550 BCE. Joseph was sold by his brothers to a caravan of Ishmaelites carrying balm and other spices down to Egypt (Genesis 37:25). Also, in another Biblical story Jacob, asked his children to collect the Tzori described there as one of special crops of Israel as a gift to Pharaoh King of Egypt. However, according to specific version, which is in my opinion a concensus among experts dealing with this issue, during the time of Jacob – early beginning of the second millennium BCE, apharsemon was not yet grown in Israel as it was brought to Israel by the Queen of Sheba as a gift to King Solomon at the end of the tenth century BCE. This unexplained Biblical story may relate to a possible mistake, for example in considering the word Tzori always as representing the Apharsemon (Pictures 6, 7, and 8). It is possible that Tzori may be at times used to describe a natural medical product. Such a disagreement with the biblical story occurred at the period of the first Temple. At a post Biblical period no such problem of lack of agreement between the accepted history of the apharsemon and the many written historical documents existed.

Present Day Investigations to Find the Apharsemon

Currently, no residue of the ancient apharsemon has been discovered, and all archaeological attempts to find it have failed (Hirschfeld 2007).

Picture 6 *Commiphora gileadensis* leaves and fruits (Taken by Lumír Hanuš on August 1, 2013 in Kibbutz Almog, Jordan Rift Valley, West Bank. Notice the different shape of the leaves between the Israeli and the Oman plants. Explanation is not yet available)

Picture 7 *Commiphora opobalsamum* male flowers (Taken by Lumír Hanuš on May 11, 2013 in Kibbutz Almog, Jordan Rift Valley, West Bank)

Picture 8 *Commiphora gileadensis* seeds (Taken by Lumír Hanuš on August 1, 2013 in Kibbutz Almog, Jordan Rift Valley, West Bank)

In one of the many archaeological projects carried out to locate residues of apharsemon, Patrich and Arubas (1990) discovered a juglet, half full of a dense liquid, in a cave near Qumran, in the Dead Sea Basin. They suggested that this oil might be made from the apharsemon. However, two chemical studies negated this suggestion; one was performed by Eizenstadt and Ashengraw and reported as an appendix in the paper by Patrich and Arubas (1990), and the other was unpublished data by S. Ben-Yehoshua and L. Hanuš. This oil, according to our data, had none of the chemical markers of the *Commiphora* species or, for that matter, of *Boswellia*. Nevertheless, Vendyl Jones, one of the initiators of the Qumran expedition, claimed in several of his reports to his financial sponsors for the lost treasures of the Holy Temple in the Qumran region that he had found the oil of the biblical apharsemon inside the juglet that Patrich had discovered. However, no data confirming this claim have been presented.

In another work of the Vendyl Jones Research Institute (Jones 1995), a hidden silo in the bedrock in a cave at Qumran was found during the 1992 excavation, which contained a reddish material that appeared to be organic in nature. Tests allegedly indicated that the reddish material was a mixture of 11 ingredients of the holy incense (*pitum haqetoret* in Hebrew) used in the Temple in Jerusalem, which also contains the oil of the apharsemon. Over 400 kg of the reddish material were removed that year from the cave. These two items are listed in the Copper Scroll, one of the Dead Sea Scrolls, which Vendyl Jones studied. In his work, he further claimed that this incense was prepared in the precise order as had been written in the Torah. However, Vendyl Jones's reports were greatly criticized by many researchers. The late Yehuda Feliks (see Amar 1998) said that the reliability of this article is dubious and the finding of the holy incense is just a fantasy. Amar (1998) also analyzed this report in detail and concluded that the silo was possibly a factory to produce soap from the local Dead Sea Basin herbs.

The precious perfume boxes: Four powder boxes made of gold and silver were given to the senior author for chemical evaluation by a famous antiques collector, one box bearing the inscription "Balsam". Analysis of the top layer of the material of all four boxes did not reveal any of the chemical markers of the *Commiphora or Boswellia* species. However, a chemical that is a known component of the aromatic gum *ladanum* from *Cistus creticus* was found in one box. The ladanum spice (*lot* or *lotem* in Hebrew) was one of the important ancient spices of Canaan and of the Israelites. Furthermore, it was one of the spices that the Ishmaelites who had purchased Joseph from his brothers carried on the backs of their camels: astragalus, balm, and ladanum ("*nechot, tzori velot*" Hebrew, (Genesis 37:25). This is the first time that both the apharsemon and ladanum are mentioned in the Bible. It was suggested that these chemical markers could be used to identify the ladanum spice (Ben-Yehoshua and Hanuš, unpublished data 2007).

Crude Drug Uses of Myrrh

Myrrh was a central factor in religious ceremonies in ancient Egypt. Plutarch wrote: "Every day they make an offering …to the Sun …of myrrh at midday." Animal sacrifices and rituals were accompanied by the burning of myrrh, to mask smells and disperse evil spirits (Babbitt 1928).

The resin *bdellium* (*b'dolach* in Hebrew) is obtained from *Commiphora africana* (A. Rich.) Engl., named the "African myrrh" by Duke (2008). This resin is mentioned in the Bible (Genesis 212) and was regarded as a costly gum. It was well known in the ancient world, and Theophrastus (*Historia Plantarum* IV: 2.1 and 2.6) (Theophrastus Eresius 2009), Pliny (*Historia Naturalis* Book 12), and Galen (*Opera Omnia* Vol. 14) all mentioned it.

Stacte (*nattaf* in Hebrew), which appears in the Bible in Exodus (30:34), probably refers to the liquid form of myrrh, a solution of myrrh resin in oil. Pliny (Book 12, Chapter 35) refers to a naturally flowing gum, called stacte, which sometimes flows from the bark of the tree without any cutting, before the actual harvest. However, Dioscorides (Osbaldeston, Book I, 73) and Theophrastus (Hort., Chapter 9) interpret stacte as distilled myrrh. Johnson (1987) suggested that stacte is the myrrh resin dissolved in oil of *Balanites aegyptiacus* (L.) Delile.

Embalming in ancient Egypt was an elaborate process, involving many different materials. The inclusion of frankincense and myrrh from Punt was symbolic as originating from the land of the Egyptian gods. Great quantities of myrrh and frankincense were employed to treat the dead body and preserve it from decay and deterioration. The antibacterial properties of these resins were important in protecting the body from putrefaction. The embalming procedure of ancient Egypt, as described by Herodotus in the fifth century BCE, used myrrh extensively, as evidenced from archaeology, where myrrh can still be smelled in newly excavated burial tombs. Myrrh was a very effective antiputre-faction and antimicrobial agent in corpses, its efficiency allowing the examination of intact mummies several thousands of years old. Tomb paintings at the tomb of Petosiris show ancient Egyptian perfumers preparing resins to perfume the air and mask the odors of the embalming process and generally the unsanitary conditions of life in those times.

The New Testament mentions myrrh in John 19:39, where Nicodemus brought about 45 kg myrrh and aloes for treating the crucified body of Jesus before his burial. This huge quantity of such an expensive material demonstrates the esteem conferred on Jesus. Matthew 2:11 relates that the Magi traveled to the birthplace of Jesus and they opened their treasures and presented him with gifts – "gold, incense, and myrrh."

Olibanum

The name given to the natural oleo gum resin of *Boswellia* was olibanum, not indicating oil from Lebanon but most probably taken from the Arabic *laben* or *alluban*, meaning "white," since the clear white drops of resin are the most valuable (Miller and Morris 1988). The Hebrew name is *levonah*, also suggesting "white" (Hebrew: *lavan*). The other name – "frankincense" has its origin either from the French crusaders ("frank" or "French incense") or from the Old French *franc encens*, meaning "pure incense". The chemical composition of olibanum or frankincense oil has been investigated by Hamm et al. (2003, 2005). Incense burning at religious ceremonies is one of the chief uses of frankincense which will be described later.

Chemical Constituents, Bioactive Compounds and Current Research

Apharsemon

It is worth mentioning, that it is not entirely certain that all the publications cited, referring to *C. opobalsamum*, describe the same plant. Since there are chemical differences between *C. opobalsamum* from Saudi Arabia or Yemen, and from Israel and from China. The exudates of Apharsemon measured there were imported to China from India. It is therefore questionable if the researchers are dealing with the same plant – *C. opobalsamum*.

Structures of typical known terpenic compounds from *C. opobalsamum* – linalool, α–terpineol, and 4-terpineol (Al-Massarany et al. 2007), α-cubebene, β-cubebene, α-copaene, β-selinene, γ-muurolene, germacrene D, δ-cadinene, β-caryophyllene, spathulenol, cembrene, and α-cadinol (Al-Massarany et al. 2007; Hanuš 2012) – are presented at Fig. 1.

The other compounds, typical for *C. opobalsamum* – 5β,10α-hydroxy-2α-methoxy-6-oxoguaia-7(11),8-dien-8,12-olide, and 2α-methoxy-6-oxogermacra-1(10),7(11)-dien-8,12-olide (Shen et al. 2008a), cycloartane-24-ene-1α,2α,3β-triol (Shen et al. 2008b), guaia-6α,7α-epoxy-4α,10α-diol (Shen et al. 2007), cycloartane-24-ene-1α,2α,3β-triol-1,2-acetonide, 1β, 8β-epoxy-2α-methoxy-6-oxogermacra-9(10), 7(11)-dien-8,12-olide, and ent-4(15)-eudesmene-1β,6α-diol (Yang and Shi 2012) – are in Fig. 2.

Chemical Constituents and Bioactive Compounds of Myrrh

There are several compounds which are typical for *C. myrrha* – curzerene, 2-hydroxyfuranodiene (Morteza-Semnani and Saeedi 2003), furanoeudesma-1,3-diene (Brieskorn and Noble 1983; Jingai and Shangwei 1996; Dolara et al. 1996), lindestrene, curzerenone, furanodiene (Brieskorn and Noble 1982; Provan et al. 1987), 2-methoxyfuranodiene, 2-acetoxyfuranodiene (Brieskorn and Noble 1983; Monti et al. 1986), germacrene B (Baser et al. 2003), germacrene D (Brieskorn and Noble 1982), and T-cadinol (Zhu et al. 2003). Their structures are in Fig. 3.

Ahmed et al. (2006) succeeded in isolation of six new compounds from the oleo-gum resin of *Commiphora myrrha*, which were identified as myrracadinol A–C and myrracalamene A–C (Fig. 4).

The resin of *C. myrrha* gave a new cycloartane-type triterpene, cycloartane-1α,2α,3β, 25-tetraol (neomyrrhaol; Fig. 5), along with four known compounds, sandaracopimaric acid, abietic acid, 2-methoxy-5-acetoxyfruranogermacr-1(10)-en-6-one, and dehydroabietic acid (Shu-Lan et al. 2009). 2-methoxy-5-acetoxyfruranogermacr-1(10)-en-6-one, and dehydroabietic acid exhibited significant aromatase inhibitory activity with IC_{50} values at 0.2 μM and 0.3 μM, respectively. The four

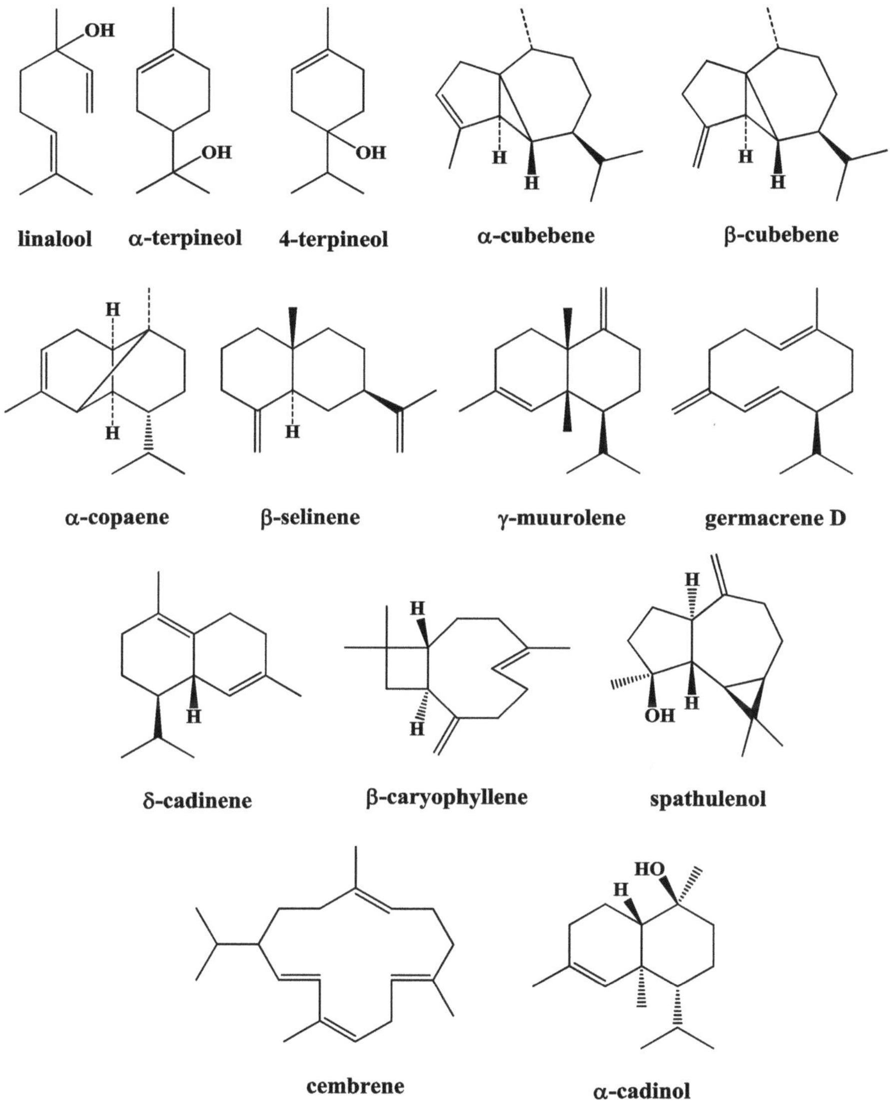

Fig. 1 Structures of typical known terpenic compounds extracted from *C. opobalsamum* (Al-Massarany et al. 2007; Hanuš 2012)

known terpenes had inhibitory effects on human umbilical vein endothelial cells growth with IC_{50} values of 0.122 μM (2), 0.125 μM (3), 0.069 μM (5).

After isolation from methanol extract, the air dried oleo-resin of *Commiphora myrrha* revealed n-dodecanyl myristate, henetriacosanyl laurate, and three new tetraterpenyl esters, myrrhatetraterpenolyl salicylate, myrrhatetraterpenolyl salicylate triacetate, and myrrhatetraterpenolyl vanillic acetate (Shuaib et al. 2013) (Fig. 6).

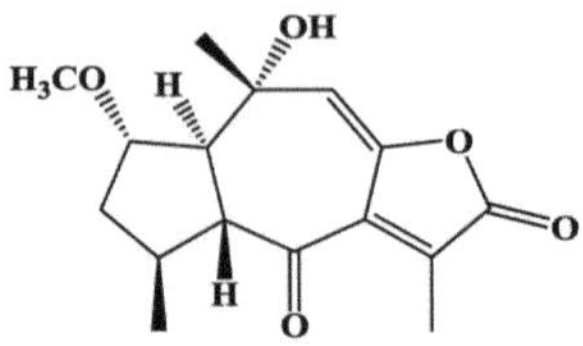

5β-10α- hydroxy-2α-methoxy-6-oxoguaia-7(11),8-dien-8,12-olide

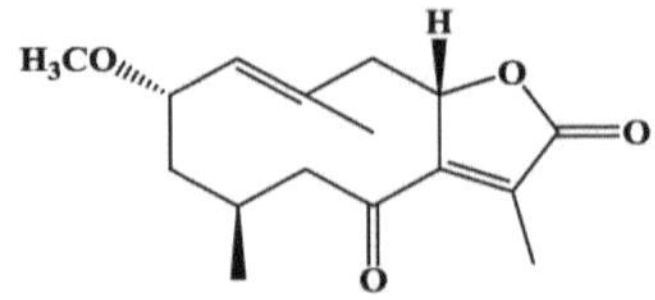

2α-methoxy-6-oxogermacra-1(10),7(11)-dien-8,12-olide

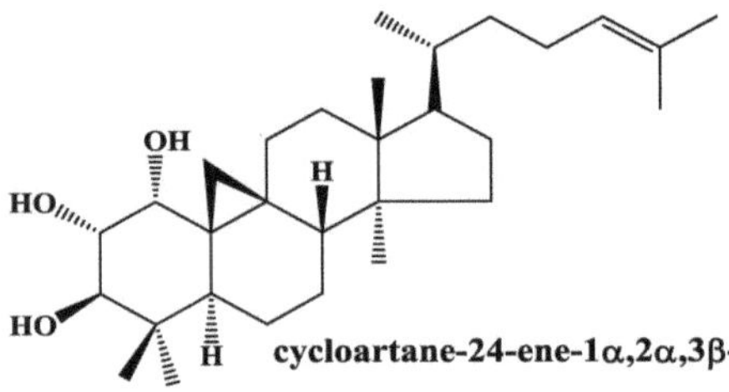

cycloartane-24-ene-1α,2α,3β-triol

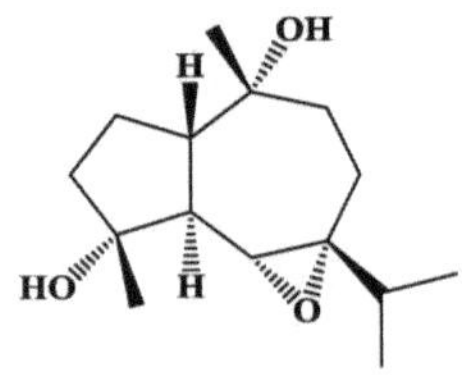

guaia-6α,7α-epoxy-4α,10α-diol

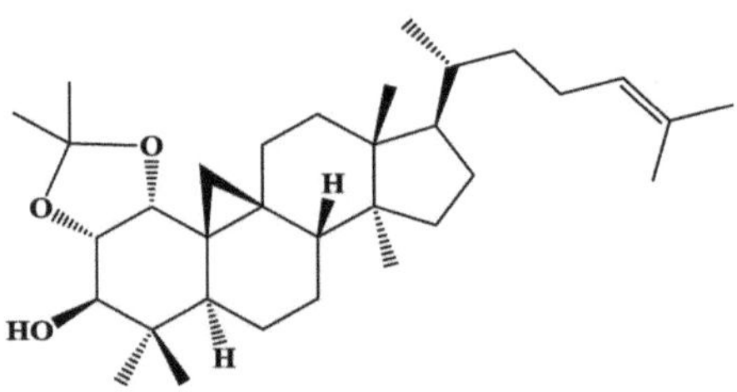

cycloartane-24-ene-1α,2α,3β-triol-1,2-acetonide

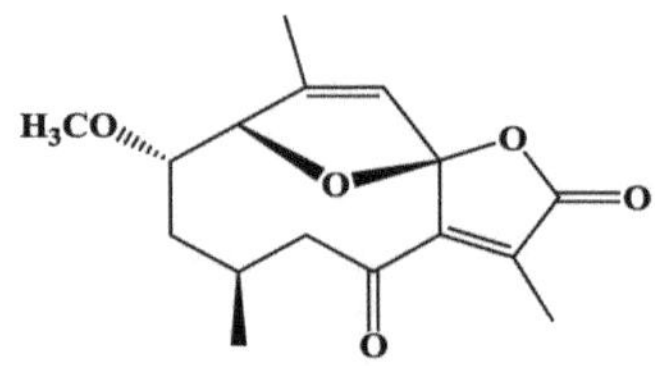

1β, 8β-epoxy-2α-methoxy-6-oxogermacra-9(10), 7(11)-dien-8,12-olide

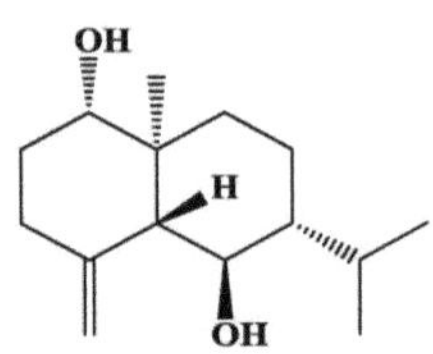

ent-4(15)-eudesmene-1β,6α-diol

Fig. 2 Structures of typical apharsemon compounds extracted from *C. opobalsamum* (Shen et al. 2007, 2008a, b; Yang and Shi 2012)

Methanolic extract of the resin of *Commiphora erythraea* gave after isolation a new cadinene sesquiterpenoid named agarsenone, which easily decompose to the agarsenolides A and B and myrrhone (Santoro et al. 2013). The isolation revealed also already known compounds, 1,10(15)-furanogermacra-dien-6-one, 1(10),4-furanodien-6-one, rel-3*R*-methoxy-4*S*-furanogermacra-1*E*,10(15)-dien-6-one, rel-2*R*-methoxy-4*R*-furanogermacra-1(10)*E*-en-6-one, dihydropyrocurzerenone, curzrenone, alismol, fura-

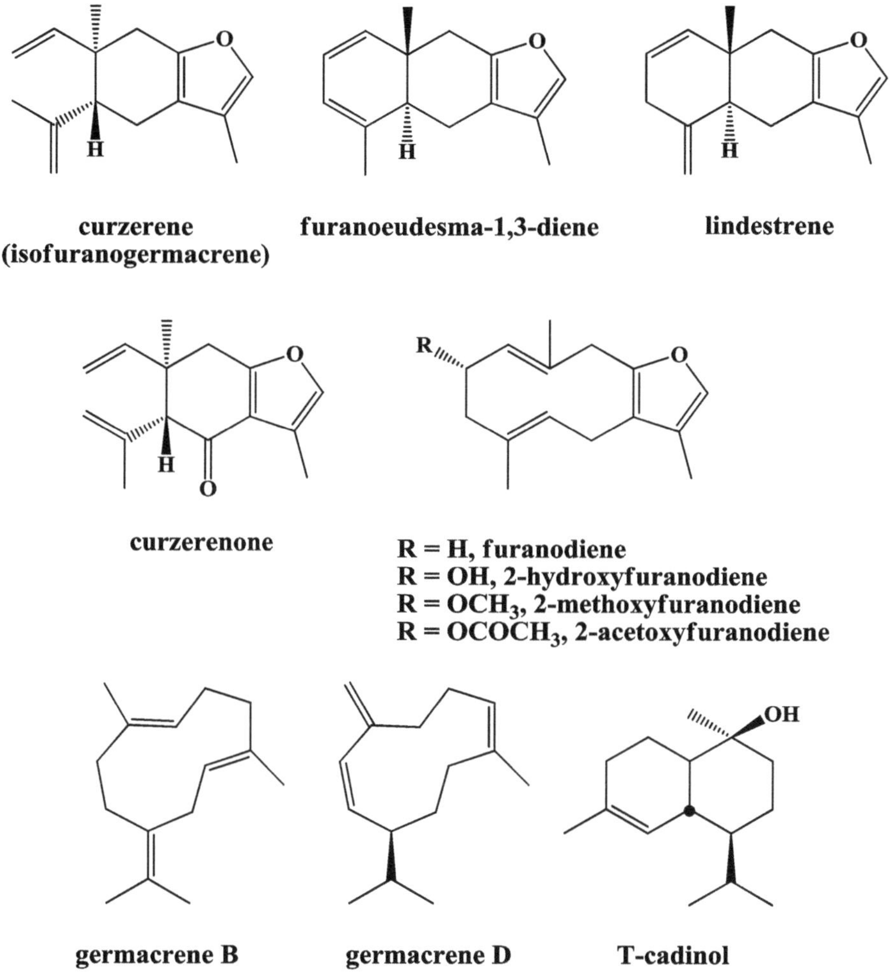

Fig. 3 Structures of typical myrrh compounds extracted from *C. myrrha* (Morteza-Semnani and Saeedi 2003; Brieskorn and Noble 1983; Jingai and Shangwei 1996; Dolara et al. 1996; Brieskorn and Noble 1982; Provan et al. 1987; Monti et al. 1986; Baser et al. 2003; Zhu et al. 2003)

noeudesma-1,4-dien-6-one and rel-1*S*,2*S*-epoxy-4*R*-furanogermaca-10(15)-dien-6-one (Fig. 8). Myrrhone, 1,10(15)-furanogermacra-dien-6-one, 1(10),4-furanodien-6-one, rel-3*R*-methoxy-4S-furanogermacra-1*E*,10(15)-dien-6-one, and rel-2*R*-methoxy-4*R*-furanogermacra-1(10)*E*-en-6-one are known for their antiradical, anti-inflammatory, and antiviral activity (Fig. 7).

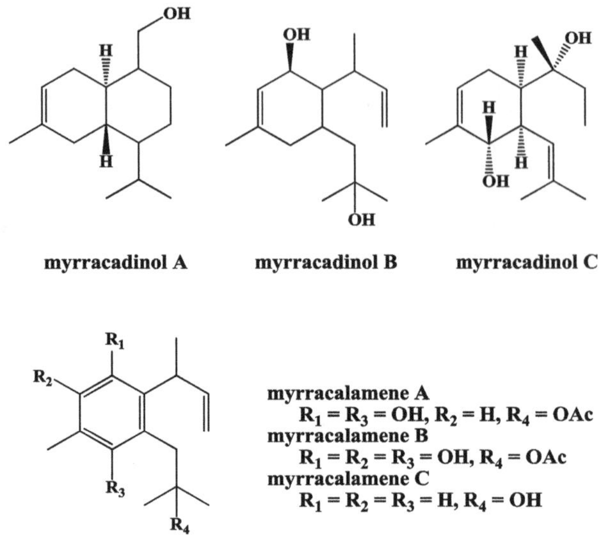

Fig. 4 Structure of new compounds isolated from *C. myrrha*

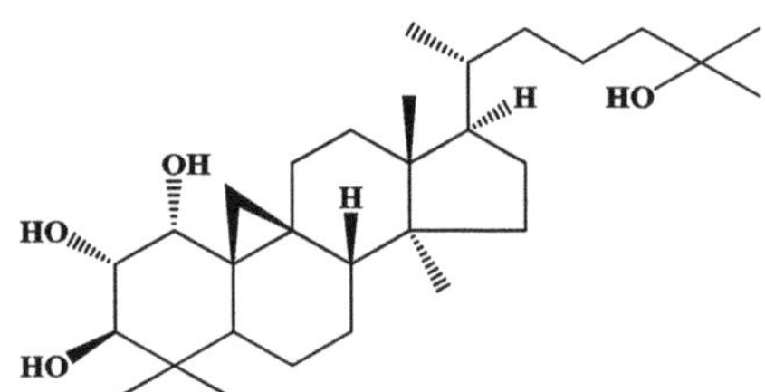

Fig. 5 Structure of compound isolated from *C. myrrha* (Shu-Lan et al. 2009)

Major Chemical Constituents and Bioactive Compounds of Olibanum

Figure 8 shows typical *Boswellia* compounds. These are α-boswellic acid, α-boswellic acid acetate, β-boswellic acid, and β-boswellic acid acetate (Fattorusso et al. 1983), incensole (Klein and Obermann 1978), incensole acetate (Basar et al. 2001), 24-noroleana-3,12-diene and 24-norursa-3,12-diene (Hanuš et al. 2007), incensole oxide and incensole oxide acetate (Hamm et al. 2005; Hanuš et al. 2007).

The analysis of the essential oil of *B. sacra* revealed the main compounds to be β-ocimene and limonene (Table 1; Al-Harrasi and Al-Saidi 2008).

Woolley et al. (2012) compared the main compounds in the essential oils of *B. sacra* and *B. carterii* with α-pinene as the main compound (Table 2).

Morikawa et al. (2010) isolated four new ursane-type triterpenes, olibanumols K, L, M, and N from *Boswellia carterii* (Fig. 9) together with 19 known triterpenes (epilupeol acetate, lup-20(30)-ene-3α,29-diol, glochidiol, lupeol, lup-20(29)-ene-

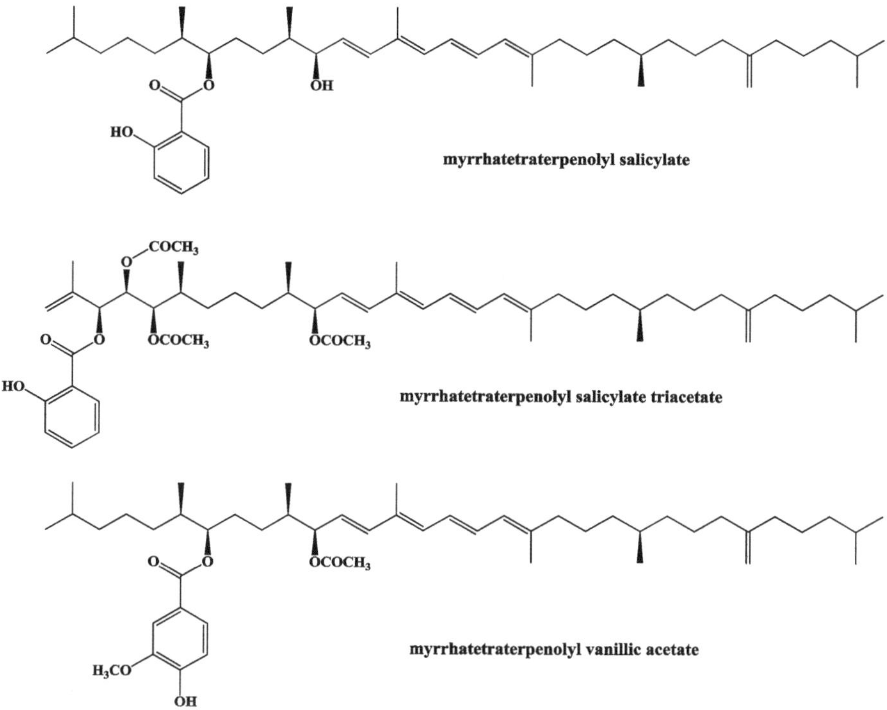

Fig. 6 Structures of compounds isolated from *C. myrrha* (Shuaib et al. 2013)

2α,3β-diol, 3β-acetoxylup-20(29)-en-11β-ol, lupenone, urs-9(11),12-dien-3β-ol, neoilexonol, neoilexonol acetate, urs-12-ene-3β,11α-diol, urs-12-ene-3α,11α-diol, dammarenediol II, dammarenediol II acetate, 3-*O*-acetyl-3β,20S,24-trihydroxydammar-25-ene, isofouquierol acetate, ocotillol acetate, 3β-hydroxymansumbin-13(17)-en-16-one, and mansumbinol).

From the nine compounds isolated from *Boswellia carterii*: 3-keto-tirucall-8,24-dien-21-oic acid and acetyl-11-keto-β-boswellic acid (Fig. 10) showed proliferation inhibition activity against Bel-7402, MCF-7, SMMC-7721, K562 and Hela (Li et al. 2010).

Wang et al. (2011) isolated from the resin of *Boswellia carterii* two new triterpenoids, 3-oxotirucalla-7,9(11),24-trien-21-oic acid (1) and 18Hα,3β,20β-ursanediol (Fig. 11) and another 15 already known compounds: α-amyrin, α-boswellic acid, β-boswellic acid, acetyl α-boswellic acid, acetyl β-boswellic acid, 9,11-dehydro-β-boswellic acid, 9,11-dehydro-α-boswellic acid, acetyl 11α-methoxy-β-boswellic acid, 11-keto-β-boswellic acid, acetyl 11-keto-β-boswellic acid, acetyl α-elemolic acid, 3β-hydroxytirucalla-8,24-dien-21-oic acid, elemonic acid, 3α-hydroxytirucalla-7,24-dien-21-oic acid, and 3α-hydroxytirucall-24-en-21-oic acid.

The Hougari Regular frankincense samples of *Boswellia sacra* were collected from the various locations in Oman and were also supplied by a trustful partner. A new boswellic acid derivative, 11α-ethoxy-β-boswellic acid and a new ursane-type triterpene,

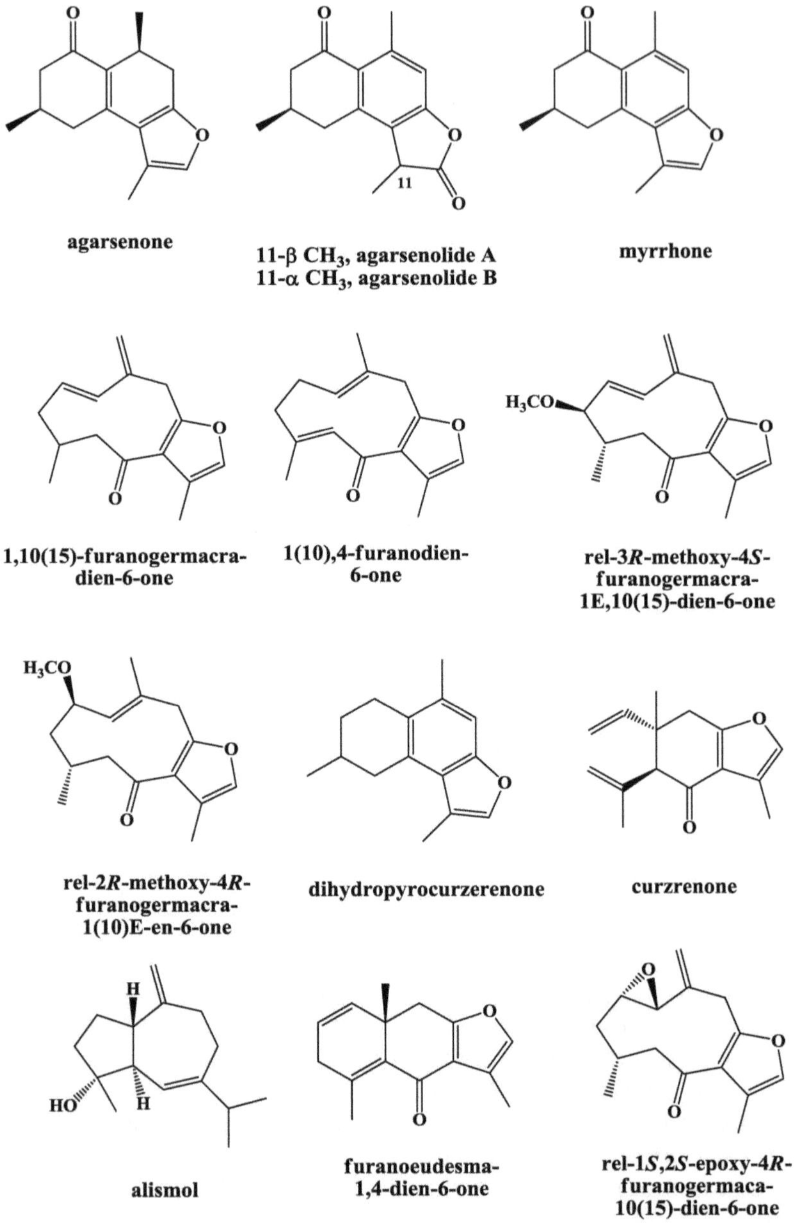

Fig. 7 Structures of compounds isolated from *Commiphora erythraea* (Santoro et al. 2013)

named nizwanone, were isolated from Omani frankincense *Boswellia sacra* Flueck. together with two known compounds papyriogenin B and rigidenol (Fig. 12), which were isolated from *Boswellia* spp. for the first time (Al-Harrasi et al. 2013).

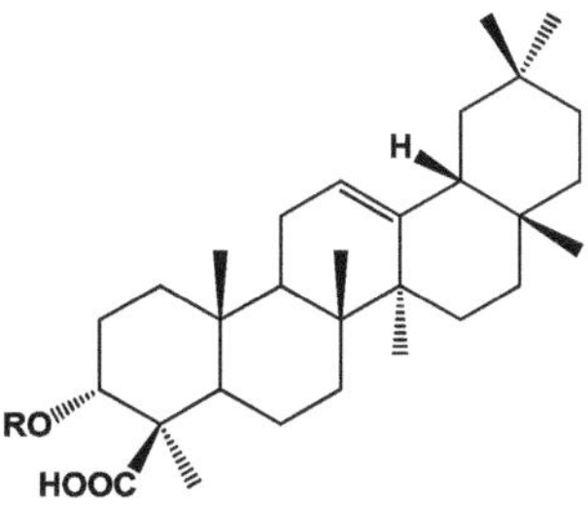

R = H, α-boswellic acid
R = CH_3CO, α-boswellic acid acetate

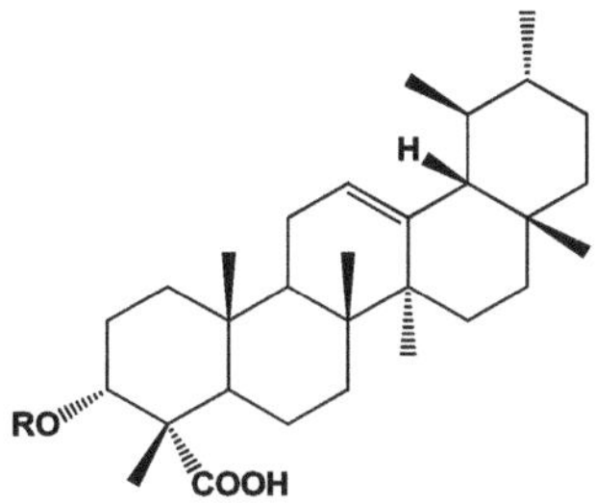

R = H, β-boswellic acid
R = CH_3CO, β-boswellic acid acetate

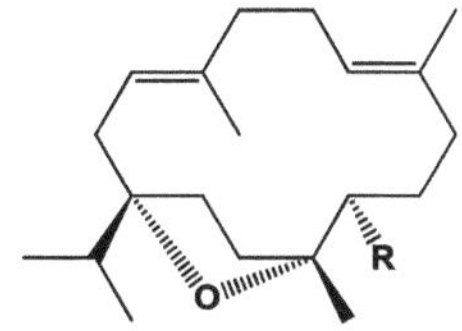

R = OH, incensole
R = $OCOCH_3$, incensole acetate

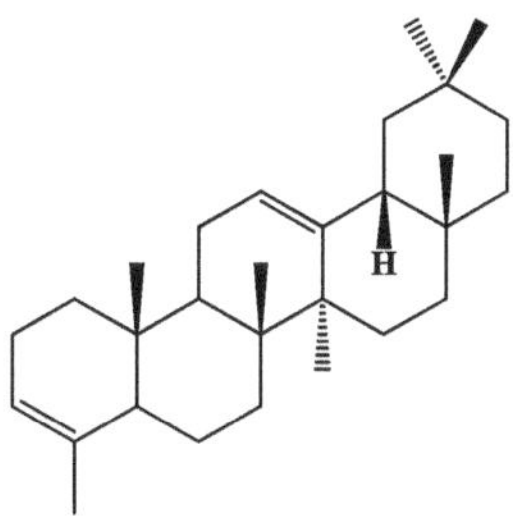

24-noroleana-3,12-diene

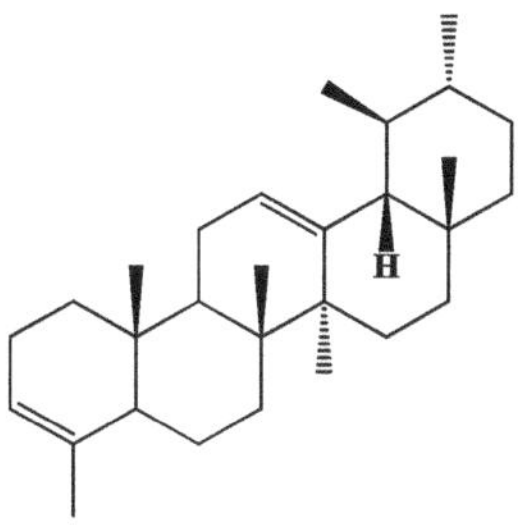

24-norursa-3,12-diene

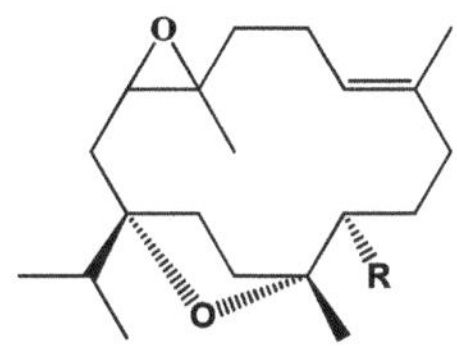

R = OH, incensole oxide
R = $OCOCH_3$, incensole oxide acetate

Fig. 8 Structures of typical *Boswellia* compounds extracted from *B. carterii* and *B. sacra* (Fattorusso et al. 1983; Klein and Obermann 1978; Basar et al. 2001; Hamm et al. 2005; Hanuš et al. 2007)

Table 1 The main terpenes identified in *Boswellia sacra*

E-β-ocimene (32.3 %)
Sabinene (5.2 %)
β-pinene 1.8 %)
Myrcene (6.9 %)
α-pinene (5.3 %)
Limonene (33.5 %)
γ-terpinene (1.0 %)
E-caryophyllene (0.9 %)

Al-Harrasi and Al-Saidi (2008)

Table 2 Comparison of the main terpenes in *B. sacra* and *B. carterii*

Boswellia sacra	*Boswellia carterii*
α-thujene (0.6 %)	α-thujene (7.9 %)
α-pinene (68.2 %)	α-pinene (37.3 %)
Camphene (2.1 %)	Camphene (0.8 %)
Sabinene (2.9 %)	Sabinene (4.9 %)
β-pinene (2.0 %)	β-pinene (1.8 %)
Myrcene (0.7 %)	Myrcene (7.3 %)
Limonene + β-phellandrene (6.2 %)	Limonene + β-phellandrene (14.4 %)

Woolley et al. (2012)

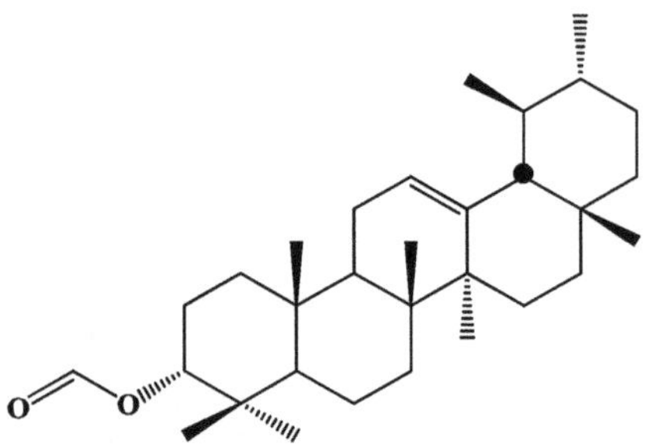

olibanumol K

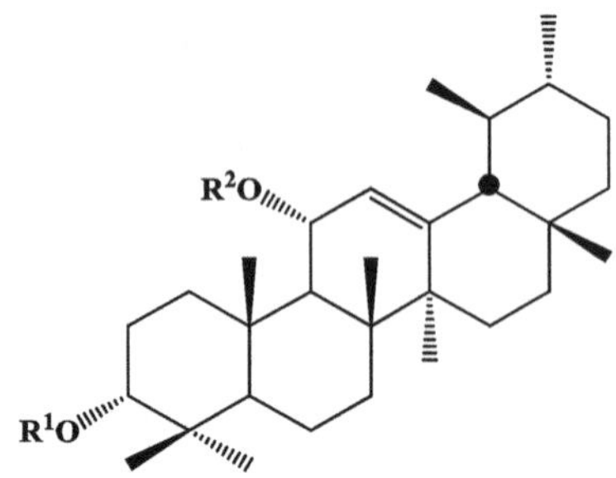

olibanumol L: R^1 = Ac, R^2 = H
olibanumol M: R^1 = H, R^2 = CH_3
olibanumol N: R^1 = Ac, R^2 = CH_3

Fig. 9 Structures of compounds isolated from *B. carterii* (Morikawa et al. 2010)

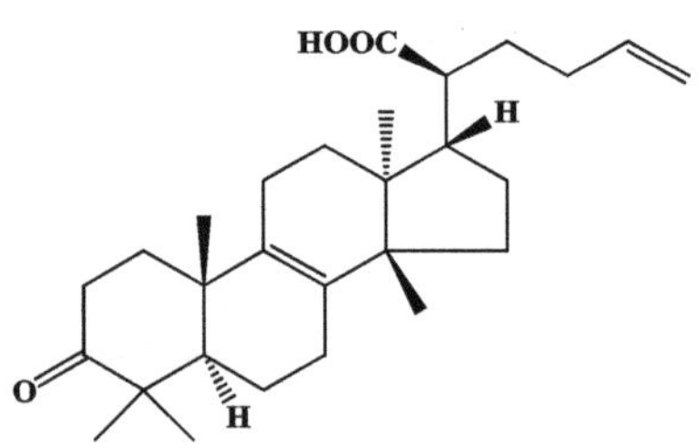

3-keto-tirucall-8,24-dien-21-oic acid

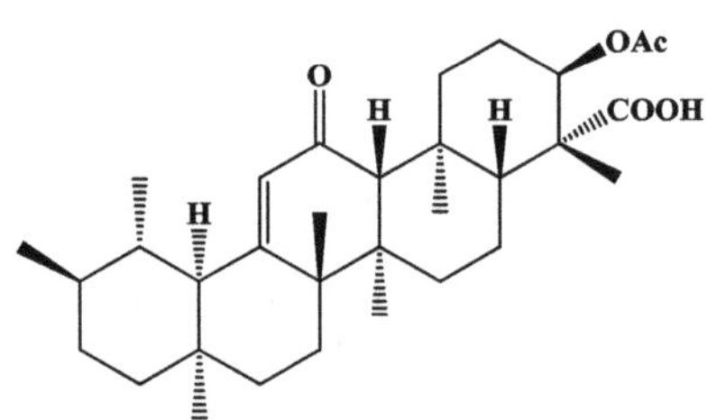

acetyl-11-keto-β-boswellic acid

Fig. 10 Structures of compounds isolated from *B. carterii* (Li et al. 2010)

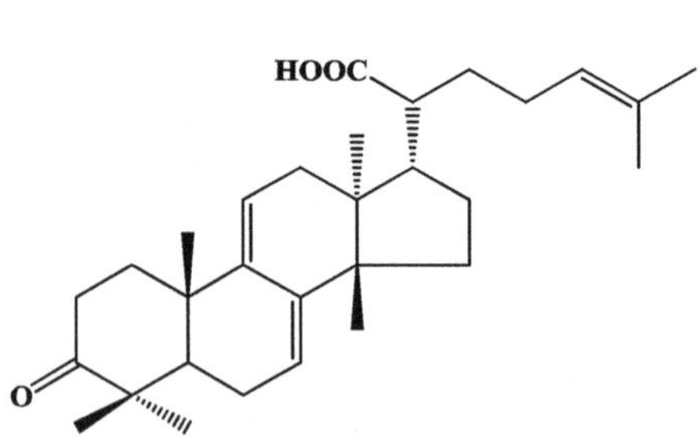

3-oxotirucalla-7,9(11),24-trien-21-oic acid

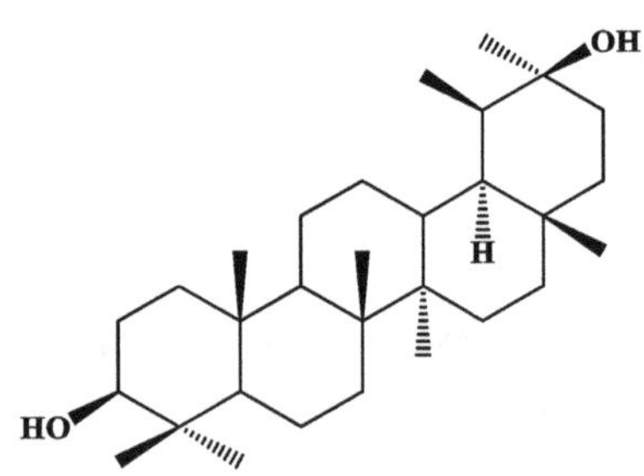

18Hα,3β,20β-ursanediol

Fig. 11 Structures of compounds isolated from *B. carterii* (Wang et al. 2011)

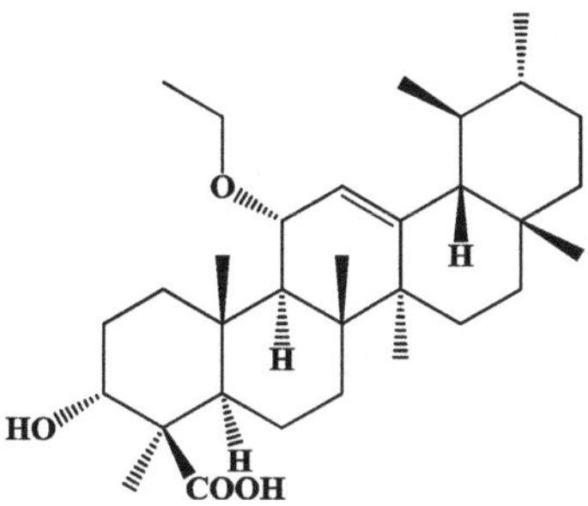

11α-ethoxy-β-boswellic acid

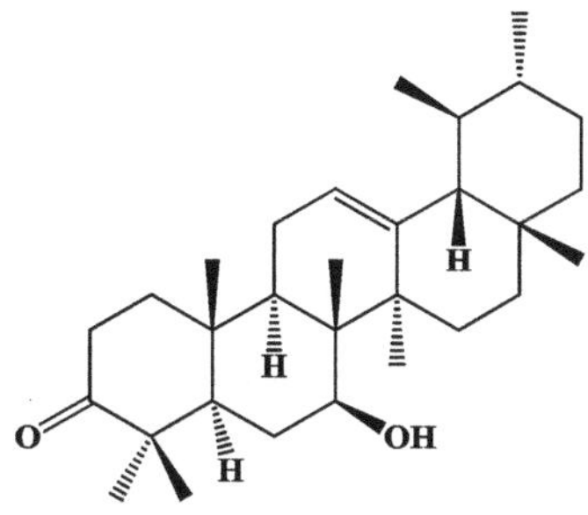

nizwanone
(7β)-7-hydroxyurs-12-en-3-one

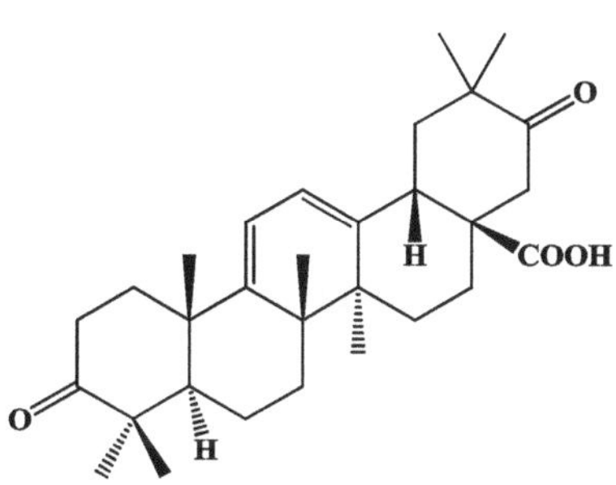

papyriogenin B
3,21-dioxo-oleana-9(11),12-dien-28-oic acid

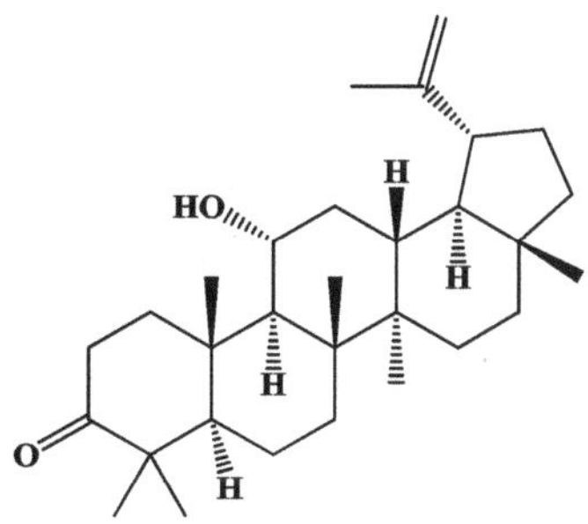

rigidenol
(11α)-11-hydroxylup-20(29)-en-3-one

Fig. 12 Structures of compounds isolated from *Boswellia sacra* (Al-Harrasi et al. 2013)

Morphological Description

Apharsemon

The below description was given by Forsskål, Peter on 1775 in the following report: Flora Aegyptiaco-Arabica, sive, Descriptiones plantarum. Ex Officina Mölleri. Hauniae. Ternate leaves, entire; one-flowered peduncle. Leaves are 3–5 foliate, terminal leaflets obovate (rarely elliptic) about 1 cm long, base attenuate, apex rounded to emarginate; margin entire, lateral leaflets fully developed, about the same size as the terminal leaflets. Leaves are alternate and ternate. Thread-like petiole. Leaflets are close together at the base; the middle ones are great and obovate; the lateral ones are ovate. All of them are glabrous, flat, entire. Description: Medium-sized tree. Branches are strongly divaricate. Bark is smooth, ash grey; when wounded, it produces a balsamic juice (resin) and odor. The plant has long and slender branchlets without spines;

Calyx: Perianth – one-leaved, campanulate, placed underneath, 4-toothed, persistent: short teeth, appressed to the corolla (Picture 6).

Peduncles in branchlets are terminal, solitary or several/many, thread-like, one-flowered

Flowers are in clusters, drupe apiculate (Picture 7).

Corolla: 4 petals, linear, erect, converging to a quadrangulate prism, obtuse, at the margin bent slightly inwards, at the base – more.

Nectary – a fleshy ring between the stamens and the ovary: forming a twisted small channel, circular, close to the ovary, yellow. On the outside there is a little elevation between each stamen.

Filaments – 8, inserted between petals and nectary, thread-like – subulate, in the middle they are curved inwards, shorter than the corolla.

Anthers are oblong, didymous, and yellow.

Ovary is upper/placed above, small.

Style is extremely short, narrow.

Stigma is obtuse, 4-angled.

Pistil is shorter than the filaments.

Pericarp – berry, ovate-acute, glabrous, 4 sutures, or rather valves, Pulp is tough, viscid. With 2 cavities on the inside, one-seeded: often with 1 cavity, empty, often united.

Seed is ovate-acute, solitary (Picture 8).

Flowers seem monoecious. In some of them the anthers were fresh, the ovary – green and the stigma – narrow. In others the anthers were withered, the ovary – dark/dark brown, grooved: the stigma – thick, hyaline, 4-angled: the seed – blackish, abortive: perhaps the balsamic juice hinders fructification; thus several capsules were big, nevertheless empty.

The tree was first seen near the caravanserai/hostel of Oude, not far from the city of Haes in Yemen.

Arabic: Abuscham.

Myrrh

The *Commiphora* species of *gileadensis and myrrh* found in Yemen and Oman share many features. They are typically small trees about 2–4 m high with a relatively stout, dark-green trunk and thin, papery, peeling bark. The genus *Commiphora*, together with *Acacia* and *Grewia*, provides scrub cover on most dry stony hills up to about 1,500 m. Most species are drought resistant. Photographs of several species, including the *C. gileadensis*, growing in Yemen and Oman are found in Al-Hubaishi and Muller-Hohenstein (1984).

Many species in the *Burseraceae* are woody perennial trees or shrubs with fragrant resins in the leaves and/or stems. From a botanical perspective, such resins are known to repel herbivores, and some resinous extracts have insect repellant and insecticidal properties (Birkett et al. 2008; El Ashry et al. 2003). These properties are used by the indigenous peoples where these plants grow naturally. In the New World, the most famous of the resins from the *Burseraceae* are forms of copal, produced from various species of *Bursera* from Mexico and Central America. Like frankincense and myrrh, copal (comes from the sturdy Bursera tree in the regions of southern Mexico) is used as incense and for ritual and medicinal purposes (Alcorn 1984).

Pliny describes the myrrh tree in this way (Book 12, Chapter 34): "The tree grows to a height of five cubits (length of the forearm, 43–53 cm) and has thorns upon it: the trunk is hard and spiral, and thicker than that of the tree, and much more so at the root than at the upper part of the plant." According to van Beek (1958) and Zohary (1982), the myrrh tree normally grows to a height of about 2 m, but, at higher elevations, it is reported to reach a height of about 5 m. The tree is in foliage for only a short time after the rainy season; during the remainder of the year, it is leafless. The leaves are small and single or often 3-foliolate, with two tiny leaflets at the base. The myrrh species is characterized by a terminal leaflet that is up to 1.5 cm long and the lateral leaflets are rudimentary and entire. The myrrh tree is completely different from the frankincense tree, except for its production of spice gum resin (Wood 1997). The myrrh tree has a peeling bark; the underbark is green and photosynthetic. The gum resin of the myrrh tree is yellow to brown red. All myrrh trees grow wild in nature and are not raised in agricultural plantations (Hepper 1992).

Olibanum

The *Boswellia* species "resembles a shrub more than a tree. Some of its species have no central trunk – the branches emerge near the ground – and it grows to a height of seven to eight feet" (Thomas 1932, quoted by van Beek 1958).

Boswellia sacra Flueckiger (syn. *B. carterii* Birdw.) is considered to be the source of the biblical frankincense (Tucker 1986). This tree grows wild, in dry regions, particularly in Somalia, and the Dhofar Valley, Oman (van Beek 1958). The Dhofar Valley, really a high plateau, is a lush, green oasis, watered by the monsoon rains, in contrast to the barren, stony desert surrounding it. It is regarded as having the optimal conditions for *Boswellia sacra* and produces the highest-grade resins, named Silver and Hojari (Ghazanfar 1994).

The Silver frankincense of the B. *sacra* gives a better scent in the dry desert air. The color of the resin and size of the tears also dictates its commercial value – the pale, large clumps are more expensive. *Boswellia sacra* also grows in the Nejd district of southern Oman, where the foggy climate produces an even more expensive resin and very slow tree growth, resulting in large white clumps of resin (Morris 1989).

B. serrata is taller than the other *Boswellia* trees and has a straight trunk. The scent of the resin extracted from *B. serrata* is quite distinct from that obtained from the other *Boswellia* trees and is heavier than the African resins, more of an orange type of scent, while the *B. sacra* resins have a lighter, lemon scent. The difference in odor between the various *Boswellia* tree resins is due to their complex sesquiterpenes (Tucker 1986).

Geographical Distribution

Apharsemon

Trees of the *Commiphora* genus are found in southern Arabia and northeast Africa and India (Hanuš et al. 2005). Like *Boswellia*, they belong to the resinous *Burseraceae* family, which occurs in tropical and subtropical areas in both the Old and New World. There are about 18 genera and 700 species in the family (Weeks and Simpson 2007).

Sometime during the Iron Age, probably at around 1000 BCE, after the Queen of Sheba gave King Solomon incense plants, the cultivation of a few wild plants producing incense had begun in the kingdom of Judea, in the region of Jericho and Ein Gedi (Josephus, see Ben-Yehoshua et al. 2012). Archaeological evidence suggesting the extraction of this special incense plant was found by Mazar et al. (1973) at Tel Goren near Ein Gedi, and some of the equipment used in the processing of this plant to obtain the spices has been discovered (Dayagi-Mendels 1989). These strange plants were brought to an arid ecological system much different than their original location at the Kingdom of Sheba. The farmers living in Jericho and Ein Gedi and their kings realized the economic potential of adapting the new plant to their land. They domesticated the wild plants, intensified their cultivation, created and developed unique production, advertising, and marketing systems that made the apharsemon a most lucrative crop.

Its other unique feature was the fact that it was cultivated by farmers around the Dead Sea Basin in Judea, unlike the other spices derived from myrrh and olibanum that were wild trees. This new plant apharsemon may be considered as a Judean-developed substitute for the classical spices, myrrh and frankincense, which were expensive imports from Sheba. Regardless of the cost of these spices, the Children of Israel were commanded in the Bible to use these spices and many others (e.g., Exodus 30, 34; Leviticus 2,1 and 24, 7). Thus, the gift of the spice plants the Queen of Sheba gave King Solomon the opportunity to try growing his own spices. This ability became especially important with his establishment of the Holy Temple in Jerusalem, replacing the old and much more humble tabernacle located at Shilo. The ritual ceremonies and animal sacrifices in the temple demanded a much larger supply of deodorant, incense, and preservatives of animal meat from rapid spoilage (Ben-Yehoshua and Rosen 2009). It made sense for Solomon to grow these plants, rather than import their products at very high prices, and to establish their cultivation around the Dead Sea Basin in a climate as similar as possible to their original one.

Facilities for the manufacture of perfumed oils, presumably apharsemon cultivated at the site, were discovered in the Dead Sea oasis of Ein Gedi at Tel Goren, dating to the late seventh and early sixth centuries BCE. The excavators believed that during the reigns of King Josiah and his successors, the oasis was a royal estate, and kings were anointed with apharsemon from the time of Josiah's reign (Mazar et al. 1966). The existence of gatherers of apharsemon at the time of the destruction of the First Temple is recorded in the Book of Jeremiah (52:16), where it is stated that Nebuzaradan, the captain of the guard, left the poorest Jews to be vinedressers (*kormim*). In the Talmud, Rabbi Joseph says that the *kormim* were actually gatherers of the apharsemon from Ein Gedi to Ramah (Tractate *Shabbath 26, 71* Talmud).

This unique apharsemon bush was cultivated in only few orchards in a very small area around the Dead Sea Basin, and nowhere else in the world. Most probably at the beginning of this domestication with the king's or ruler's command, the relevant protocols were developed for cultivation and for the efficient, complicated methods of resin extraction. Today these special plants are no longer cultivated. Their products are harvested only in the wild from plants growing in special ecological niches in Ethiopia, Somalia, Arabia, and India and possibly a few other neighboring countries (Hepper 1969).

Groom said that in medieval times, there was the Makkah balsam tree and its main center of production was in a region Hijaz, particularly at ak Arg - about halfway between Makkah and Madinah (Groom 1981). Miller and Cope (1996) reported several other locations of growth of the *C. opobalsamum* in Arabia. Indeed the resin of this plant, called in Arabia balasan, was used pharmaceutically throughout the medieval period until the eighteenth century CE (Duncan 1804).

Only after the identification of the *Commiphora opobalsamum* by Forsskal and Linaeous did observations appear that this plant is often associated with *C. myrrha*, growing on dry stony hills in the Tihama foothills in Yemen. It grows up to an elevation of 1,200 m and also on the stony slopes south of the Hays Mountains but has not been found north of Jebel Ash Sharafayn.

Although Pliny states that the Romans themselves did not see the plant that produces frankincense and myrrh (Bostock Book 12, Chapter 31), descriptions by contemporary Greek and Roman historians provided information on these plants. At that time, the source of the incense was from trees that grew wild in southern Arabia and from the kingdom of Sheba, first cited in the biblical description of the visit of the Queen of Sheba to King Solomon (I Kings 10:1–2; II Chronicles 9:1). This nation, Sheba, is in the list of the sons of Joktan (Genesis 10:26–29), and it is interesting that the name of Abraham's last wife was Ketura, meaning "incense" (Genesis 25:1). Furthermore, the names of the children of Ketura are the names of some of the Arab tribes in Arabia: Sheba, Dedan, Midyan, and Aifa (Genesis 25:2–4). The children of Ishmael, the first son of Hagar and Abraham, were Bashmath and Mibsam (Genesis 25:13), meaning, in Hebrew, "spice" (the Hebrew word *bosem* being the root basis for these two names).

The earliest Greek accounts of the Sabaeans and other south Arabian people are of the third century BCE (Groom 1981). Eratosthenes (276–194 BCE), quoted in Strabo XV 4.2 (Jones 1924), indicated that the extreme south of Arabia, opposite Ethiopia, is inhabited by four great nations: the Minaeans on the Red Sea, whose chief city was Carna; the adjacent Sabaeans, whose capital was Mariaba (biblical

Mariab); the Catabanes; and, farther east, the people of Hadramut, with their city Sabota. The Catabanes produced frankincense and Hadramut myrrh, and there was a trade in these and other spices with merchants who made the journey from Aelana (Elath, on the Gulf of Akaba) to Minaea in 70 days. The Gabaeans (Pliny's Gebanitae Book 12, Chapter 32) took 49 days to go to Hadramut (Artemidorus, 100 BCE, quoted in Strabo-Jones 1924, XVI: 4:4). The Minaeans formed a political and linguistic island in the Sabaean country. Pliny states (Book 12, Chapters 30, 51) that frankincense was collected at Sabota (the capital of Hadramut) and exported only through the Gebanites, whose kings received custom dues on it (Pliny, Book 12, Chapter 32).

Strabo provides a similar account of the wealth and trade of the Sabaeans and their capital, Mariaba, adding that each tribe received the wares and passed them on to its neighbors as far as Syria and Mesopotamia (Jones 1924-XVI: 4:19). The Sabaeans also had colonies in Africa. Abyssinia probably was settled by the Sabeans from south Arabia, as indicated by the similar language and writing. This interrelation between the Kingdom of Sheba and the Horrn of Africa also contributed to the spice trade, as the plants were grown in both areas (Groom 1981).

The source of these important ancient spices was not commonly known in antiquity, and the Arabians involved preferred to keep this information secret. This led to confusion among classical writers such as Theophrastus, Artimedorus (as related by Strabo), and Diodorus Siculus (of Sicily), a first-century Greek historian, who maintained that frankincense grew in the land of the Sabaeans (Van Beek 1958).

In actuality, frankincense grew in the Horn of Africa (Somaliland) and farther east in Arabia, in the region of Dhofar, Oman. The Minaeans and other peoples of the Arabian Peninsula, such as the Qedarites, the Gerrhaeans, and the Nabateans, maintained control over the inland trade routes to the Mediterranean and particularly to Egypt. The trade was never the monopoly of one people. According to Strabo: "Those tribes who live close to one another receive in continuous succession the load of spices and deliver them to their next neighbors as far as Syria and Mesopotamia" (Jones 1924, Book XVI).

Biblical citations allude to Sheban trade in incense and perfumes, gold and precious stones, ivory, ebony, and costly garments (Ezekiel 27:15, 20, 22; Job 6:19). These passages attest to the wealth and importance of Saba (Sheba) from the days of Solomon to those of Cyrus.

Myrrh

Classical sources refer to myrrh as growing in the Ma'in, Hadhramaut, Qataban, and other areas of southern Arabia, and these sources, as well as modern investigation, indicate that the production of myrrh was confined to these areas. However, the current growing areas of the myrrh district are centered in the west and central part of Somaliland (Van Beek 1958) and also in India.

Olibanum

Van Beek (1958) concludes that the geographical distribution of the frankincense tree is governed by definite rainfall patterns and soil factors.

Boswellia carterii is also referred to as olibanum and dragon's blood. Some authorities regard this species as the biblical frankincense and the same species as the B. sacra, but there is some dispute over this.

Boswellia papyrifera Hochst grows in Ethiopia and Sudan but is not marketed in the western world. The resin is transparent and oilier than the other resins.

Boswellia serrata Triana & Planch. (Roxb.) Colebr is the *Indian frankincense*, considered by some to be of inferior quality. The golden brown color resin is soft and hardens slowly; it is mainly burned as incense but also used in Ayurvedic medicine (Miller and Morris 1988). The *B. serrata* is taller than the other *Boswellia* trees and has a straight trunk. The scent of the resin extracted from *B. serrata* is quite distinct from that obtained from the other *Boswellia* trees and is heavier than the African resins, more of an orange type of scent, while the *B. sacra* resins have a lighter, lemon scent. The difference in odor between the various *Boswellia* tree resins is due to their complex sesquiterpenes (Tucker 1986).

By the eleventh century BCE, the demand for the resin of the *Boswellia* trees was well developed, resulting in the improvement of overland routes. Historic records of trade in biblical times, and earlier, link this specific shrub with the trade routes. Thesiger (1959) writes: The civilizations of Arabia for 1,500 years had depended for their prosperity on frankincense gathered on the mountains of Dhaufar.

The price of the *B. sacra* varies according to the grade, the most expensive being the Hojari frankincense locally available in Oman. The scent of the Hojari is greatly appreciated in the damp air of Europe, although, to the Arabian dealers.

Recently, the ancient site of Punt, which was the target for the Hatshepsut in her search for spices, has been identified as Eritrea and eastern Ethiopia, based on work of Nathaniel Domino and Gillian Leigh Moritz of the University of California, Santa Cruz, with oxygen isotope tests carried out on the fur of two ancient Egyptian mummified baboons imported by Hatshepsut and compared to baboons found in other countries. The isotope values in baboons in Somalia, Yemen, and Mozambique did not match. It was estimated that the mummified baboons dated from about 3,500 years ago, when Hatshepsut's fleet sailed to Punt and brought them back as pets (Bressan 2013).

Present Day Production in Africa

Teketay (2003) reviewed the frankincense and myrrh resources of Ethiopia suggesting these resources could contribute towards the conservation and management of frankincense as well as local ecosystems. However, a study of *B. papyrifera* in northern Ethiopia in 2002 by Gebrehiwot et al. gave a depressing picture of the decline of this species, despite what had been a flourishing market for extracted incense, and the large demand from churches in Ethiopia and Europe. Human encroachment, and unrestricted grazing and harvesting had greatly reduced the population of this tree.

The Forest Ecology and Forest Management Group, at Wageningen University, the Netherlands, inaugurated a project (2006–2010) to promote natural regeneration of *B. papyrifera* in Eritrea, the semiarid areas of Ethiopia, which is more than half of total land area and became an independent state. In 2007, the estimated mean frankincense annual yield was 127 kg/ha-for closed forest land and 85 kg/ha for open forest land. The price for exported frankincense was estimated at $53 for closed and $39/ha-year for open sites, with rural households earning about 74 % of this for tapping and collecting the resin. This benefit was considered to be superior to alternative land use (Wageningen University). Scholarly historian Thieret (1996) suggests that total yearly production of myrrh is perhaps 500 t, and frankincense, 1,000 t. Recently, U.S. imports ran 5–20 t. The UK imports about 30 t frankincense each year, one perfume manufacturer alone consuming 5 t annually.

Most frankincense comes from Somalia (following bananas and cattle as leading exports) where it provides work for some 10,000 Somali families, and some is gathered in Arabia. The actual export amounted to 1,000–1,500 t annually. Most of the frankincense is marketed in Saudi Arabia, Yemen, and Egypt, which are the major markets, with lesser quantities marketed in other countries.

Despite the evidence that small holders would be able to derive obvious economic benefits from frankincense, constraints have been revealed. Local people have been producing and trading frankincense for centuries to diversify their income sources. However, production of frankincense varies considerably among the producing dryland regions of the country. *Boswellia papyrifera* comprised 51 % of the species composition of the vegetation of the district with a potential of 254 kg/ha/year. The total annual frankincense production potential of the district was estimated to be 79 thousand tonnes. The question is why the export of the Somali amounts only to around 1,500 t. Almost all inhabitants do not benefit economically from the species due to cultural influence, unattractiveness of income from frankincense compared to other economic activities in the area, property tenure, government policy on incense production, poor knowledge on frankincense production, and unawareness of the potential of frankincense as a source of income. The absence of direct economic benefits for the local people from the woodlands is triggering widespread degradation mostly from human induced fire, improper forest use, and agricultural land expansion (Lemenih et al. 2007).

Currently, replanting projects in the wild are underway in Ethiopia and Somalia. However local scientists in these countries report that the future of these projects seemed far from being assured. Examination of various databases of exports from countries that used to sell these spices, such as Oman, Yemen, Ethiopia, Somaliland [Index Mundi, the FAO and USDA], shows that spices are such a minute item in the export lists that they do not appear at all.

Recent studies have indicated that frankincense tree populations are declining due to over-exploitation, since heavily tapped trees have been found to produce seeds that germinate at only 16 %, while seeds of trees that had not been tapped germinate at more than 80 % (Howes 1946; Bergstrom et al. 1982; Miller and Morris 1988; Ghazanfar 1994).

Ecological Requirements

The ecology of the geographical origins of these three medical plants is rather varied. These three plants are growing in the wild in several new locations, but the tendency is to relate these plants to their origin at the Southern Arabian Peninsula and to the African Horn: Somalia, Kenya, Tanzanya and Ethiopia and possibly also India.

The Apharsemon has moved to Judea and was cultivated there for a long period of about 1,500 years. Then it moved to Matariya in Egypt for a period of several hundred years. Consequently it became extinct, but its past reputation and taxonomical research identified several botanically related plants in various parts of the Arabian Peninsula, whose exact taxonomical identification is still waiting for more research.

Olibanum and myrrh trees are considered unusual for their ability to grow in environments so unforgiving that they sometimes appear to grow directly out of solid rock. The means of initial attachment to the stone is not known but is accomplished by a bulbous disk-like swelling of the trunk of the frankincense.

This disk-like growth at the base of the tree prevents it from being torn away from the rock during the violent storms that frequent the region they grow in. This feature is slight or absent in trees grown in rocky soil or gravel. Each species has its particular characteristics and quality, depending on its growth environment, its harvesting procedures and type of resin produced. Van Beek (1958) concludes that the geographical distribution of the frankincense tree is governed by definite rainfall patterns and soil factors. There are various and many grades of the resin extracted from the *Boswellia* trees, related to the exact climatic conditions prevalent where they are grown, often the most deprived soils producing the highest quality resins. A soil containing limestone and dry conditions is the preferred growing environment for the *Boswellia* (Bergstrom et al. 1982, quoted in SEPASAL).

Harvest Practice

Apharsemon

The writings of Pliny, Josephus and others on apharsemon contain details on the special techniques regarding resin harvest of these special plants (Rosen and Ben-Yehoshua 2007; Ben-Yehoshua and Rosen 2009). Pliny noted that the producers of apharsemon possessed special techniques to extract the apharsemon exudates:

> … an incision is made in it with a piece of glass or a stone, or with a knife made of bone – it strongly dislike having its vital parts wounded with steel and die off at once, though it can

> stand having superfluous branches pruned with it. The hand of the operator making the incision must be poised under skilful control, to avoid inflicting a wound going below the bark.

Josephus (*Wars of the Jews* 2, 6; 6) mentioned these stone knives (Rosen and Ben-Yehoshua 2007).

Traditional tapping methods, used where the resin is collected from the wild plants, employ crude incisions by axing, which injure the cambium and shorten the tree's life span. In the twenty-first century the axe is still used to extract the resin from both frankincense and myrrh, a crude practice that endangers these trees (Ben-Yehoshua and Rosen 2009). Thus, we conclude that those Judean farmers had solved this injury problem by developing a special tool kit from glass, stone, and bone, to perform harvest operations on the bush. Iron tools were relegated to pruning. There is current support for Pliny's observation of the "the lethal effect of iron on the bush." (Book 12, Chapter 54). Iron tools imbedding in the bark, cambium, and the sap overloads them with soluble iron.

Myrrh

Harvest of myrrh is similar to that of frankincense. Resin exudes from the branches after an incision is made in the bark of the trunk or of branches, allowing the pale yellow liquid gum to run out and accumulate on a mat or container next to the trunk. As the gum hardens it turns a reddish brown color, in the shape of tears. Pliny wrote of a "bundle of tears," the form in which the myrrh is marketed (Book 12, Chapter 32).

Olibanum

Both Theophrastus (372–287 BCE) (Hort 1916) and Pliny (23–79 CE) (Bostock 1855) reported on methods of harvesting the resin, which have hardly changed. The trees were wild plants grown in a few isolated locations and no actual cultivation practices were reported. The resin is harvested by wounding the trunk or big branches by scraping about 2 cm of the bark with a tool (*mengaff*), which results in the resin exuding on to the trunk. This resin hardens into clumps in the shape of tears, as it dries. After 2 weeks the harvester returns to the tree and collects the accumulated resin, which is left to harden for a few weeks before being brought to market. The first tears of clear resin collected are the best quality – the extracts running down the tree or onto the ground are not so fine, as mentioned by Theophrastus. The aroma of the tears is prized for their healing abilities and are also said to be more pleasing to the gods.

Early records of harvesting in the Oman described slaves gathering the resin since it was unpleasant work, in hostile conditions. Later, Pliny added to his records that the harvesting was carried out by a small group of elite natives, the privilege being an inherited one and jealously guarded. This group was celibate during the harvest time and ordered to avoid pollution either by contact with women or dead bodies (Book 12, Chapter 30). Herodotus relates the tall tale of frankincense that the local tribesmen passed off to gullible foreigners: "When they gather frankincense, they burn storax (the gum which is brought into Greece by the Phoenicians) in order to raise a smoke to drive off the flying snakes; these snakes, the same which attempt to invade Egypt, are small in size and of various colors, and great numbers of them keep guard over all the trees which bear the frankincense, and the only way to get rid of them is by smoking them out with storax" (Rawlinson 1859 Book 3:107).

Usually there are two harvests each year, giving the tree ample time to recover, thus ensuring high quality resin. Tapping is done two to three times a year. High quality resin can be visually discerned through its level of opacity.

There is a rich folklore surrounding the harvesting and use of the Dhofar frankincense, and full details are documented in SEPASAL. Daily life, and particularly harvesting in the Dhofar Valley of the Oman, is accompanied by ancient and intricate rituals relating to the wealth attained by the sale of the frankincense and its daily use for medicinal, religious, and social purposes, especially rites of passage, such as childbirth, weddings, and funerals. Childbirth in the Dhofar Valley is accompanied by the burning of incense during the 40-day period following the birth and for treating the mother to prevent infections and other problems. Weddings, religious celebrations, the reception of guests to the home in Dhofar, and formal occasions are accompanied by a welcoming incense-burning ceremony, promoting social harmony and peace. The women of Dhofar use the incense to smooth and oil their hair and also to sweeten their breath. The soot of the burning resin was collected and used for eye makeup. The soot from the incense burner was used in Arab communities to mark tattoos on the skin after piercing.

History, Traditional Use, Ritual and Cosmetic

Archaeology and History

Frankincense and myrrh were available in the biblical period only in limited parts of southern Arabia and the Horn of Africa. Due to the high demand for these spices, trade routes were developed to carry this precious burden over long distances through many countries to their foreign markets (Keay 2006).

In ancient times, the resin of the *Commiphora gileadensis*, the apharsemon, was used to treat many ailments. Although sources of supply of the resin were limited, it was in great demand in the classical and biblical world due to its medical benefits. The increased use of drugs of herbal origin in medicine instead of employing surgery was encouraged in Mesopotamia, perhaps because the Code of Hammurabi threatened amputation if the surgeon was unsuccessful and found responsible (Rosengarten 1970).

Spices and Medical Plants in Ancient Egypt

The ancient Egyptians used spices for their religious ceremonies that they purchased from the Land of Punt, long thought to be in the Horn of Africa (Kitchen 1993). At the beginning of the third millennia BCE, pharaohs went to great lengths to obtain spices, particularly myrrh, from other climes, since they were not grown locally. References to the importation of myrrh to Egypt from Punt appear as early as the fifth dynasty ca. 2800 BCE under King Sahure and King Isesi; later there were expeditions under Mentuhotep III in 2100 BCE and under Amenenhat II and the Sesostris dynasty. Since the price of these spices was exorbitant, the Queen Pharaoh Hatshepsut organized an expedition to Punt about 1500 BCE to investigate the option of importing the spice plants into Egypt The famous depictions (Picture 9) of the expedition of Queen Hatshepsut (1473–1458 BCE) are ships loaded with many treasures are depicted in the Temple in Thebes. In the mortuary temple at Deir el-Bahri, detailed descriptions of trade between Punt and the Egyptians are carved on the wall. On one of the reliefs, the boats from Egypt have arrived at Punt and are stopped on the beaches. "Here we see a flotilla of five large ships sailing from their Red Sea port, the arrival at Punt where the inhabitants lived in grass huts built on piles, the Egyptians offering the trade-goods of all such African adventures ever since – strings of beads, axes, and weapons – and the triumphant return with gold, ivory, apes, and precious myrrh-trees, their root-balls carefully protected by baskets for transplanting in Thebes" (Aldred 1961). One ship has 31 young trees that some scholars believed to be frankincense in tubs (Hepper 1969; Zohary 1982; Dayagi-Mendels 1989; Dominy et al. 2010).

However, Groom (1981) believed them to be myrrh as, according to his opinion, depictions of trees at that period were mainly schematic, presenting an image rather than a specific plant, and he referred also to the opinion of most previous experts that these trees were myrrh. Some scholars, however, find the trees on the Punt reliefs too conventionally drawn to be of any help in identifying them (Nielsen 1986).

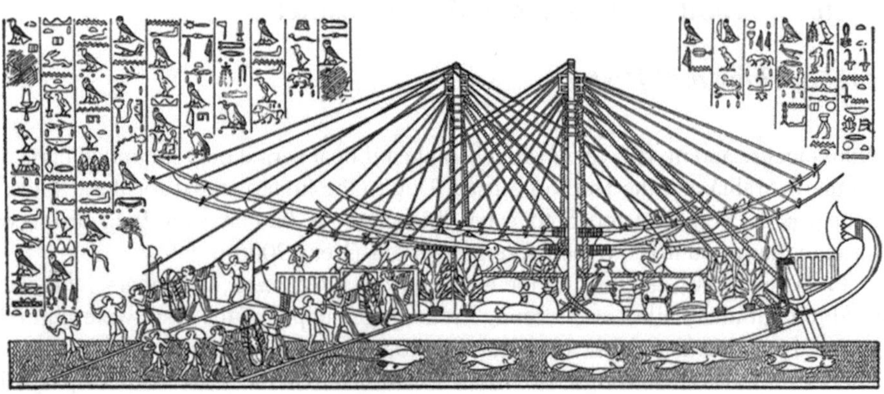

Picture 9 Queen Hatshepsut's expedition to the Land of Punt. It was primarily a trading expedition, for Punt, or God's Land, produced myrrh, frankincense, and fragrant ointments that the Egyptians used for religious purposes and cosmetics (From J. Dümichen "Die Flotte einer ägyptischen Königin aus dem 17. Jahrhundert vor unserer Zeitrechnung" 1868)

According to George Rawlinson (1897), the Egyptians entered the incense forests and either cut down the trees for their exuded resin or dug them up. Specimens were carried to the seashore and placed upright in tubs on the ships' decks, screened from sun by an awning. The day of transplanting in Egypt concluded with general festivity and rejoicing. Seldom is any single event of ancient history so profusely illustrated as this expedition, but there is no documentation for the growth of myrrh or frankincense in Egypt following this import.

Spices, an important part of Egyptian life, were used extensively on a daily basis. The Egyptian word for myrrh, bal, signified a sweeping out of impurities, indicating that it was considered to have medicinal and, ultimately, spiritual properties (Schoff 1922). Ancient Egyptians regularly scented their homes and were commanded to perfume themselves every Friday (Ziegler 1932). Idols were regularly anointed with perfumes, and incense became an important element in religious ceremonies; prayers were believed to be transported to the gods by the smoke of incense rising upward (Ziegler 1932). Every large Egyptian temple contained facilities for producing and storing perfumes (Brun 2000). The Egyptians ground the charred resin into a powder called *kohl*, which was used to make the distinctive black eyeliner seen on many females and males too in Egyptian art.

A growing body of archaeological evidence indicates that the volume of trade between Arabia and the surrounding areas accelerated during the Assyrian Empire. Assyrian documents record a growing interaction with the peoples of the Arabian Peninsula due to Assyrian attempts to control and capitalize on trade emanating from southern Arabia during the fifth century BCE.

Spices in Judea

Archaeological evidence of trade between southern Arabia and the Mediterranean coast has been found as early as the eighth century BCE in Tel Beer Sheva and Arad in Judea and includes the first appearance of alabaster containers and small limestone incense altars (Singer-Avitz 1996, 1999). The containers were a preferred means of storing and transporting raw incense resins, according to the Roman writer Pliny (Bostock 1855, Book 36, Chapter 60). New archaeological findings also indicate commercial relationships between southern Arabia and Judea, along the Incense Road. Much commercial activity existed in the Beer Sheva Basin, serving this trade during the seventh century BCE. In Tel Beer Sheva, several covers used for sealing the alabaster containers were found, as well as a stone object bearing the inscription of Cohen "priest" in a South Arabian language (Singer-Avitz 1999). At Kuntillet Ajrud, located on the Incense Road from Eilat to Gaza, Ayalon (1995) found drawings and inscriptions in two buildings and a large assemblage of Judaean and Israelite tools on sites along this incense road. These were dated to the end of the ninth century BCE. Singer-Avitz (1996) describes an altar, dated to the eighth century BCE, excavated at Tel Beer Sheva, decorated with a one-humped camel. This trade was greatly expanded at the end of the eighth century BCE under the Assyrian kingdom; its track was through the Edomite Mountains and the south of Judea, where security could be controlled. The Assyrians established several fortifications and commercial centers there, such as Ein Hatzeva south of the Dead Sea, Botzera near Petra, Tell el-Kheleifeh (Ezion Geber) at the northern end of the Red Sea, and other sites along the Mediterranean Sea near Gaza (Finkelstein and Silverman 2006). A broken ceramic seal (7 × 8 cm) found in Bethel with the south Arabian inscription *Chamin Hashaliach*, in south Arabian letters of that period, was estimated to date from the ninth century BCE (Van Beek and Jamme 1958). The archaeologists (Hestrin and Dayagi-Mendels 1979; Dayagi-Mendels 1989) suggested that the seal meant Chamin the messenger.

Apharsemon

Apharsemon as a Spice and a Medical Plant

Apharsemon (*tzori Gilead* in Hebrew) is described in the Bible as the gift that the Queen of Sheba gave to King Solomon. In Judea, it was grown around the Dead Sea for about 1,500 years and achieved fame due to its aroma and medicinal properties.

From the Book of Nehemiah 3:8, it is evident that the apothecaries (*roqeah* in Hebrew) who mixed spice substances were organized into guilds similar to those known in earlier periods at Ugarit (Neufeld 1971). In the First Temple period (957–587 BCE), incense was widely used in domestic settings to provide pleasant scents in homes, as insecticides, and as protection against disease (Neufeld 1971).

Apharsemon was one of the several components of the special incense that was used twice daily in the Holy Temple in Jerusalem. Rabbi Shimon Ben-Gamliel president of the Jewish High Court – Sanhedrin said that the apharsemon is the resin that exudes from the trees of *kataf* (ph; *Yoma 41:74* Jerusalem Talmud).

"Secret of the Town"

Ben-Yehoshua and Rosen (2009) discussed the relation of the apharsemon to the "secret of the town" mentioned in the Aramaic inscription on the floor of the sixth century CE synagogue at Ein Gedi (Barag et al. 1981).

Anyone causing a controversy between a man and his friend, or whoever slanders his friend before the Gentiles or whoever steals the property of his friend, or whoever reveals the secret of the town to the gentiles – He whose eyes range through the whole earth and Who sees hidden things. He will set his face on that man and on his seed and will uproot him from under the heavens and all the people said: Amen and Amen Selah (Levine 1981).

These authors support the suggestion of Feliks (1981) that the secret involved special technologies of producing and extracting the apharsemon resin. These authors suggested the existence of a special guild of the Judaean farmers that were involved in the production of this special lucrative crop.

Keeping secret the production of the apharsemon must have been a major factor in sustaining the economy and thus the survival of "a very large Jewish village"

Wolf 1971). The complete agrotechnical and legal history of the Judaean balsam since the Iron Age and during Babylonian, Persian, and Ptolemaic rule is unknown.

History of Apharsemon Production

Only a small part of our knowledge concerning the agrotechnology of Judaean balsam production comes from Jewish sources. Talmudic literature did not intend to cover this subject, and if it was supposed to be a secret, as little as possible of it would have been promulgated. Much of our meager knowledge of this agrotechnology comes from Greek and Roman authors. In his record of odiferous trees, Pliny, writing in the second half of the first century CE, supplies us with the most detailed descriptions about apharsemon, describing it as more preferable than any other odor and saying that it was "a plant that has only been bestowed by Nature upon the land of Judea" (Book 12, Chapter 54). He describes at some length the different grades of apharsemon and the care required to harvest the precious resin. Pliny noted that the high price motivated the production of fraudulent merchandise, describing such frauds and their detection, including the adulteration of the resin with hypericon produced at Petra. Josephus comments: "This country bears that apharsemon, which is the most precious drug that there is, and grows there alone" (Ant. XV. 4.2).

It is likely that, Queen Cleopatra, aware of the background of secrecy, after receiving the ownership of the apharsemon in Judea from her lover Anthony, hired King Herod to continue caring for this crop. Thus, despite the hostility between Cleopatra and Herod, she wisely selected him as she knew that as the previous owner he knew how to handle this guild of Judaean farmers that she was anxious to control.

Around this period, the cultivation of apharsemon was introduced farther south at The 'En Boqeq. Here archaeologists have uncovered an *officina* (a workshop) that the excavators believe was used to produce perfumed oils during most of the first and early second century CE (Fischer et al. 2000). The 'En Boqeq *officina* may have constituted part of the Roman imperial *fiscus* in the early Roman period (Cotton and Eck 1997). The Jewish revolt and the subsequent conquest of Judea by the Roman legions under Vespasian and his son Titus marked a traumatic period that affected the production of apharsemon.

According to Pliny, during the revolt, the Jews tried to destroy the trees in order to hurt Roman economic interests. The Jews vented their rage upon this shrub just as they were in the habit of doing against their own lives and persons, while the Romans protected it (Pliny, Book 12, Chapter 54), resulting in conflict in defense of a shrub. The economic and national importance of controlling the apharsemon was aptly demonstrated in the Roman triumph staged following the suppression of the great revolt. Pliny reports that apharsemon plants were paraded in the triumphant victory procession (Book 12, Chapter 54). The Roman rulers were keen to take control of this lucrative source of revenue, and the royal Judaean plantations were confiscated by Rome after Herods' death (Cotton 2001). However realizing the very complicated growth of this crop involved in the secrets that the Judaeans embrace of this special crop the Roman rulers applied a wise policy allowing the Judaeans to keep raising the crop themselves and maintaining their secrets while the Roman administration took over the marketing of this crop so that they completely controlled this crop and even supported the expansion of the areas of this Aparsemon into new locations. It appears that the Jewish Guild that controlled the secrets of growing this crop were able to select these new growers in order to maintain their embrace of the secrets. Thus although the control of the Roman administration was apparently absolute about the marketing still the background of the secrets attached to this special lucrative crop was not broken.

Apharsemon and the Dead Sea Basin

All observers, beginning with Theophrastus and Pliny, stated that this special crop was raised only in the Dead Sea Basin. A casual note of Diodorus Siculus (Oldfather 1935) raised the important question of whether Judea was the sole producer of the apharsemon during ancient periods. He reported that in 300 BCE, myrrh, apharsemon, and frankincense were especially important trade items going through Petra. Was the apharsemon sold in Petra the resin produced around the Dead Sea Basin? In that period, no other spice was called apharsemon. New evidence about this came from Arabia. Pliny did not know of Arabian apharsemon (Groom 1981). Further, it has been suggested that the Arabs of classical times did not regard the apharsemon as worth exploiting at all (referring to the plants that were, in our opinion, the ancestors of the apharsemon), continuing "We do not, at present, know enough about its gum" (Groom 1981).

An outstanding element of this production system was the possession of unique agrotechnical knowledge accumulated by long and intensive study of the environment. Due to the unique climatic conditions required for apharsemon production, similar knowledge could not have been accumulated in other parts of the Roman and Byzantine empires that lacked such a specific climate. The free farmers of Ein Gedi organized in well-established groups and guilds preserved this knowledge and kept trade secrets. Their communities, formed or disbanded at the pleasure of their master, could not have formed such a long-lasting, closed rural society of well-established, affluent manufacturing specialists. Free or semi free specialists organized in guilds are known from several economic environments. The producers at Ein Gedi established the oldest continuous center of cultivated apharsemon production that ever existed in a natural and human environment that was often extremely hostile. Several times these communities were almost completely destroyed. Still, this special crop survived for over 1,000 years developing several systems of new, improved apharsemon cultivars as well as new production techniques (Rosen and Ben-Yehoshua 2007).

Apharsemon was cultivated exclusively in royal gardens in Judea at Jericho and Ein Gedi (Whiston 1737, Josephus *Ant.* XIV.4.1, XV.4.2) and was cultivated during the First Temple period. Pliny refers to the great expense of small quantities of the extracted resin in the time of Alexander the Great. The resin must have been an important source of income for the Hasmoneans – the ruling Jewish dynasty in Judea from mid second century, 164 BCE to 40 BCE – and their plantations were highly coveted. Herod was forced to pay rent for his own plantations for 10 years to the Egyptian Ptolemaic queen Cleopatra under arrangements demanded by his friend Mark Anthony (Josephus Ant. XV.4.1; *Wars* I.18.5). Following the death of Cleopatra and her lover, Herod become one of the wealthiest men in the Roman Empire, and his monopoly on the cultivation, processing, and marketing of this valued substance was one of the sources of his wealth (Erickson-Gini 2007). Herod utilized this great wealth in one of the most ambitious building programs of any ancient monarch. In addition to his many desert palaces, the greatest one being the Northern Palace at Masada, Herod built whole cities in Judea and abroad. His most ambitious project was the construction of the Jewish Temple in Jerusalem. According to the Jewish historian Flavius Josephus (Ant. XV.11.1), Herod funded its construction at his own personal expense (Whiston 1737).

In Aden's trade statistics (Yemen) from the nineteenth and twentieth centuries, apharsemon gum, although technically a bdellium, appears to have been accounted for as "myrrh. As another product of *Commiphora*, it would seem most likely that in classical times, too, it was among the many types of myrrh (Groom 1981).

Diodorus Siculus also discussed the use of apharsemon wood (Book 19, Chapter 98), which indicates that the plant he discussed is the true apharsemon, since the wood of no other spice plant was utilized. This suggests strongly that the source of this apharsemon is the Dead Sea area, as we know from many other sources that only the apharsemon had several products, one of which was the wood taken from the xylem (Ben-Yehoshua and Rosen 2009). Pliny provides the prices of the different spices: strengthened also the option that this resin was the Judaean apharsemon. Those prices, as quoted by Pliny, are to be 300–1,000 dinars for a pint of the apharsemon, as compared to the price 2–6 dinars for frankincense and 11 for myrrh (Book 12). It is highly unlikely that an Arabian spice would fetch such a high price and not to be discussed in any report of that period. Furthermore, it is known that Pliny and all other historians of this Greek and Roman period spoke often of the special apharsemon and its very high price, which was much higher than all other spices. Many also said that the plant grew only in the Dead Sea Basin.

In conclusion, we may discount the possibility that the apharsemon used in Petra was imported with the myrrh and frankincense from Arabia. It suggests that the Nabatean merchants in Petra had also purchased or processed there apharsemon from the Dead Sea despite or, possibly, because of its high price. Presumably, apharsemon continued to be cultivated by the Jews, at least at the Dead Sea oasis of Ein Gedi, as late as the sixth century CE.

Decline of the Aphasemon Production

Ein Gedi was often destroyed between the inception of its special agriculture during the Iron Age and its end in the sixth century CE. However, it was always revived because doing so was economically sensible. The economic crisis of the third century CE undoubtedly affected the demand for apharsemon oil by the late Roman period. Records show that at least one rabbi, Shimon ben Eliezer, was required to explain the nature of apharsemon to his students: “Apharsemon (*tzori*) is merely the sap of resinous trees” (Shabbath 26a). References to mundane uses of apharsemon in the late Roman period – for example, for kindling Sabbath lamps (Shabbath 25b, 26a) – may be another indication that there was no longer a robust demand for this substance as there had been in earlier times.

The last destruction, by marauders or an early breakout of the Justinian pandemic that occurred between 541 and 750 CE (Ziegler 1932), or both, occurred prior to the conquest by Islam. The Islamic-Arab conquest flooded Middle Eastern markets with products of Arabia and the East, such as dates and Arabian balsam; previously free trade in such goods was hindered by political borders. At the same time, a trade barrier was created between Ein Gedi and the traditional markets in Byzantium and Europe. Consequently, the economic advantages of the oasis of Ein Gedi vanished. The well-advertised, deeply ingrained brands – “Ein Gedi,” “Jericho,” “Gaza,” “Ashkelon,” “Palaestina.” and “Holy Land” – lost much of their value and their customers. Maintaining a productive plantation system on the border of the desert was expensive. Thus the special agrotechnical system of Ein Gedi vanished. Only time will tell if it can be revived.

However, evidence (Lev 2003) of the purchase of apharsemon oil in Jerusalem by the Bishop Wilibald at the beginning of the eighth century testifies to the fact that the region of Palaestina remained a source for apharsemon oil, either because the trees continued to grow there or because the region served as a place of transit for this trade.

Apharsemon Production in Matariyya, Egypt

Another center of production of the apharsemon was developed later in the Coptic Church garden in Matariyya, Egypt (Milwright 2003). The dating of this place is not well documented, but the pricing of the resin was double its weight in gold.

In spite of the glorified aura that once surrounded apharsemon oil in ancient times, its production and medicinal uses nearly ceased altogether. The end of apharsemon production in Judea is probably related to the Arab conquest in the seventh century. The markets of this new Arab-controlled country were opened to the myrrh and olibanum from Arabia, which probably were sold at much lower prices. The earlier lucrative markets of Rome and Byzantium were now closed to the Judaean producers of the apharsemon. Furthermore, the Arab rulers expected these farmers to produce the newly introduced crop of cane sugar.

With time, the established reputation of the Judaean apharsemon, the great past demand for its products, and its high prices, alerted the Arabians to consider whether this apharsemon is indeed related to the ancestor plants of *Commiphora* growing wild in the kingdom of Sheba as the Bible suggested, but they did not accept this theory. Thus, they started to harvest and sell the balm of these ancestor plants and introduced this balm as a new product of their spice trade, in addition to the myrrh and olibanum that they had monopolized. In all likelihood, this ancient successful trade item was the reason why other balms of Gilead are now sold in several parts of the world. The Arabians also sold the Mecca balsam that was produced from a resin extracted from, a close relative, or the ancestor of the source of apharsemon.

Many corporations have utilized the name "apharsemon" for their own products. Thus, a healing compound (a balm) made from the resinous gum of the North American tree species *Populus candicans* is sold as apharsemon. However, the Remingon edition of the *Dispensory of the United States of America* (Remington and Wood 1918) defines the Judaean balsam of Gilead and the Mecca balsam as the resinous juice of *Commiphora opobalsamum*. Poplar buds (tears) are often, incorrectly, called apharsemon buds. At present, several corporations sell products under the name of apharsemon and claim that all the best fortunes would happen to consumers of their product; these commercial medications are extracted from other trees, and their sellers do not suggest any relation to the apharsemon. Persimmon is the name given to another Japanese fruit tree called in Japan *kaki* – *Diospyros kaki* – probably trying to make use of the fame of this ancient biblical plant which was then extinct.

Apharsemon oil or balsam of Mecca is still used as incense and in the preparation of perfumes. Its world supply is limited, and it is inexpensive (Hill 1952; Uphof 1968).

Myrrh

Religious Rituals, Myrrh and the Embalming

Pliny wrote that "there commeth not so much incense of one whole year's yield in Saba [the land of Saba or Sheba was the provider of the spice] as the Emperor Nero spent in one day when he burnt the corpse of his wife Poppea" (Book 12, Chapter 41). According to Tacitus (Church and Broadribb 1876), her body was filled with fragrant spices before the cremation. Strabo the Roman geographer wrote of Alexander: "His army used spikenard (*Valerianaceae*) and myrrh for tent covering and bedding, thus at the same time enjoying sweet odor and a more healthy air" (Dalby 2002). In the Roman Empire, myrrh was valued highly as a perfume and exotic fragrance and was burned as incense.

The ancient Egyptian perfume *kyphi* was an important material used for religious and medical purposes; frankincense and myrrh were among its 16 ingredients, and it was mixed according to a special prescription, accompanied by readings from sacred writings. Kyphi was first mentioned in the *Pyramid Texts*, a collection of religious texts from the time of the Old Kingdom, which describe the afterlife of the ancient Egyptians and, especially, the divine pharaoh (Faulkner 1969). The Papyrus Harris I, found in a tomb and purchased for the British Museum in 1855, records a delivery of ingredients for the manufacture of kyphi in the temples of Ramses III. Detailed instructions for its preparation decorate the walls in the temples at Edfu and Dendera. Dioscorides' *Herbal* Book One, Aromatics I, 24 (Osbaldeston 2000) presents the first description of kyphi in Greek. Galen wrote a poem on the medicinal uses of kyphi, and Plutarch notes that the mixture of kyphi could be used as a medicine and an ointment.

Myrrh was also mixed into wine (Fabius Dorsennus, writer of plays, quoted by Pliny Book 13, Chapter 5, referred to myrrh-wine in his play *Acharistio*) and served it at feasts in wine cups named *vasa murrina*, made of fluorspar from the eastern land of Parthia. The delicate fluorspar was reinforced with myrrh resin, giving the wine the taste of myrrh. Martial (*Epigrams XIV:113*) (Bohn 1897) states: "If you drink from it hot, the vase myrrhina suits the ardent Falernian and gives the wine abetter flavor." In the nineteenth century, some alcoholic drinks containing myrrh were commercialized: Becherovka, invented in 1807 by Josef Becher, is an herbal spirit drink made in the Czech Republic. Its 32 herbs including myrrh make it a popular remedy for digestive problems. Fernet Branca, invented in 1845 (Fratelli Branca Company, Milan), contains 40 different herbs, including myrrh. This drink is very popular in South and North America, where it is served neat, with ice, or mixed with other drinks; it is also promoted as a home herbal remedy for indigestion problems and colic.

Olibanum

Incense burning at religious ceremonies is one of the chief uses of frankincense. Frankincense (*levonah* in Hebrew) was an ingredient in the grain offering (Leviticus 2:1) and the showbread (Leviticus 24:6–8), while liquid myrrh or *stacte*, cinnamon, and cassia were prepared with olive oil "according to the art of the perfumer" to make the anointing oil (Exodus 30:22–30). Perfumers were employed in the palaces of the early Israelite kings (I Samuel 8:13).

The Bible has many references to incense that accompanied the sacrificial rites in the Temple in Jerusalem:

Exodus 30:34–5: "and the Lord said unto Moses: Take unto thee sweet spices… these sweet spices with pure frankincense"
Leviticus 2:1: "and he shall pour oil upon it and put frankincense thereon"
Leviticus 2:15: "and thou shall put oil upon it and lay frankincense thereon"
Leviticus 2:16: "and the priest shall burn …with all the frankincense thereof"
Leviticus 6:15: "and all the frankincense which is upon the meat offering"
Leviticus 24:7: "and thou shall put pure frankincense upon each row"
Numbers 5:15: "nor put frankincense thereon for it is an offering of jealousy"
I Chronicles 9:29: "all the instruments of the sanctuary…and the frankincense"
Nehemiah 5:15: "and they laid the meat offerings, the frankincense" Malachi 1:11: "and in every place incense shall be offered unto my name"
Psalms 141: "Let my prayer be set before thee as incense"

The sacrificial altar dating from the eighth century BCE excavated at Tel Dan, and exhibited at the Skirball Museum in Jerusalem, shows the marks of soot from the incense burned at the ceremonies: I Kings, 12:28–30: "Whereupon the King made two calves of gold. He set the one in Beth-el and the other he put in Dan …and he offered upon the altar and burnt incense."

In the New Testament, in the lament over the final fall of Babylon, Revelations 18:13, mourns: "there is no one left to buy her goods …spice, incense, myrrh, frankincense." Frankincense was burned to accompany prayer:

Luke 1:10: "the crowded congregation was praying at the actual time of the incense burning" Revelations 5:8: "golden bowls full of incense, which are theprayers of the saints"

Revelations 8:3: "and the smoke of the incense rose up before God mingled with the prayers of the saints"

The growth of Christianity depressed the market for frankincense during the fourth century, but the Roman Catholic Church later adopted the use of incense for religious services (Howes 1946). It was also thought that the white smoke carried the prayers up to heaven (Armenian Orthodox). By the Middle Ages, frankincense was incorporated into regulated use, with detailed instructions on its use (Catholic Encyclopedia).

Olibanum and Folklore

The growth of Islam curtailed the use of frankincense in the Middle East, since Islam does not require the burning of incense in religious rites and ceremonies. However, the aroma of frankincense is said to represent life, and the Judaic, Christian, and Islamic faiths have often used frankincense mixed with oils to anoint newborn infants and individuals moving into a new phase in their spiritual lives.

In Yemen, the home is fumigated and perfumed by burning the incense on a special implement. Pots and jars are mended and cleaned by pouring the resin inside, where it hardens in the cracks, making the article watertight. A frankincense candle is burned in the house during the night to give light and also keep evil spirits away – perhaps commercial rivals, seeking to steal the precious harvest.

The bark of the *Boswellia* tree was used as a dye for the cotton gowns worn daily; the bark was cooked until a red-brown color was obtained, and then the garment was lowered into it. It was also used for dyeing leather, and the red-brown color was very popular. Frankincense can also be added to coffee to give a "spicy" flavor. Frankincense is a staple household medicinal for dental problems, swellings, bronchitis and coughs. It is claimed that memory can be enhanced by soaking some incense with iron in water overnight and drinking it in the morning of exams. Perhaps this has a calming effect, overriding panic (Hepper 1992).

Frankincense has been long associated with the phoenix, a mythical and mysterious bird. The Roman poet Ovid (43 BCE–18 CE), exiled for his uninhibited verse, describes the phoenix in this way: "The Assyrians call it the Phoenix. It does not live on fruit or flowers, but on frankincense and odoriferous gums. When it has lived five hundred years, it builds itself a nest in the branches of an oak, or on the top of a palm tree. In this it collects cinnamon, and spikenard, and myrrh" (Melville 1998). This is repeated by Pliny: "In Arabia he is held a sacred bird, dedicated unto the Sunne: that he liveth 660 years [modern texts have 540 years]: and when he growth old, and begins to decay, he builds himself a nest with the twigs and branches of the Canell or Cinnamon, and Frankincense trees: and when he hath filled it with all sort of sweet Aromaticall spices, yee yieldeth up." (Book 10, Chapter 2).

Pliny (Book 12, Chapter 32) describes that the olibanum trees grew in isolated and inhospitable areas, and their harvest was surrounded by myths and fables, mainly to deter rivals eager to enter into the trade and share the enormous profits. Because of the dangerous routes for delivery of the harvest, the merchants were selected carefully, mostly from the nomadic tribes of Arabia and Nabatea who were familiar with the terrain and its perils.

Frankincense is used in perfumery and aromatherapy. Olibanum essential oil is obtained by steam distillation of the dry resin, some of the smell of the olibanum smoke resulting from the products of pyrolysis.

At present, frankincense is in demand as a component in some perfumes and colognes and particularly also in the currently fashionable aromatherapy procedures, promoting serenity and well-being. The "perfume amouage" is based on frankincense and produced in the Oman, using a combination of fragrant resins (Hepper 1969). Frankincense is also used in soaps, powders, and creams, especially for the treatment of skin problems and for softening (Rees 1995).

Modern Research Based on Traditional Medicine Uses

Apharsemon

In traditional and popular medicine, apharsamon has long been considered both as a panacea and an important component in the treatment of a wide range of ailments. Muntner (1971), quoting Asaph Harofeh (Asaph the Physician, nineth or tenth century BCE from his "Sefer Harefuot"), states the apharsemon was used against "evil vapors of the stomach and also as an antidote against poisons". Lev (2003), in reviewing medieval medicine, mentions that, in addition to its antidotal properties, apharsemon was also used for pain relief, dissolving urinary tract stones and curing infertility.

Hooper (1937) conducted a comprehensive survey of "Useful plants and drugs of Iran and Iraq" in 1934, scouring the local markets and examining the uses of his finds there. For *Commiphora opobalsamum* Kunth, locally known as "balasan", he concluded that "the fruits are carminative, stomachic, expectorant and stimulant; in Isfahan, for shivering and colds". These observations are supported by Feliks (1968) and Uphof (1968).

According to Lev (2003), apharsemon was considered efficacious against the venom of all kinds of serpents, beneficial to the eyesight, disperses films on the eyes, assuages hardness of breathing, acts as an emollient, prevents the blood from coagulating, acts as a detergent on ulcers, and is beneficial for diseases of the ears, headache, trembling, spasms, and ruptures. Taken in milk, it is an antidote to the poison of aconite, and used as a liniment upon the onset of the shivering fits in fevers, modifying their violence. However, it should be used sparingly, since it is very caustic, and, if not used in moderation, is apt to augment the malady.

For general and external use, apharsemon is recommended as a painkiller (Budge 1913; Brunet 1933; Said 1973; Malandin 1986); an antiinflammatory (Budge 1913; Said 1973; Majno 1975; Haefeli-Till 1977; Qataya 1981) and for treating general diseases in the body (Greenhill 1705). It reduces the temperature of fevers (Greenhill 1705; ibn al-Baytar 1874; Stannard 1966; Temkin 1973; Malandin 1986) but is also recommended for raising the body temperature (Adams 1844; ibn al-Baytar 1874; Budge 1913; Meyerhoff and Sobhy 1932; Brunet 1933; Gunther 1933; Cahen 1947; Levey 1966a; Temkin 1973; Said 1973; Qataya 1981; Sayyid 1985; Malandin 1986). It is considered beneficial for lethargy and tiredness (Qataya 1981); for cases of paralysis, spasms, and dizziness (ibn al-Baytar 1874; Budge 1913; Said 1973; Haefeli-Till 1977; Qataya 1981). For skin conditions, apharsemon was applied as an antidote to stings, bites, and allergies (Postlethwayt 1766; Langkaevel 1868; ibn al-Baytar 1874; Jones and Omerod 1918; Siddiqui 1928; Gunther 1933; Levey 1966b; Watson 1966; Rosner 1984); as an astringent; and to heal wounds, prevent rotting and putrefaction, and heal skin blemishes and warts (Starkey and Pitt 1678; Greenhill 1705; Postlethwayt 1766; Low 1924; Kühn 1964; Said 1973; Rubin 1974).

Apharsemon was administered: for internal problems and for bladder and digestive problems (Budge 1913; Said 1973; DeFenoyl and Sauneron 1979; Qataya 1981); as a diuretic (Budge 1913; Gunther 1933; Haefeli-Till 1977; Qataya 1981; Malandin 1986); for relief from flatulence (Budge 1913; Said 1973; DeFenoyl and Sauneron 1979); for removing obstructions and hardness of the liver (Pagel 1893; Budge 1913); as a laxative (Aldredge 1996); for cataracts (Hassler 1843; Adams 1844; ibn al-Baytar 1874; Budge 1913; Meyerhoff 1928; Brunet 1933; Wood 1936; Kühn 1964; Levey 1973; Nielsen 1974; Rubin 1974); to relieve cough, breathing difficulties, pneumonia, and asthma (Gunther 1933); as a disinfectant in protection against infection (Greenhill 1705; DeFenoyl and Sauneron 1979); and in the preparation of medicines and mixtures (Costeo and Mongio 1608; Starkey and Pitt 1678; Bonwicke 1725; Spencer 1938; Kühn 1964; Watson 1966; Anawati 1987). Furthermore, it was considered effective as an abortifacient (Gunther 1933), for menstrual problems (DeFenoyl and Sauneron 1979), and to encourage sexual potency (Levey 1966a).

As present reviewers have faith in these species as medical plants whose products shall be developed in the close future, we bring forth reports that evoke some alarm about the use of these plants: *Balsamodendrum gileadense* (spelling is original and unusual), was suspected of producing allergic effects (Bardel 1935). This plant is the source of Balm of Mecca (Genders 1972) which was a favourite beauty preparation in the Middle East but could produce alarmingly violent reactions on the face (Montagu 1717, cited by Woods and Calnan 1976).

Balm of Gilead is likely to be a source of confusion because a number of other plants and plant products may be referred to under the same name. For example, the shrubby *Cedronella canariensis* Willd. ex Webb (syns *Cedronella triphylla* Moench, *Dracocephalum canariense* L.; fam. Labiatae) is popularly known as balm of Gilead. Also, the oleoresin from *Abies balsamea* Mill., fam. Pinaceae, which is usually called Canada balsam, may also be known as balm of Gilead, as may the resinous matter coating the buds of certain *Populus* L. species (fam. Salicaceae). Stuart (1979) asserts that the balm of Gilead of commerce is now derived from *Populus candicans* Aiton,

P. balsamifera L. and possibly other *Populus* species, for example *P. tremuloides* Michaux. Probably some of these changes are due to commercial reasons of available markets and easier marketing due to the high reputation that the Balm of Gilead provides. The hypotensive effect of aqueous extract from the branches of the *C. opobalsamum* tree on blood pressure and heart rate in rats is due to the activation of muscarinic cholinergic receptors (Abdul-Ghani and Amin 1997). However, it is surprising that this author found plants of *Commiphora opobalsamum* growing wild in the mountains of Ramallah, Palestine, since, according to past data, these plants do not grow at such relatively high altitudes in climates different from those of the Dead Sea Basin. Possibly another species is involved in that article (probably *Lycium* spp. ?).

Extracts of *C. opobalsamum* exhibited anti-inflammatory, analgesic and diuretic activities in rats and mice, hepatoprotective ability and ulcer protective effects (Al-Howiriny et al. 2004, 2005).

The essential oil from the fresh aerial parts exhibited anti-microbial activity against *Bacillus subtilis*, *Staphylococcus aureus*, *Candida glabrata*, *C. krusei*, *Cryptococcus neoformans* and *Mycobacterium intracellulare* (Al-Massarany et al. 2007).

The chemical composition and biological evaluation of the essential oil of *Commiphora opobalsamum* was examined for its identified constituents using essential oil samples from the stored aerial parts, fresh aerial parts and fresh flowering tops, obtained by hydrodistillation. Following a photochemical investigation of the aerial part of *C. opobalsamum* growing in Saudi Arabia, six compounds were isolated and identified: friedelin, canophyllal, oleanonic acid, mearnsetin, quercetin, and syringic acid. Extracts and isolated compounds were preliminary assayed in vitro for antimicrobial, antimalarial, antitumor, anti-inflammatory, antioxidant and estrogenic activity. The ethyl acetate extract was moderately active against *Staphylococcus aureus*, *Pseudomonas aeruginosa*, and *Plasmodium falciparum*, while the petroleum ether and chloroform extracts inhibited COX-2 at 5 and 10 μg mL^{-1}, respectively. Mearnsetin and quercetin exhibited antioxidant activity and syringic acid showed moderate antimalarial, anticandidal, and antimycobacterial activity (Abbas et al. 2007).

The antiproliferative effect of resinous exudates of *Commiphora opobalsamum* was examined on human prostate cancer cells and secondary metabolites were isolated (Shen et al. 2007): cycloartane-24-en-1α,2α,3β-triol, octadecane-1,2*S*,3*S*,4*R*-tetrol, 1-*O*-α-*L*-rhamnopyranoside, eudesmane-1β,5α,11-triol, and guaia-6α,7α-epoxy-4α,10α-diol (Fig. 14) along with six known sesquiterpenoids (guaianediol, myrrhone, dihydropyrocurzerenone, 2-methoxy-5-acetoxy-furanogermacr-1(10)-en-6-one, (1(10)*E*,2*R*,4*R*)-2-methoxy-8,12-epoxygermacra-1(10),7,11-trien-6-one, and curzerenone) (Fig. 13).

Two sesquiterpenoids – [(1(10)*E*,2*R*,4*R*)]-2-methoxy-8,12-epoxygemacra-1(10),7,11-trien-6-one and 2-methoxy-5-acetoxyfuranogermacr-1(10)-en-6-one (Fig. 14) were isolated from the resinous exudates of *C. opobalsamum* by Kai et al. (2008). With the help of MTT assay, they evaluated the cytotoxicity of these compounds against the colorectal cancer cell line, hormone-independent human prostate cell lines and immortalized liver normal cell line. An inhibitory effect on the proliferation of hormone-independent prostate cancer cell lines in a concentration-dependent manner was proved for both compounds.

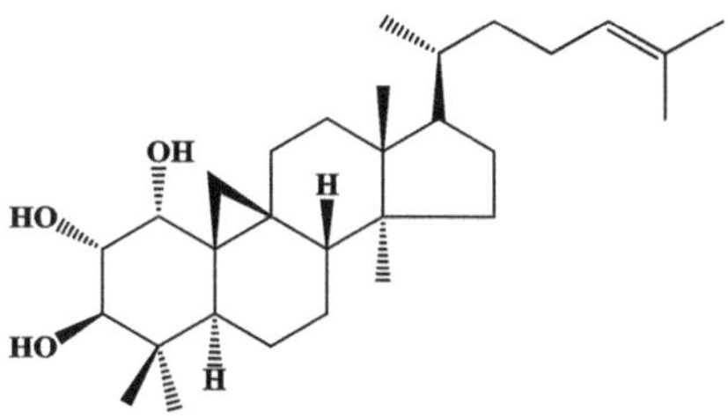

cycloartane-24-en-1α,2α,3β-triol

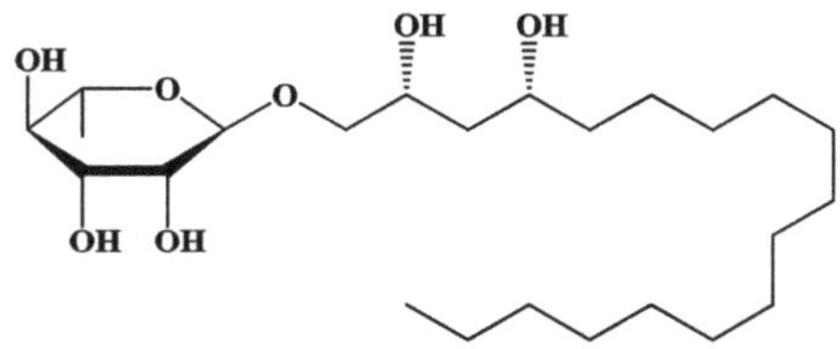

octadecane-1,2*S*,3*S*,4*R*-tetrol 1-*O*-α-L-rhamnopyranoside

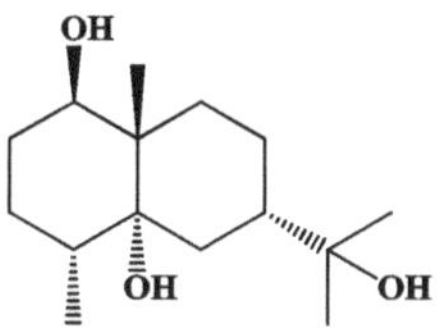

eudesmane-1β,5α,11-triol

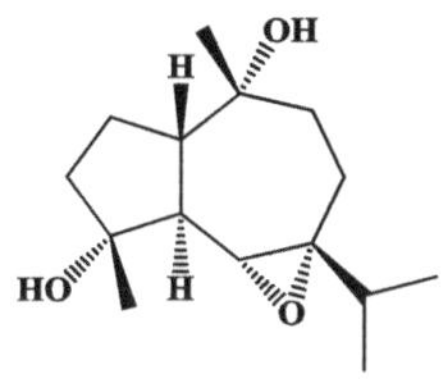

guaia-6α,7α-epoxy-4α,10α-diol

Fig. 13 Structures of compounds isolated from *C. opobalsamum* (Shen et al. 2007)

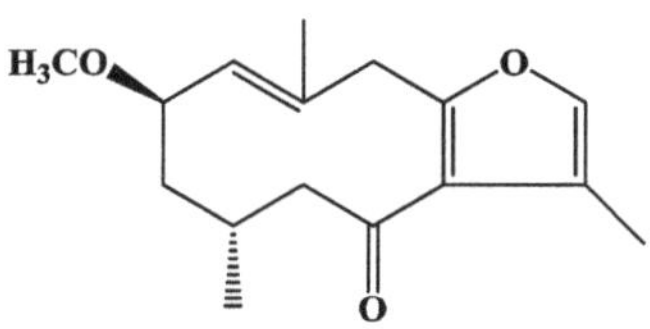

[(1(10)*E*,2*R*,4*R*)]-2-methoxy-8,12-epoxygemacra-1(10),7,11-trien-6-one

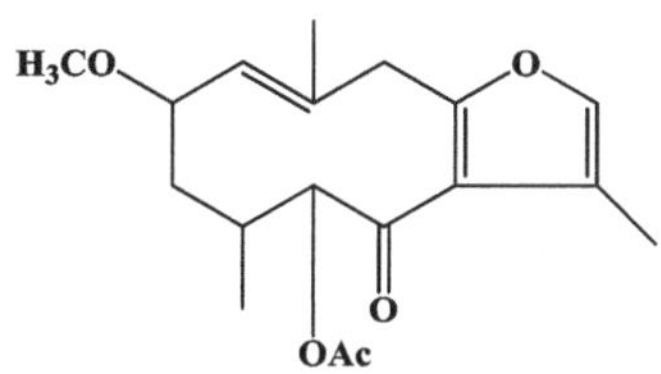

2-methoxy-5-acetoxyfuranogermacr-1(10)-en-6-one

Fig. 14 Structures of compounds isolated from *C. opobalsamum* (Kai et al. 2008)

Studies of Iluz et al. (2010) have shown that *C. gileadensis* possesses antibacterial activities that validate its usage in the local treatment of wound infections. The results showed that the thick milky, viscous C. gileadensis sap significantly inhibited *B. cereus* growth. Investigations of the antimicrobial activity of *C. gileadensis* showed the plant sap blocking of *Pseudomonas aeruginosa* lectins

Inhibition of Cancer Cell Proliferation by a *C. Gileadensis* Extract

Anticancer activity against lung and blood cancer cell lines was found by studying extracts from leaf and stem extracts and from the resin of *C. gileadensis*. Apharsemon extracts were investigated for anticancerous activity against tumor cell lines (Amiel et al. 2011, 2012; Ben-Yehoshua et al. 2013). The effect *C. gileadensis* extract was assessed against two Limphoma cancer cell lines: mouse (BS-24-1) and human (MoFir) using ethanol based stem extracts.

The results obtained from ethanol-based extracts and from essential oils of apharsemon indicated that β-caryophyllene is a key component – 20.12 % – in the essential oil extracted from the apharsemon. β-Caryophyllene can be found also in spice blends, citrus flavors, soaps, detergents, creams, and lotions, as well as in a variety of plants, food and beverage products, and it is known for its anti-inflammatory, local anaesthetic, and antifungal properties. Beta caryophyllene was found to be a potent cytotoxic compound over two cell lines of Lymphoma cancer. Apharsemon – *Commiphora gileadensis* stem extracts and essential oil of this plant have an antiproliferative proapoptotic effect against tumor cells and not against normal cells. β-caryophyllene caused a potent induction of apoptosis accompanied by DNA ladder and caspase-3 catalytic activity in tumor cell lines. In summary, apharsemon – *C. gileadensis* plant contain an apoptosis inducer that acts, in a selective manner, against tumor cell lines and not against normal cells (Amiel et al. 2011, 2012; Ben-Yehoshua et al. 2013).

Amiel, E., N. Dudai, T. Rabinsky, S. Rachmilevitch, R. Ofir, and S. Ben-Yehoshua. US patent No. 20140030289 A1 (WO2012104845 A1). Compositions Comprising Beta–Caryophyllene and Methods of Utilizing the Same.

Yang and Shi (2012) isolated several new compounds from a resin of *C. opobalsamum* imported from India. The compounds were tested for their cytotoxic activity and were identified as cycloartane-24-ene-1*S*,3*R*-diol, cycloartane-23-ene-1S′,3*R*′, 25-triol,cycloartane-24-en-1α,2α,3β-triol-1,2-acetonide,1β,8β-epoxy-2α-methoxy-6-oxogermacra-9(10),7(11)-dien-8,12-olide, 1β,8β-epoxy-2α-methoxy-12α-hydroxy-6-oxogermacra-9(10),7(11)-dien-8,12-olide,1β,8β-epoxy-2α-methoxy-12β-hydroxy-6-oxogermacra-9(10),7(11)-dien-8,12-olide,2α-methoxy-8α-hydroxy-6-oxogermacra-1(10),7(11)-dien-8,12-olide, guaia-4β,7β,10α-trihydroxy-5-ene, commipholinone, 1β,4β-epoxy-eudesmane-11-ol (Fig. 15), and five already known compounds. Guaia-4β,7β,10α-trihydroxy-5-ene showed the most potential cytotoxic activity aganist HeLa and HepG2 cancer cell lines.

Myrrh

Traditional medicine uses the resins of the *Commiphora* species for the treatment of inflammation, arthritis, obesity, microbial infection, wound, pain, fractures, tumor and gastrointestinal diseases (Shen et al. 2012).

A special formulation of myrrh is the mithridatum, a poison antidote containing myrrh, named after King Mithradates (first century BCE) who was terrified of being poisoned or falling sick, and took a dose of strengthening antidote daily (Milwright 2003). The mithridatum is based on a mixture described by Theophrastus,

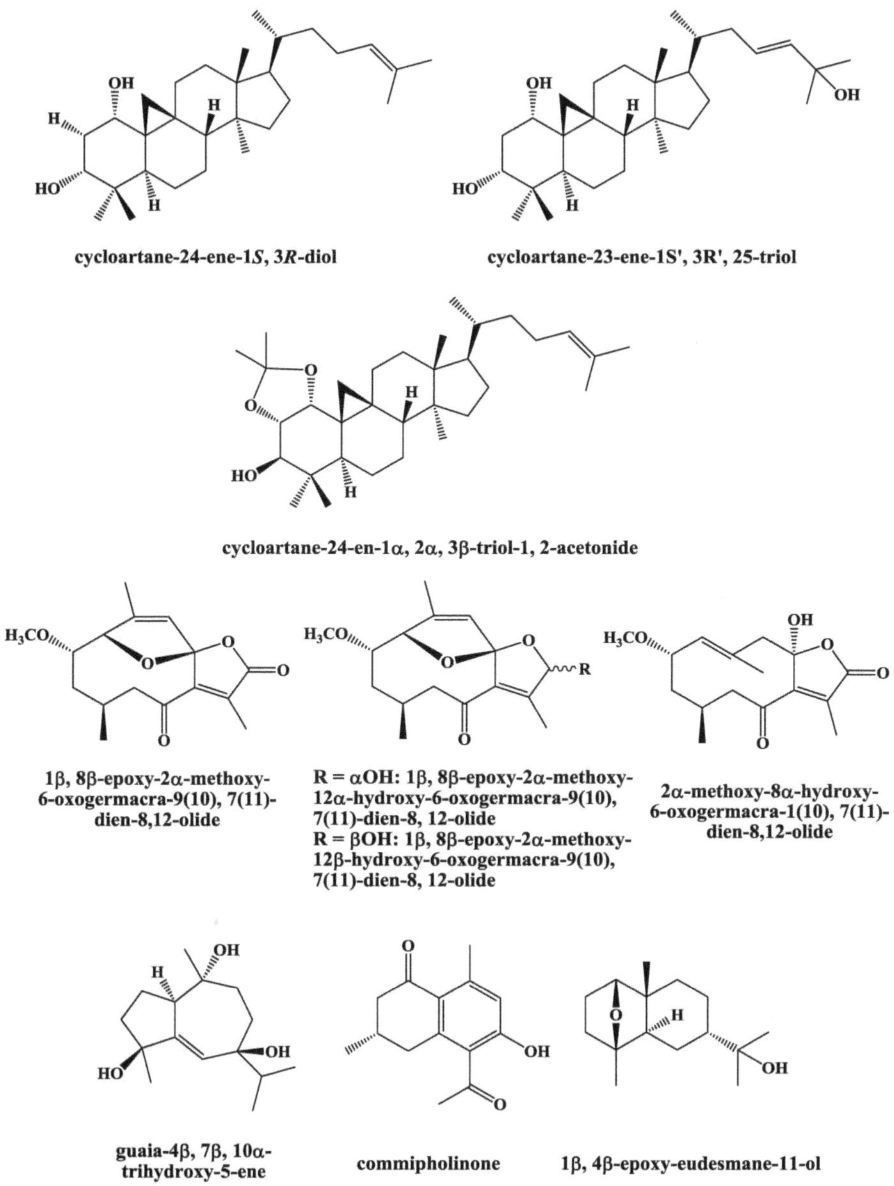

Fig. 15 Structures of compounds isolated from *C. opobalsamum* (Yang and Shi 2012)

called megalium, a sweet-smelling potion containing myrrh to relieve wounds. Plutarch mentioned a similar mixture named Egyptian kyphi, and Ptolemy's doctor, Zopyrus, detailed a combination of megalium and Egyptian kyphi in a letter to Mithridates. With some more additions, the combination became mithridatum, especially recommended for recovery from serious falls as well as an antidote to food poisoning. A century later, Galen created his own version and named it theriac, recommended for all internal indispositions. This formulation was continuously developed throughout the ages, and by 1659, mithridatum contained 63 ingredients (*Pharmacopoeia Londonensis*). It is interesting to note that Jerusalem was a major site of its production during all these ages, possibly due to the presence of many ingredients in plants nearby and also the city's reputation of holiness, which could contribute to the marketing of mithridatum (Lev 2003).

Research in Egypt and Saudi Arabia into parasitic diseases (mainly schistosomiasis – bilharzias – but also fascioliasis and monieziasis) has yielded a new treatment, based on an alcohol extract from the *C. molmol* plant. Schistosomiasis is a parasitic infection attacking millions of people, mainly in Africa and China, but also in other countries; the infection is debilitating and sometimes fatal, attacking and damaging the kidney and liver. Research has focused on an extract from *C. molmol*, mirazid. Despite some side effects, a complete cure has been realized, not only for the bilharzia parasite but for several others, in both humans and animals. This cure, coupled with projects to increase production of the *C. molmol* tree, could possibly be an effective and economically feasible solution to help eradicating the incidence of parasitic diseases in developing countries (Badria et al. 2001; AboMadyan et al. 2004; Hamed and Hetta 2005; Southgate et al. 2005). However, data on the control of these parasitic diseases has been controversial, and Fenwick and Webster (2006) reported that myrrh is ineffective against schistosomiasis.

Additional conflicting reports are found in the literature about the efficacy of Mirazid as an antischistosomal drug. Contrary to the other positive studies this study gave negative effectiveness as antischistosomal drug (Ramzy et al. 2010).

In the twenty-first century, natural alternative medicine for many ailments has increased in popularity in the Western world, and myrrh has been revisited. The stringent and antiseptic properties of myrrh have been promoted as useful for cleansing and healing wounds, including bedsores. Traditionally, myrrh was used as a dressing for skin ulcers and sores. Myrrh is also a common ingredient in therapies for tonsillitis and sore gums; several commercially available tooth pastes contain *Commiphora myrrha* extract. It is considered useful as an ingredient in cough mixture, and as an effective and speedy expectorant in cases of catarrh and bronchitis. In small doses myrrh can promote digestion, but larger doses cause excessive sweating and gastric heat. Myrrh is often prescribed in combination with iron and aloes, for anemia, in connection with "female disorders". Since myrrh is a stimulant, especially to the mucous tissues, it can provoke prostration, nausea, and vomiting, if taken in excess. Myrrh is not taken alone as an internal medicine but is approved by the USDA as flavoring, fragrance, or stabilizing ingredient in beverages, cosmetics, drugs, and foods.

As is the case with other traditional herbal treatments, recent research has centered on the examination and evaluation of the unique properties of the resin, concluding that myrrh has considerable antimicrobial activity and is medicinally used in a variety of cures (El Ashry et al. 2003).

Gugulipid has been used for controlling or preventing cognitive dysfunction, hyperglycemia and some infective conditions of the skin (Pratap et al. 2002).

Gugulipid – the active ingredient of *C. mukul* (gugulipid), is widely used to treat hyperlipidemia, and positive effects were noted after examination (Cui et al. 2003).

Hypolipidemic activity of the phytosteroid extract from *C. mukul* was also observed (Urizar and Moore 2003). This hyperlipidemic agent represents a possible mechanism for the guggulsterone-mediated hypolipidemic effect (Deng et al. 2007). The chemistry and pharmacological activity of guggul derived from *C. wightii* was related to the isolates of material claimed to be efficacious for rheumatism, arthritis, hyperlipidemia, obesity, inflammation, atherosclerosis, wrinkles, and acne (Anurekha and Gupta 2006).

Extracts of *Commiphora mukul* have proved to be useful in the treatment of allergic and non-allergic inflammation concerning skin and external mucosae, in the symptomatic treatment of benign prostatic hypertrophy, and in the treatment of acne (Bombardelli and Spelta 1991).

In 1996, a team of chemists and pharmacologists at the University of Florence in Italy reported that two compounds of myrrh have pain-relieving properties. The researchers initially observed that mice injected with a myrrh solution were slower than a control group in reacting to the heat of a metal plate. They tested three main compounds of myrrh and found that two of them – furanoeudesma-1,3-diene and curzerene – have pronounced analgesic effects. Additional tests suggested that these compounds interact with the opioid receptors in mice brains to decrease the sensation of pain (Freese 1996).

Analysis and evaluation have been carried out to examine the characteristics of the various derivatives of *C. myrrha*. Eight sesquiterpene fractions were extracted, purified, and characterized from *C. molmol* (Dolara et al. 2000), and local anesthetic, antifungal, and antibacterial properties were recorded. The analgesic effects of myrrh had been examined by Dolara et al. already in 1996.

Toxicity studies on *C. molmol* were conducted in mice to determine external morphological, biochemical, and hematological changes, but no significant difference in mortality compared to controls was detected (Rao et al. 2001).

Myrrhanol A (isolated from *C. mukul*) displayed potent anti-inflammatory effects and is regarded as a plausible candidate for a potent anti-inflammatory agent (Fig. 16, Kimura et al. 2001).

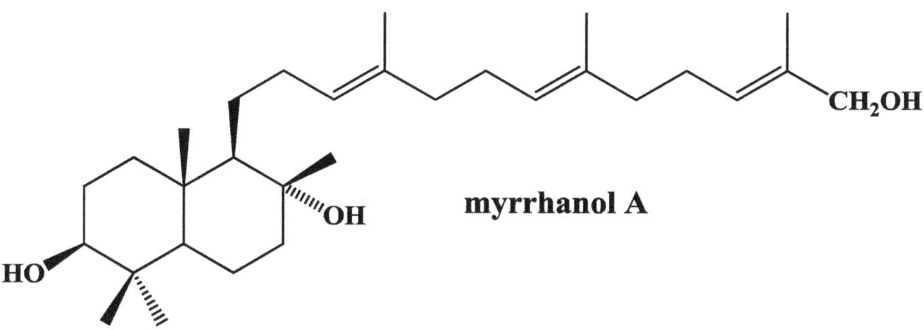

Fig. 16 Structure of myrrhanol A isolated from *C. mukul* (Kimura et al. 2001)

The active ingredients responsible for the maintenance of healthy cholesterol levels are the guggulsterones, specifically guggulsterone E and Z. These resins reduce serum lipids and cholesterol in the bloodstream, thus helping to avoid stroke (Owsley and Chiang 2003).

A randomized controlled trial performed in 2003 to study the short-term safety and efficacy of the extract concluded that cholesterol levels were not improved (Szapary et al. 2003).

Twelve Chinese medicinal herbs, including *C. myrrha*, were examined for their anticancer activity and positive results suggest that further studies were warranted (Shoemaker et al. 2005).

Other effects of myrrh resin extracts include studies on diabetes wounds (Lotfy et al. 2006), gingivitis inflammation, and general anti-inflammatory and antibacterial properties (Tipton et al. 2006; El Ashry et al. 2003).

Essential oil of *C. myrrha* (0.0001–5 %) was examined in cosmetic composition used for skin firmness, improvement of contractile properties and maturation of fibroblasts and myofibroblast, angiogenesis stimulation, weight reduction, against leg pain and circulation problems (Aurrens and Laperdrix 2008).

Compounds with antibacterial activity against *Staphylococcus aureus* strains, several *Salmonella enterica serovar Typhimurium* strains and two *K. pneumoniae* strains were isolated from oleo resin of C. molmol. The compounds were identified as mansumbinone, 3,4-seco-mansumbinoic acid, β-elemene and T-cadinol (Fig. 17, Rahman et al. 2008).

The substance responsible for toxic effect on the fourth stage larvae of Culex pipien in the essential oil of *Commiphora molmol* was found to be dl-limonene (Habeeb et al. 2009).

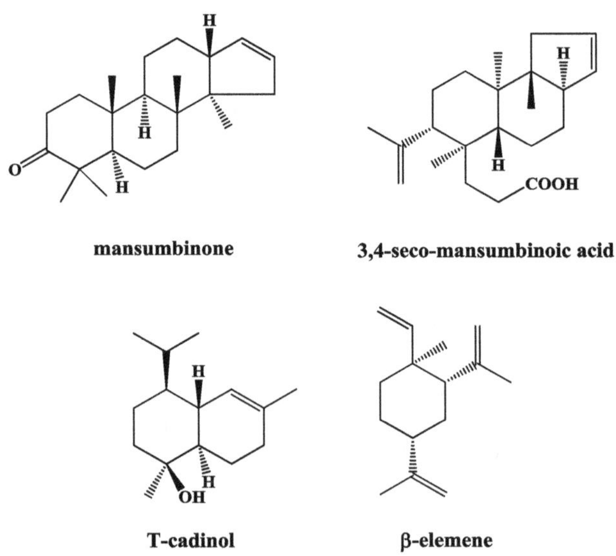

Fig. 17 Structures of compounds isolated from *C. molmol* (Rahman et al. 2008)

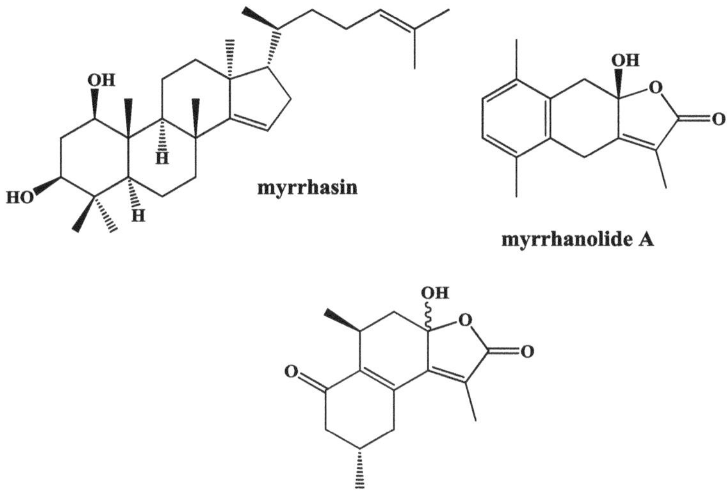

8α-OH: myrrhanolide B
8β-OH: myrrhanolide C

Fig. 18 Structures of compounds isolated from *C. myrrha* (Shen et al. 2009)

Shen et al. (2009) succeeded in isolating four new compounds from the resinous exudates of *Commiphora myrrha*: myrrhasin and myrrhanolide A–C (Fig. 18), as well as nine known sesquiterpenoids. Cytotoxic activities of these compounds were tested against the PC3 and DU145 human prostate tumor cell lines with the help of the MTT assay. Myrrhanolide A and the mixture of myrrhanolide B and C exhibited weak cytoxicity against the PC3 cell line with IC_{50} values of 38.3–46.0 μM, respectively.

Protection of *Commiphora molmol* emulsion against against PbAc-induced hepatic oxidative damage and immunotoxicity was successfully studied by Ashry et al. (2010). They found that myrrh treatments resulted in elevated leukocyte levels even in the absence of injury throughout healing (Haffor 2010). Masking of the bitter and disagreeable taste of the essential oil of *C. molmol can* improve patient acceptance and compliance (Etman et al. 2011). A stable MEO (myrrh essential oil) emulsion with an acceptable taste was formulated using Cremophore as an emulsifier and a combination of anise oil, peppermint oil, and glycerol as flavoring agents.

Plant extract of *C. molmol* was found to be efficient against *Trichomonas vaginalis* infection (El-Sherbiny and El-Sherbiny 2011).

Three new compounds were recently isolated from the resinous exudates of *C. myrrha*, named commiterpenes A–C (Fig. 19) (Xu et al. 2011a). These cadinane sesquiterpenes revealed moderate neuroprotective effects against 1-methyl-4-phenylpyridinium -induced neuronal cell death in human dopaminergic neuroblastoma SH-SY5Y cells (CRL-2266).

The resinous exudates of *Commiphora myrrha* were used for isolation of ten new furanosesquiterpenoids, myrrhterpenoids A–J (Fig. 20, Xu et al. 2011b). Their structures and relative configurations were elucidated by spectroscopic methods and

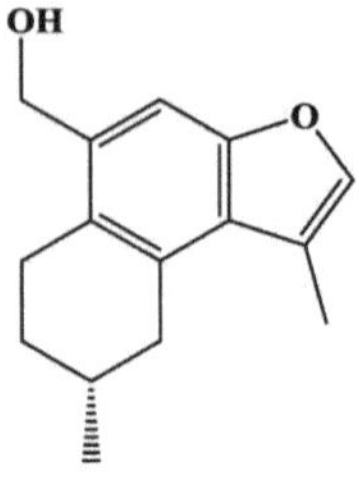

commiterpene A

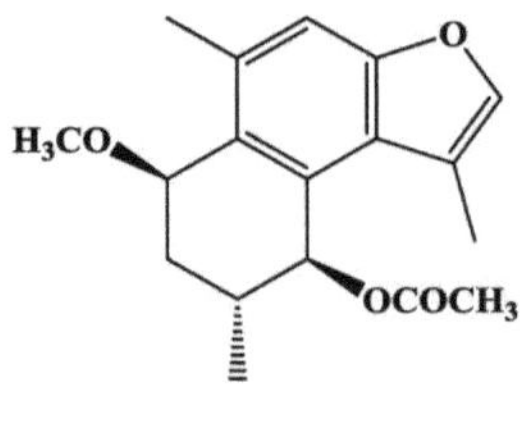

commiterpene B

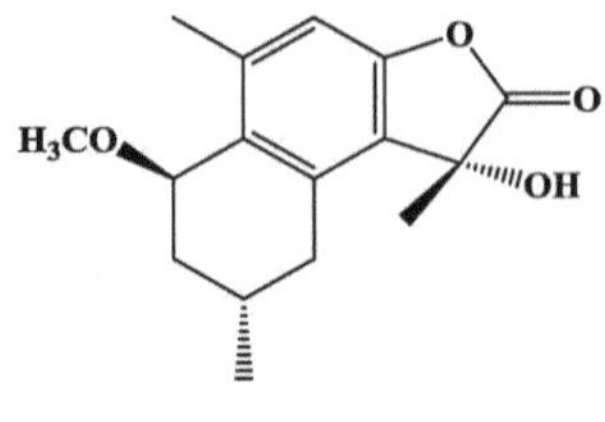

commiterpene C

Fig. 19 Structures of compounds isolated from *C. myrrha* (Xu et al. 2011a)

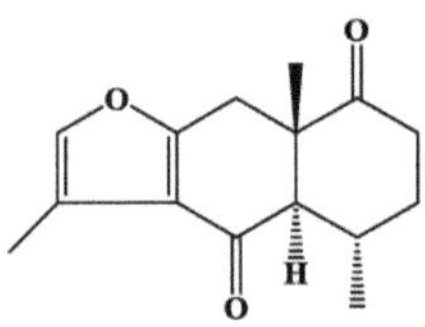

myrrhterpenoid A

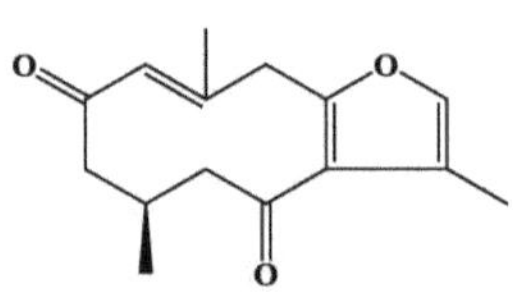

myrrhterpenoid B

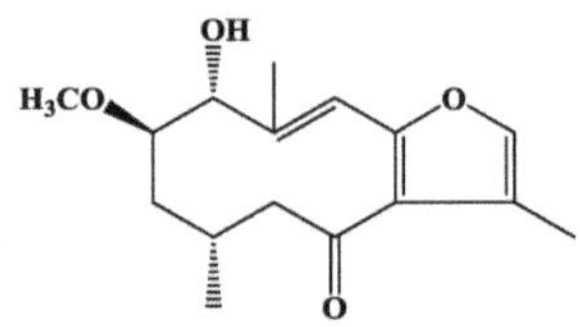

myrrhterpenoid C

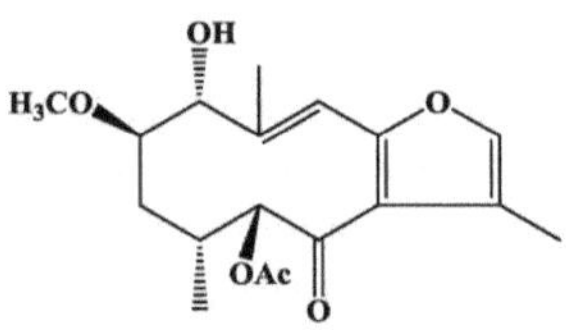

myrrhterpenoid D

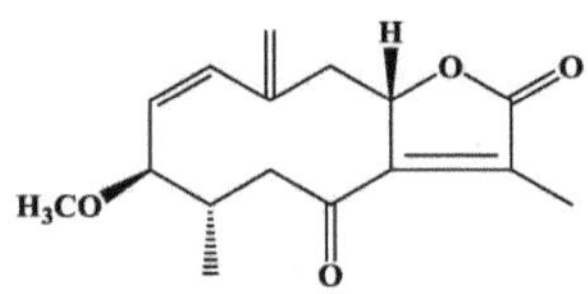

myrrhterpenoid E

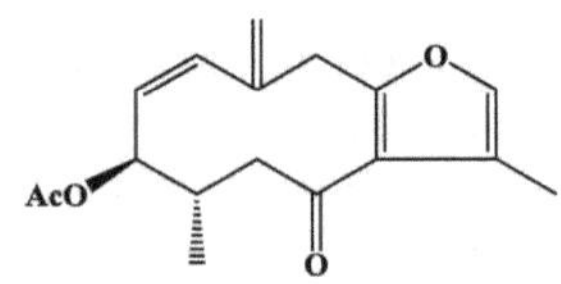

myrrhterpenoid F

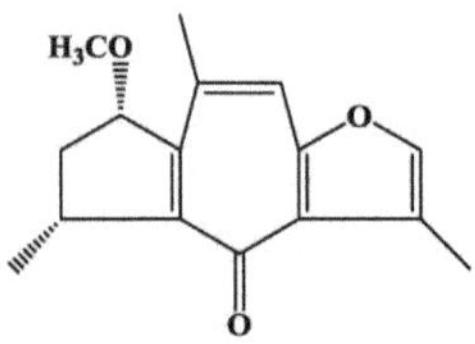

myrrhterpenoid G

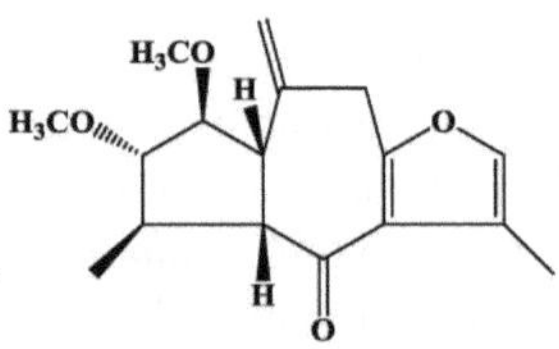

myrrhterpenoid H

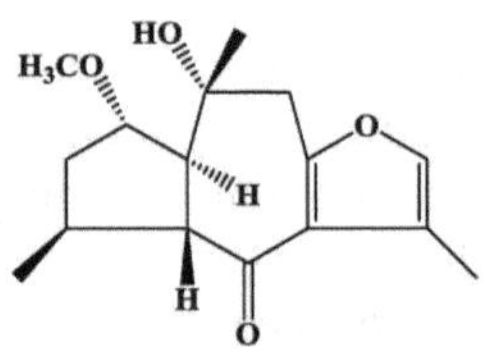

myrrhterpenoid I

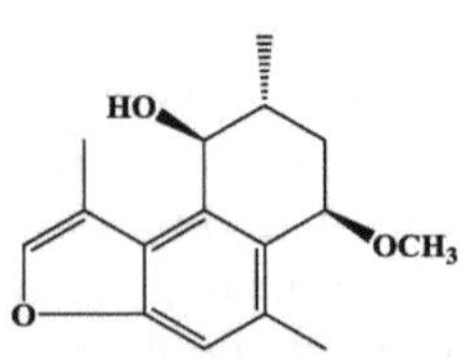

(6*R*,8*R*,9*S*)-*rel*-(+)-6,7,8,9-tetrahydro-6-methoxy-1,5,8-trimethyl-naphtho [2,1-*b*]furan-9-ol

Fig. 20 Structures of compounds isolated from *C. myrrha* (Xu et al. 2011b)

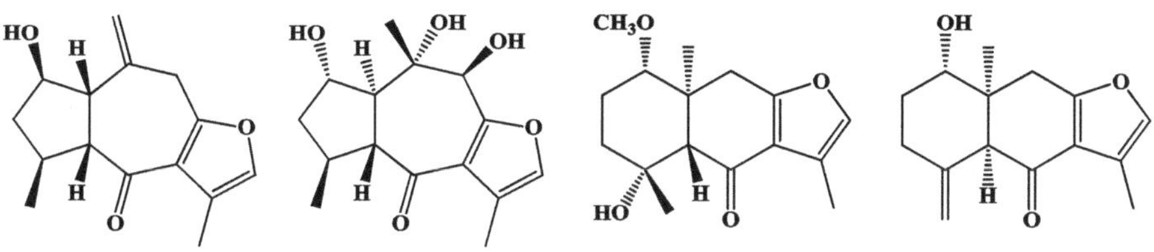

Fig. 21 Structures of compounds isolated from *C. myrrha* (Xu et al. 012)

by the ChemDraw 3D modeling using MM2. All isolated furanosesquiterpenes showed neuroprotective effects against MPP^+-induced neuronal cell death in SH-SY5Y cells.

One year later this scientific group isolated four new sesquiterpenes from *C. myrrha* resin: myrrhterpenoids K–N (Fig. 21) (Xu et al. 2012). Myrrhterpenoids K and N revealed neuroprotective effects against 1-methyl-4-phenylpyridinium -induced neuronal cell death in dopaminergic neuroblastoma SH-SY5Y cells.

Anti-inflammatory and analgesic activity of *C. myrrha* extracts were proved by Shulan et al. (2011). The authors identified seven main compounds in these extracts: 2-methoxy-8,12-epoxygermacra-1(10),7,11-trine-6-one, 2-methoxy-5-acetoxy-furanogermacr-1(10)-en-6-one, myrrhone, sandaracopimaric acid, abietic acid, dehydroabietic acid, and mansumbinone. Later Su et al. (2012) studied myrrh, frankincense and their combination for the treatment of inflammatory pain. The combined extract exhibited significant anti-inflammatory and analgesic activities. Twelve potentially active compounds in both extracts were identified – 2-methoxy-8,12-epoxygermacra-1(10),7,11-trine-6-one, 7-methoxy-3,6,9-trimethyl-6,6α,7,8,9,9α-hexahydroazuleno[4,5-β]furan-4(5H)-one, 2*R*-methoxy-4*R*-furanogermacr-1(10)E-en-6-one, 2-acetoxyfuranodiene, 3,17-dihydroxy-3β-pregn-5-en-20-one, 1,2,3-trihydroxyurs-12-en-28-oic acid, 3-keto-tirucall-7,24-dien-21-oic acid, 3-hydroxytirucall-8,24-dien-21-oic-acid, 3-keto-tirucall-8,24-dien-21-oic acid, acetyl-11-keto-β-boswellic acid, 9,11-dehydro-β-boswellic acid, and α-boswellic acid.

De Rapper et al. (2012) studied the in vitro antimicrobial activity of three essential oil samples of frankincense (*Boswellia rivae*, *Boswellia neglecta* and *Boswellia papyrifera*) and two essential oil samples of myrrh and sweet myrrh (*Commiphora guidotti* and *Commiphora myrrha*). In this study several microorganisms were used, namely *Staphylococcus aureus*, *Bacillus cereus*, *Pseudomonas aeruginosa*, *Escherichia coli*, *C. albicans* and *Cryptococcus neoformans*. Antimicrobial activity was improved for 80 % for frankincense and myrrh essential oil samples, with the best one just for *C. myrrha* oil.

C. molmol did not improve the biochemical parameters of the hepatocarcinogenesis in rats (El-Shahat et al. 2012).

Al-Abdalall (2013) showed that fresh aqueous extracts of *Commiphora myrrha* and *Commiphora molmol* inhibit pathogenic bacteria: *Micrococcus luteus*, *Neisseria*

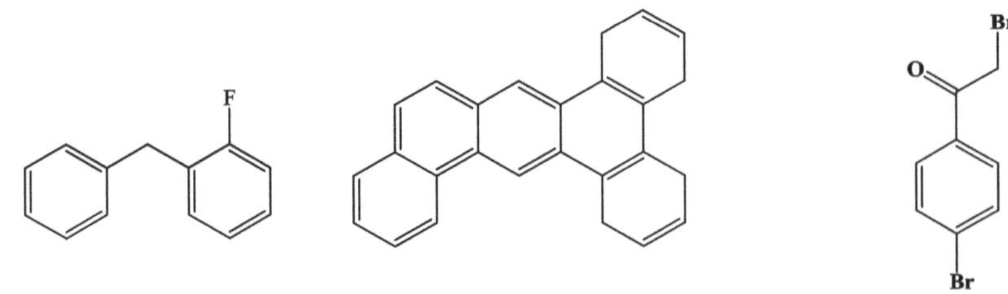

1-benzyl-2-fluorobenzene **tribenzo-1,2,3,4,5,6-anthracene** **2-bromo-1-(4-bromophenyl) ethanone**

Fig. 22 Structures of compounds isolated from *C. myrrha* and *C. molmol* (Al-Abdalall 2013)

siccca, *Proteus mirabilis* and *Pseudomonas aeruginosa*. With the longer storage time of the extracts the inhibition of bacterial growth decreased. Three compounds with inhibitive effect were identified, namely 1-benzyl-2-fluorobenzene, tribenzo-1,2,3,4,5,6-anthracene, and 2-bromo-1-(4-bromophenyl)ethanone (Fig. 22).

El-Mekkawy et al. (2013) isolated from the hexane extract the oleogum resin of *Commiphora wightii* (guggul) three new compounds, epi-mukulin, (Z)-Δ1,2 -dehydroguggulsterone and Δ6,7-dehydro-20-hydroxygugglsterone. They isolated also six known compounds, diasesartemin, (+)-epi-magnolin, (+)-diayangambin, gugglsterol I, (E)-guggulsterone and (Z)-guggulsterone (Fig. 23). α-Glucosidase inhibitory effects of the isolated compounds were evaluated calorimetrically. The hexan soluble fraction showed significant α-glucosidase inhibitory effect. Under the assay conditions, diasesartemin was found to be more potent than the positive control, acarbose, a known α-glucosidase inhibitor.

Commiphora myrrha essential oil showed remarkable in-vitro cytotoxic activity against MCF-7 breast cancer cells (MCF-7 cell line is a plural effusion from human mammary gland adenocarcinoma) using a MTT-based cytotoxicity assay (Simmons et al. 2013).

Patents

Majeed, M., V. Badmaev, K.R. Bammi, B. R. Kumar, S. Prakash, and S. Natarajan. 2002. US Patent No. 20020061929. Composition and method containing products extracted from *Commiphora* sp. for prevention and treatment of abnormal cell growth and proliferation in inflammation, neoplasia, and cardiovascular disease.

Two ferulic acid esters from *Commiphora wightii* were used for the prevention and treatment of abnormal cell growth and proliferation in inflammation, neoplasia, and cardiovascular disease.

Jindal, K. C., C. B. Rao, M. Ramanathan, and B. Suresh. 2005. Patent WO2004069262. Herbal composition comprising *Commiphora mukul*, *Allium sativum* and *Cucuma longa*.

The treatment and/or propylaxis of hypercholesterolemia, atherosclerosis, hyperlipidemia, and hypertension in mammals uses a herbal composition comprising *Commiphora mukul*, *Allium sativum*, and *Curcuma longa*.

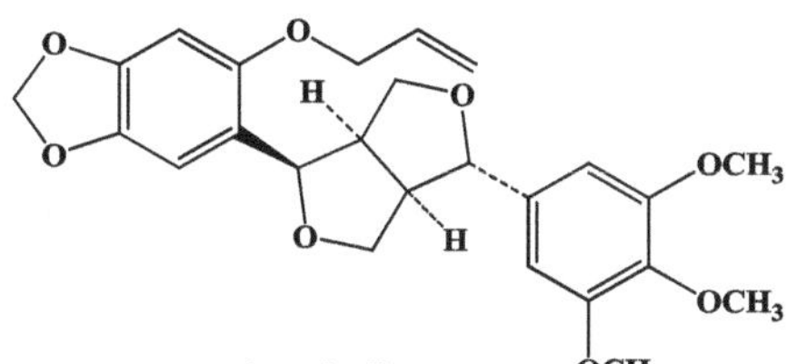

epi-mukulin

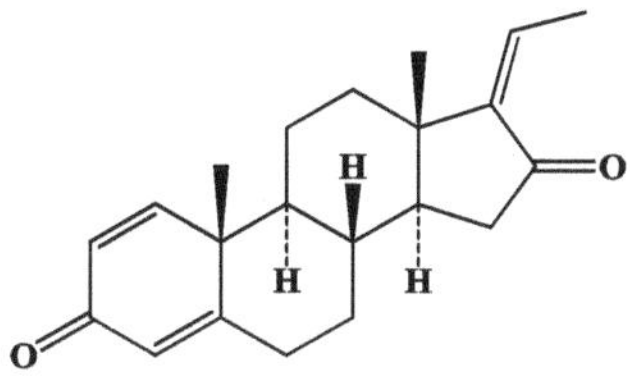

(*Z*)-$\Delta^{1,2}$- dehydroguggulsterone

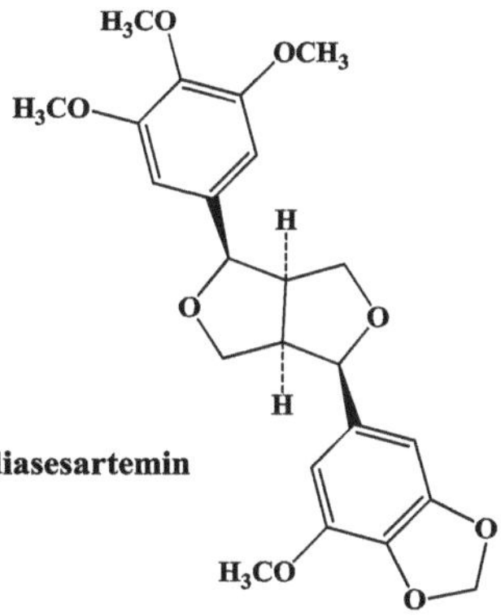

diasesartemin

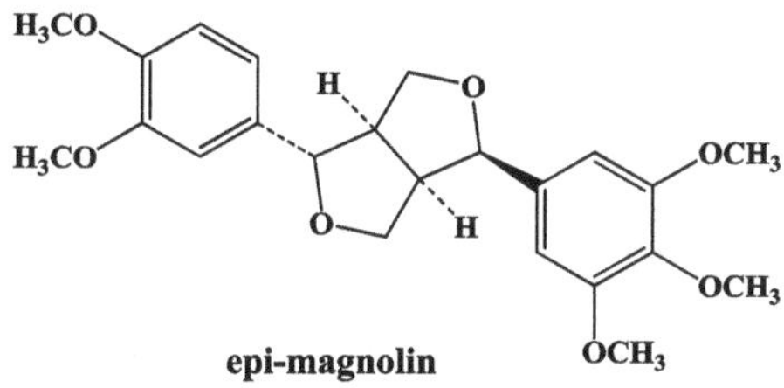

epi-magnolin

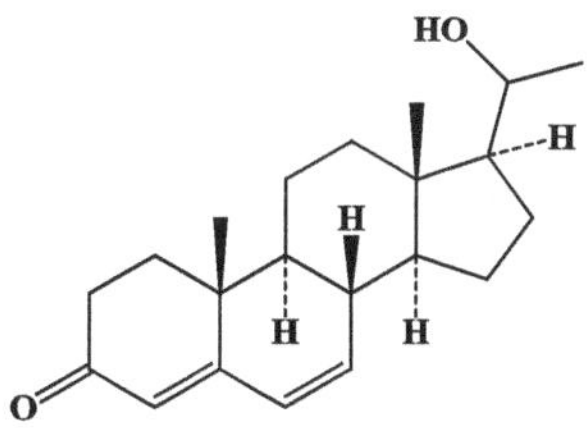

$\Delta^{6,7}$-dehydro-20-hydroxyguggulsterone

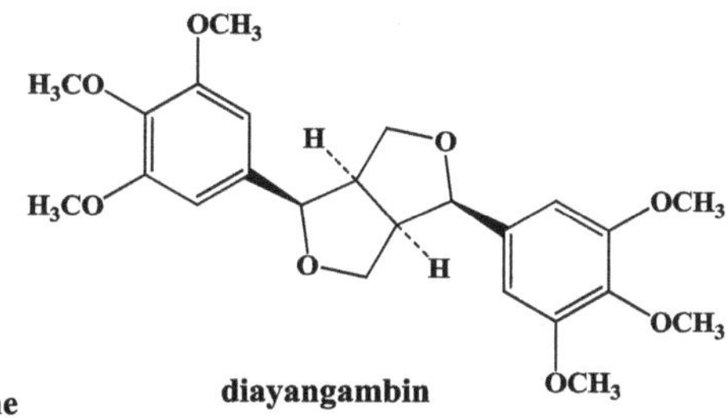

diayangambin

Fig. 23 Structures of compounds isolated from *Commiphora wightii* (El-Mekkawy et al. 2013)

McCook, J.P., J.M. Corey, P.L. Dorogi, J.S. Bajor, H.E. Knaggs, B.A. Lange, E. Sharpe, and E. Tallman. 1997. US Patent No. 5690948. Antisebum and antioxidant compositions containing gugulipid and alcoholic fraction thereof.

Gugulipid (registered and marketed by Sabinsa Corporation) is a standardized extract prepared from the oleogum resin (gum resin) of *C. mukul*, an ingredient in traditional Ayurvedic medicine. The product is claimed to be beneficial for lowering serum cholesterol. Gugulipid (from *C. mukul* or *C. wightii*) may be used in the control of oily skin conditions and protecting the skin from free radical damage.

Pratap, R., R. Pal, S. Singh, G. Shankar, C. Nath, H.K. Singh, D. Raina, A.K. Srivastava, A.K. Rastogi, P.S.R. Murthy, S. Srivastava, O.P. Astana, N. Singh, and N. Nand. 2002. US Patent No. 2003099729. Novel uses of gugulipid as cognition enhancer, antihypergly-cocemic, and for dermal conditions.

Use of an extract from the stem of *Commiphora molmol* (Family *Burseraceae*) comprising additional or alternatively a volatile oil (containing heerabolene, cadinene, elemol, eugenol, cuminaldehyde, numerous furanosesquiterpenes including furanodiene, furanodienone, curzerenone, lindestrene, 2-methoxyfuranodiene and other derivatives), resin (including α-, β-and γ-commiphoric acids, commiphorinic acid, heeraboresene, α- and β-heerabomyrrhols and commiferin) or gum (composed of arabinose, galactose, xylose and 4-0-methylglucuronic acid) and particularly oleogum-resin (Myrrh) in the preparation of a medicament for the treatment of schistosomiasis. Myrrh contains approximately 7–17 % of volatile oil, 25–40 % of resin, 57–61 % of gum and some 3–4 % of impurities. Pharmaceutical compounds are provided together with a pharmaceutically acceptable carrier diluent or excipient.

Olibanum

Major aim of recent research is the therapeutic possibilities of the oleo-gum-resins of *Boswellia*. Boswellic acids, a group of medicinally important compounds, were reviewed and referenced in 276 studies. The studies emphasized anti-inflammatory properties and anticancer potential (Shah et al. 2009). Boswellic acids were found to be effective through topical application in inflammatory disorders (Singh et al. 2008).

The ancient Egyptian medical practices relied strongly on faith and belief in mystical and magical treatments, combined with practical medicinal herbs The Egyptian Ebers Papyrus (1500 BCE), which contains 876 prescriptions, states: Magic is effective together with medicine. Medicine is effective together with magic" (Wreszinski 1912). The Ebers Papyrus cites the use of frankincense in cases of throat and larynx infections, stopping bleeding, reducing phlegm, asthmatic attacks, and stopping vomiting. Pliny (Book 25, Chapter 82) mentioned frankincense as an antidote to hemlock (*Conium maculatum*). The SEPASAL database cites at least 20 different systems of medical disorders for which frankincense has been or is still used as a remedy.

Ibn Sina (Avicenna) in his *Canon of Medicine* of the tenth century (Jahier and Noureddine 1956) recommended frankincense for tumors, ulcers, vomiting, dysentery, and fevers.

Use of frankincense in China was first mentioned in the sixth century CE in the *Mingyi Bielu* (Needham and Lu 1974). Frankincense was called *fanhunxiang* and was used in memorial ceremonies. The prefix "fan-," which means "foreign" or "devil," should be interpreted to mean that the substance was imported. Frankincense is used in herbal medicine in a similar way to myrrh, to quicken the blood circulation and relieve pain. However, unlike myrrh, frankincense also acts on *qi* (the physical life force). An ancient Chinese prescription (*Qi Li San*) is prescribed for all injuries and is made up of dragon's blood, catechu, myrrh, frankincense, carthamus, cinnabar, musk, and borneol. This ointment is the base for Yunnan Bai Yao, a popular remedy today, reputedly carried by the Vietcong during the Vietnam War to stop bleeding from wounds, with apparently amazing success (Yunnan Baiyao Company).

The psychoactivity of *Boswellia* was recognized in ancient times in the Near East and Europe. In India, the traditional Ayurvedic medical systems refer to the use of the gum extracted from *Boswellia serrata*, which is recommended for arthritic and inflammatory conditions, gastric disorders, pulmonary diseases, and skin ailments. It also is reported to have a strong action on the nervous system. Yoga tradition uses frankincense oil for massage and stimulation in arthritic conditions (Miller and Morris 1988).

Patents

There are a number of recent patents involving frankincense and its derivatives, indicating a wide range of suggested applications in medicine. Relevant patents are described next.

Etzel, R. 1997. US Patent 5720975. Use of incense in the treatment of Alzheimer's disease.

This patent concerns the use of incense in the treatment of Alzheimer's disease, citing the production of a medicament composed of olibanum and boswellic acid combined with physiologically acceptable salts.

Simmet, T., and H.P.T. Ammon. 1999. US Patent 6174876. Use of Boswellic acid for treating brain tumors.

Weisman B. 1999. US Patent No. 5888514. Natural composition for treating bone or joint inflammation.

Inventor uses extracts of *Boswellia serrata* and boswellic acid among other materials for treating bone or joint inflammation.

Duranton, A., and O. De Lacharriere. 2002. US Patent 6465421. An application for modulating body/cranial hair growth, using boswellic acid as a possible ingredient.

Inventors presented an application for modulating body/cranial hair growth using boswellic acid as a possible ingredient.

Meybeck, A., and A. Zanvit. 2004. US Patent 20040166178. 3-O-acetyl-11-ketoboswellic acid for relaxing the skin.

The present invention relates to the use of 3-O-acetyl-11-ketoboswellic acid (AKBA), a plant extract of *Boswellia serrata*, as an agent to soften lines and/or relax the skin.

This patent relates to the use of 3-O-acetyl-11-ketoboswellic acid (AKBA)), a plant extract of *Boswellia serrata*, as an agent to soften lines and/or relax the skinand reduce wrinkles.

Ali, A. Y., and I. D. Bowen. 2007. US Patent 20070264363. Molluscicidal and Anti-Barnacle Compounds.

This patent, by inventors A. Ali and I. D. Bowen, relates to the use of plant material of the *Burseraceae* as a terrestrial molluscicidal and/or molloscrepel-lant agent.

Shanahan-Prendergast, E. 2004. US Patent Application No. 20040092583. A treatment for inhibiting neoplastic lesions using incensole and/or furanogermacrens.

Inventor describes a treatment for inhibiting neoplastic lesions using incensole and/or furanogermacrens. The invention discloses the use of incensole and/or furanogermacrens, derivative metabolites, and precursors thereof in the treatment of neoplasia, particularly resistant neoplasia, and immunodysregulatory disorders.

Ammon, H.P.T., and H. Safayhi. 2005. US Patent Application No. 20050209169. Use of boswellic acid and its derivatives for inhibiting normal and increased leucocyte elastase or plasmin activity.

Inventors describe the use of boswellic acid and its derivatives for inhibiting normaand increased leucocyte elastase or plasmin activity, for treatment especially in the case of pulmonary emphysema, acute respiratory distress syndrome, shock lung, cystic fibrosis, chronic bronchitis, glumerulonephritis, and rheumatoid arthritis.

Hwa, J.Y. 2007. US Patent 7223423. Skin treatment composition.

A skin treatment composition, comprising an effective combination of ingredients selected from cumin, cloves, peach kernel, olibanum, eagle wood, giant hyssop, almond, and pachira macracarpa is provided. The composition can be used as a skin cleanser, as a deodorant, and to treat a wide array of skin problems, including signs of aging, such as wrinkles, and skin sagging, dark spots, skin infections, skin irritation, cuts, scarring, acne, cold sores, chapped lips, and varicose veins. The composition is preferably formulated in water, as a mud pack, or bath preparation. Other suitable delivery systems may be included. The composition can further be used for hair treatment, to thicken hair roots, and to prevent hair loss. The composition can also be used for vaginal treatment, including for yeast and other infections, vaginal discharge, and to promote recovery after a vaginal birth and episiotomy. Such treatment is especially effective when the composition is used as a bath preparation. The composition further has a soothing and relaxing effect especially when used in a bath.

Qazi, G.N., S.C. Taneja, J. Singh, A.K. Saxena, V.K. Sethi, B.A. Shah, B.K. Kapahi, S.S. Andotra, A. Kumar, S. Bhushan, F. Malik, D.M. Mondhe, S. Muthiah, S. Singh, M. Verma, and S.K. Singh. 2009. US Patent Application 20090298938. Use of semi synthetic analogues of boswellic acids for anti-cancer activity.

Jauch, J. 2002. German Patent No. 085921. A simple method for the synthesis of Boswellic acids and derivatives thereof.

Inventor J. Jauch relates to a method for producing a pure boswellic acid from a boswellic acid mixture, comprising these steps: acetylation by a suitable acetylation reagent or deacetylation by a suitable deacetylation reagent and oxidation by a suitable oxidation reagent or reduction by a suitable reduction reagent.

Striggow, F., W. Schmidt, and T. Mack. 2004. Patent EP 04721524. Use of incense or hydrogenation products for preventing and/or treating a cerebral ischemia, and/or cerebral traumatic lesion, and/or Alzheimer's disease.

This patent relates to the use of incense or hydrogenation products for preventing and/or treating cerebral ischemia and/or cerebral traumatic lesion and/or Alzheimer's disease.

Taneja, S.C., V.K. Sethi, K.L. Dhar, and R.S. Kapil. 1997. Patent No. 5629351. Boswellic acid compositions and preparation thereof.

Disclosed herein is a novel fraction comprising a mixture of boswellic acids, wherein the fraction exhibits anti-inflammatory and antiulcerogenic activities. Also disclosed is a novel boswellic acid compound exhibiting anti-inflammatory, antiathritic and antiulcero-genic activities. Also disclosed is a process for isolating a boswellic acid fraction and individual boswellic acids therefrom.

Recent Additional Research

Boswellic acid extracted from *B. serrata*, in an experimental model of irritable bowel syndrome, reduced inflammation after administration. The conclusion was that the anti-inflammatory actions of the *Boswellia* extract may be due in part to AKBA, e.g. acetyl-11-keto-boswellic acid (Krieglstein et al. 2001).

In cases of chronic colitis, a gum resin from *B. serrata* was shown to be an effective treatment, with minimal side effects (Gupta et al. 2001).

Compared to indomethacin, AKBA significantly inhibited angiogenesis (Singh et al. 2007) and was found to have antiproliferative and apoptotic effects on metastases in human HT-29 cells (Lui et al. 2002).

Chemicals derived from, among others, *Boswellia* plants used as mixed formulations are potent in curing inflammatory diseases (Darshan and Doreswamy 2004).

Incensole obtained from the dried bark of *Boswellia dalzielii*, a species growing in West Africa, contained strong antimicrobial and antioxidant activity, but incensole itself was only moderately active (Alemika et al. 2004). Similar results were obtained with olibanum from *B. carterii* and *B. sacra* (Hamm et al. 2003, 2005) while *B. serrata* contained an unidentified sesquiterpene. Broad-spectrum inhibition against bacteria and fungi was obtained with *B. dalzielii* (Adelakun et al. 2001).

In research on aging-associated abnormalities in mice, it was suggested that acetyl-11-keto-boswellic acid (AKBA) may provide a new therapeutic innovation for the treatment of aging-related brain disorders, such as Alzheimer's disease and different motor dysfunctions with adequate gastrointestinal tolerability (Bishnoi et al. 2005).

Boswellic acid and its derivatives were used for inhibiting normal and increased leucocyte elastase or plasmin activity (Ammon and Safayhi 2005).

Gum resin extracts of *Boswellia* species experimentally tried in animal models and studies in human subjects confirmed their potential for the treatment of not only inflammations but also of cancer (Poeckel and Werz 2006).

Ammon (2006) concluded that oleogum resins from *Boswellia* species have efficacy in some autoimmune diseases, including rheumatoid arthritis, Crohn's disease, ulcerative colitis, and bronchial asthma. Side effects were not severe when compared to modern drugs used for the treatment of these diseases.

B. serrata gum resin extract prevents diarrhea and normalizes intestinal motility, which explains the clinical efficacy of this Ayurvedic remedy in reducing diarrhea in rodents with inflammatory bowel disease (Borrelli et al. 2006).

The *Boswellia* resin is the natural defense of the tree reacting to the trauma of a wound; the polyphenols present in the gum offer protection against fungus and pests. As a skin treatment for dyshidrosis and related skin disorders and a wide array of skin problems, including signs of aging, such as wrinkles, skin sagging, dark spots, skin infections, skin irritation, cuts, scarring, acne, cold sores, chapped lips, and varicose veins, olibanum and boswellic acid are combined with other materials for supposedly safe and effective therapy (Hwa 2007).

The anti-inflammatory action of the boswellic acids is similar to that of the conventional nonsteroidal anti-inflammatory drugs (NSAID). However the NSAID often cause joint damage by inhibiting glycosaminoglycan synthesis, whereas boswellic acids do not have this undesirable action, making them a potential choice for long-term treatment. The crude methanolic extract and the isolated pure compound are capable of carrying out a natural anti-inflammatory activity at sites where chronic inflammation is present (Gayathri et al. 2007).

Boswellia serrata is the subject of many research studies in India, where this species grows. A double-blind, randomized, placebo-controlled study of the efficacy and safety of 5-Loxin (an enriched *Boswellia serrata* extract) for treatment of osteoarthritis of the knee was performed on 75 patients. 5-Loxin was observed to reduce pain and improve physical functioning and considered safe for human consumption (Sengupta et al. 2007).

A photochemical study by Sharma et al. (2007) concluded that *B. serrata* is a potent and safe alternative to conventional NSAIDs

Research on the effects of incense on humans at the Hebrew University of Jerusalem indicated that one of the major active ingredients of frankincense is incensole acetate (Moussaieff et al. 2007, 2008). The researchers demonstrated that incensole acetate lowers anxiety and causes antidepressive-like behavior in mice.

The effect of AKBA from *B. carterii* on the development of atherosclerotic lesions showed a significant reduction in the expression of several proathero-genic genes, NF-kB activity, and lesion size in treated mice (Cuaz-Perolin et al. 2008).

The inhibitory effect of AKBA-containing drugs on prostate cancer cells showed that this material could be used for the development of novel therapeutic chemicals (Yuan et al. 2008).

Boswellic acids from *B. serrata* gave a protective effect on gastric ulcers in rats (Singh et al. 2008).

Boswellic acids were found to be effective through topical application in inflammatory disorders (Singh et al. 2008).

Studies carried out by Shah et al. (2009) emphasized the anti-inflammatory properties and anticancer potential.

Six new terpenes, olibanumols A, B, C, H, I, and J (Fig. 24), were isolated from an extract of exuded gum-resin of *Boswellia carterii*, together with seven known terpenoids (3,6-dihydroxy-*p*-menth-1-ene, *p*-menth-1-en-4α,6β-diol, (−)-*trans-sobrerol*, *p*-menth-4-en-1,2-diol, *p*-menth-5-en-1,2-diol, isofouquierol, and epilupeol) (Yoshikawa et al. 2009). The extract showed anti-inflammatory effect. Olibanumols A, H, and I, and isofouquierol, revealed inhibitory effects on production of NO in LPS-activated macrophages.

Frankincense oil appears to distinguish cancerous (bladder transitional cell carcinoma J82) from normal bladder cells (urothelial cells) and suppress cancer cell viability (Frank et al. 2009)

Three purified tetracyclic triterpenoids: 3-oxo-tirucallic acid, 3-α-acetoxy-tirucallic acid, and 3-β-acetoxy-tirucallic acid (Fig. 25) from the oleogum resin of *Boswellia carterii* induced apoptosis in prostate cancer, but not in nontumorigenic cells. Thus, tirucallic acid derivatives represent a new class of Akt inhibitors with antitumor properties (Estrada et al. 2010).

Van Vuuren et al. (2010) compared nine samples of *B. carterii* and two samples of *B. sacra* for the main terpenes (see Table 3) and evaluated their antimicrobial

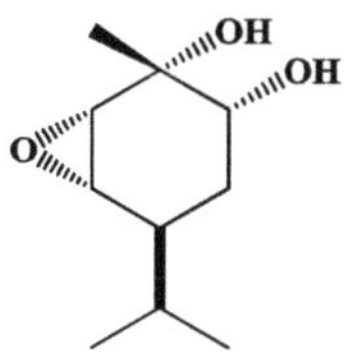
olibanumol A

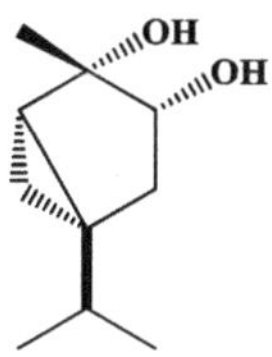
olibanumol B

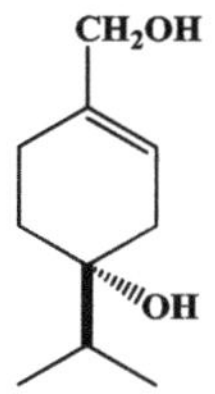
olibanumol C

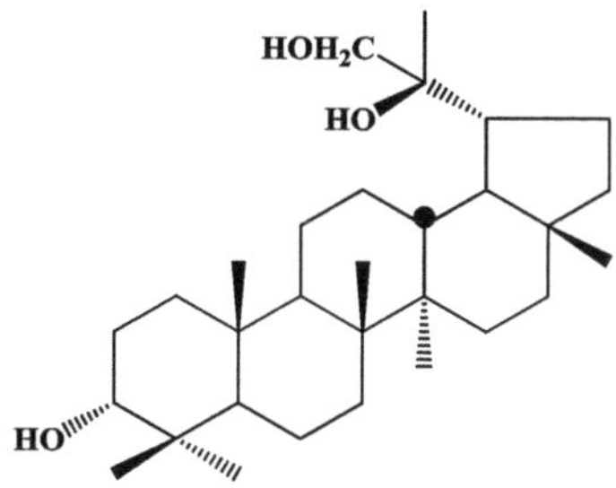
olibanumol H

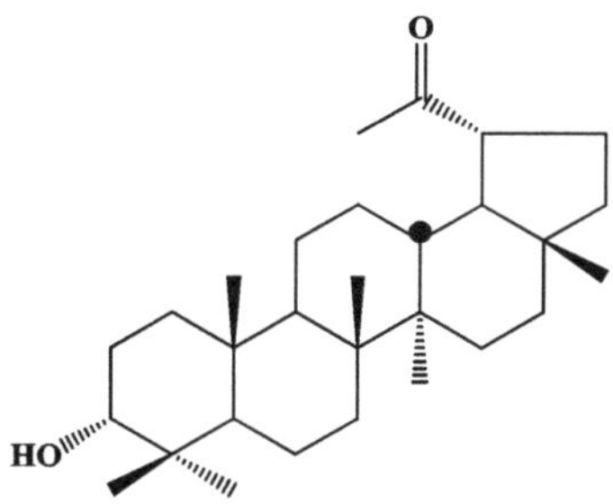
olibanumol I

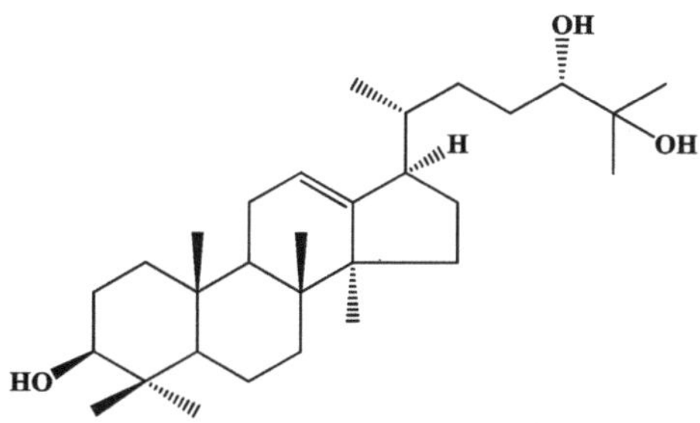
olibanumol J

Fig. 24 Structures of compounds isolated from *B. carterii* (Yoshikawa et al. 2009)

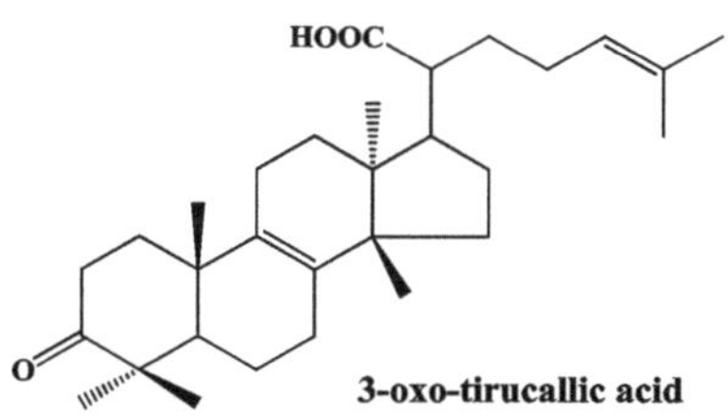
3-oxo-tirucallic acid

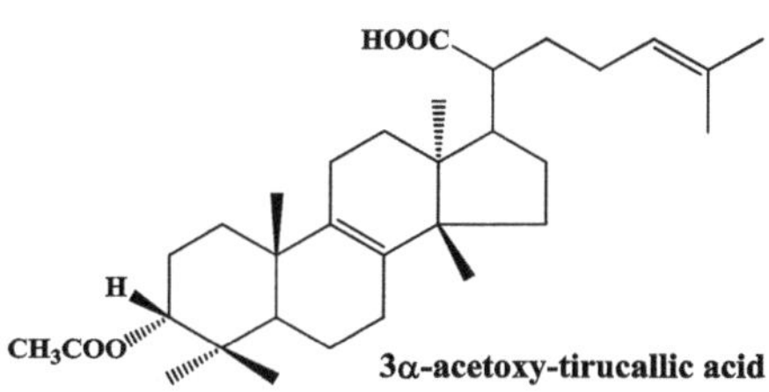
3α-acetoxy-tirucallic acid

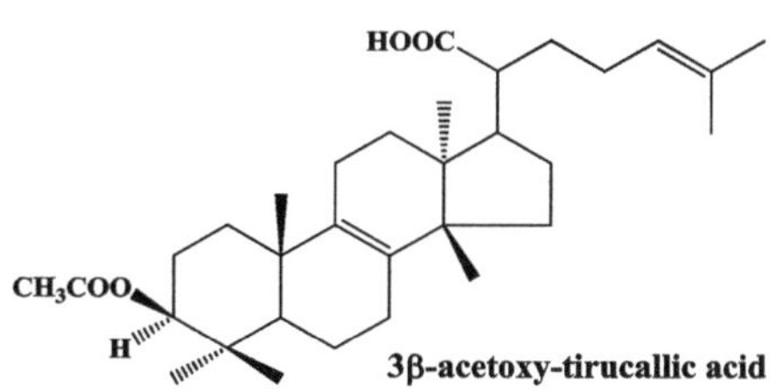
3β-acetoxy-tirucallic acid

Fig. 25 Structures of compounds isolated from *B. carterii* (Estrada et al. 2010)

Table 3 Comparison of *B. carterii* and *B. sacra* samples for the main terpenes

Boswellia carterii	*Boswellia sacra*
α-pinene (4.8–40.4 %)	α-pinene (18.3–22.5 %)
α-thujene (1.6–52.4 %)	α-thujene (3.9–11.2 %)
Limonene (2.6–20.4 %)	Limonene (11.2–13.1 %)
p-cymene (3.6–6.2 %)	*p*-cymene (4.7–5.9 %)
β-caryophyllene (0.3–10.5 %)	β-caryophyllene (7.2–7.6 %)

Van Vuuren et al. (2010)

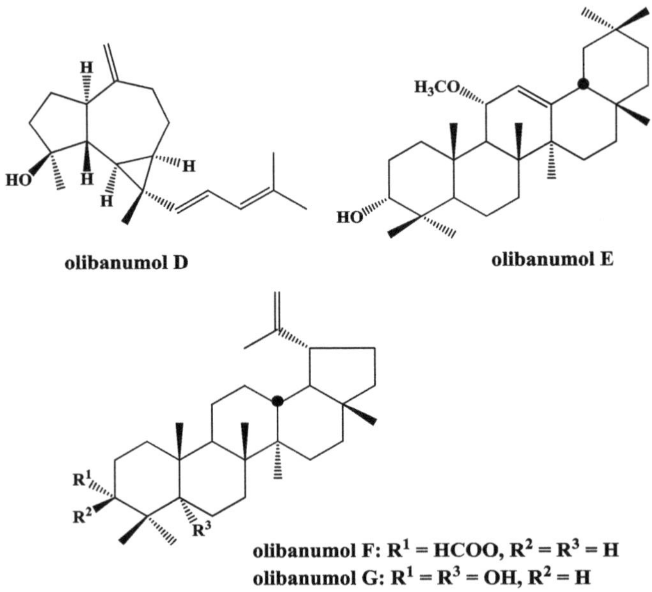

Fig. 26 Structures of compounds isolated from *B. carterii* (Morikawa et al. 2011)

activity against *S. aureus*, *B. cereus*, *E. coli*, *P. vulgaris*, and *C. albicans*. The essential oils exhibited good to poor in vitro antimicrobial activity.

Four new compounds, olibanumol D–G (Fig. 26), were isolated from the exuded gum-resin from *Boswellia carterii* (Morikawa et al. 2011). Olibanumols D and E showed significant inhibitory effects on nitric oxide (NO) production in lipopolysaccharide (LPS)-activated mouse peritoneal macrophages.

Boswellia sacra essential oil has tumor cell-specific cytotoxicity in multiple cancer cell types. Suhail et al. (2011) proved that this essential oil may induce breast cancer cell-specific cytotoxicity and may be effective for advanced breast cancer.

Four different oleo-gum resins from *Boswellia sacra* from the Dhofar region of Oman, with the presence of α-pinene as the principal constituent (Table 4), revealed pronounced activity against Gram-positive (*Bacillus subtilis*, *Micrococcus luteus*, *Staphylococcus aureus*) and Gram-negative (*Pseudomonas aeruginosa*, *Klebsiella pneumoniae*, *Enterobacter aerogenes*, *Escherichia coli*, *Proteus vulgaris*) bacteria

Table 4 The main terpenes in the *Boswellia sacra* samples from Oman

α-pinene (46.8–76.0 %)
Camphene (1.4–2.4 %)
β-pinene (1.3–2.0 %)
Myrcene (1.0–8.9 %)
α-phellandrene (0.6–3.2 %)
p-cymene (1.6–2.7 %)
Limonene (1.7–15.9 %)
p-mentha-1,5-dien-8-ol (1.1–3.4 %)
(*E*)-caryophyllene (0.5–1.5 %)
β-elemene (0.9–2.6 %)
β-selinene (1.2–1.8 %)
α-selinene (0.5–1.1 %)

Al-Saidi et al. (2012)

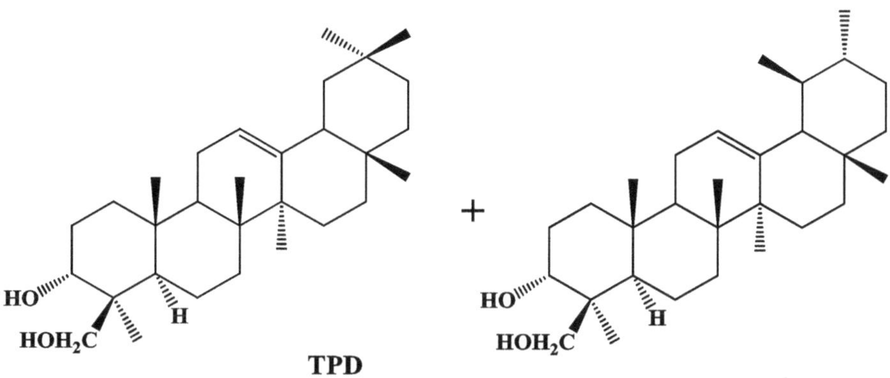

Fig. 27 Two isomers of a new pentacyclic triterpenediol from *Boswellia serrata* (Bhushan et al. 2013)

(Al-Saidi et al. 2012). The main compounds in essential oil of these four *Boswellia sacra* oleo-gum resins are shown in the following table:

Frankincense essential oil from *Boswellia sacra* suppresses viability and induces apoptosis of a panel of human pancreatic cancer cell lines (Xiao et al. 2012). Thus these authors suggested that this essential oil might be a useful alternative therapeutic agent for treating patients with pancreatic adenocarcinoma, an aggressive cancer with poor prognosis.

A novel anticancer compound was isolated from the gum of *Boswellia serrata* (Bhushan et al. 2013, Fig. 27). This pentacyclic triterpenediol (TPD) has significant cytotoxic and apoptotic potential in a large number of human cancer cell lines (Sarcoma-180 solid tumor bearing mice).

In order to assess the effect of *B. carterii* smoke, after 1 week acclimatization period, rats were randomly divided into three groups, groups 1, 2 and 3 with each group consisting of 11 animals. Each group of rats was housed separately from the other groups to avoid the cross exposure to incense smoke. Rats in group 1 served as control and were kept in normal fresh air, while rats in group 2 and 3 were exposed to *B. papyrifera* and *B. carterii* smoke, respectively, in a smoking chamber. The rats were exposed daily to the smoke emanating from the burning of 4 g of each incense material for 4 months. Smoke exposure durations lasted for 30–40 min/day. The dose and duration of incense exposure followed in this study was based on the optimized conditions from our previous studies. At the end of exposure duration, animals were killed by cervical dislocation. *B. carterii* smoke affects the process of spermatogenesis and sperm parameters in male albino rats and indicates the detrimental effects of these incense materials on the human reproductive system. Sperm analysis in rats exposed to the *B. carterii* smoke exhibited a significant decrease in their sperm count. This fact shows that incense smoke could affect men who are constantly exposed indoors to this incense (Mukhtar et al. 2013).

Ethyl alcohol extract of the gum resin of Boswellia carterii from China revealed after isolation nine new prenylaromadendrane-type diterpenes named boscarterol A–I. All compounds (except compounds D and G) exhibited moderate hepatoprotective activity against D-galactosamine-induced toxicity in HL-7702 cells (Wang et al. 2013) (Fig. 28).

Conclusions

The ancient resinous plants that produce frankincense and myrrh are, at present, in a fragile condition economically and culturally, and their use is declining. This spice industry, which earned enormous sums of money during the classical period and acquired for Arabia a great fortune, has dwindled down to 1,500 t of export, collected by nomadic people from wild trees mainly in Somalia, for a very meager financial value. The major question is whether this industry can be revived at all. Perhaps the modern world is no longer interested in these exotic spices. Our opinion is that a renewal of interest is emerging, not only because of ongoing exotic and religious appeal but due to the medical potential of several ingredients of these ancient spices. The thrust of recent research suggests efficacy of extracts of these spices and supports the documented prescriptions of the famous physicians throughout history, such as Galen, Avicenna, and Maimonides.

The search for new molecules has turned to ethnobotany and ethno-pharmacognosy as guides to lead chemists toward different sources and classes of compounds (Ben Yehoshua and Mercier 2005; Gurib-Fakim 2006). Fabre (2003) conducted a study to analyze the Roman pharmacopoeia of spices "with reference to modern criteria." He concluded that a new discipline is ready to emerge: archeo-pharmacology, aiming towards a drug research based on ancient texts. It remains to be seen if ancient herbal medicines will make a reentry in the twenty-first century.

Some attempts have been made to reestablish the roles of frankincense and myrrh in modern natural medicines, and this could promote a revival in interest. If proven efficacious, the medicinal use of these species could provide a new source of income in the disadvantaged societies where these spices grow wild. For example, in Ethiopia,

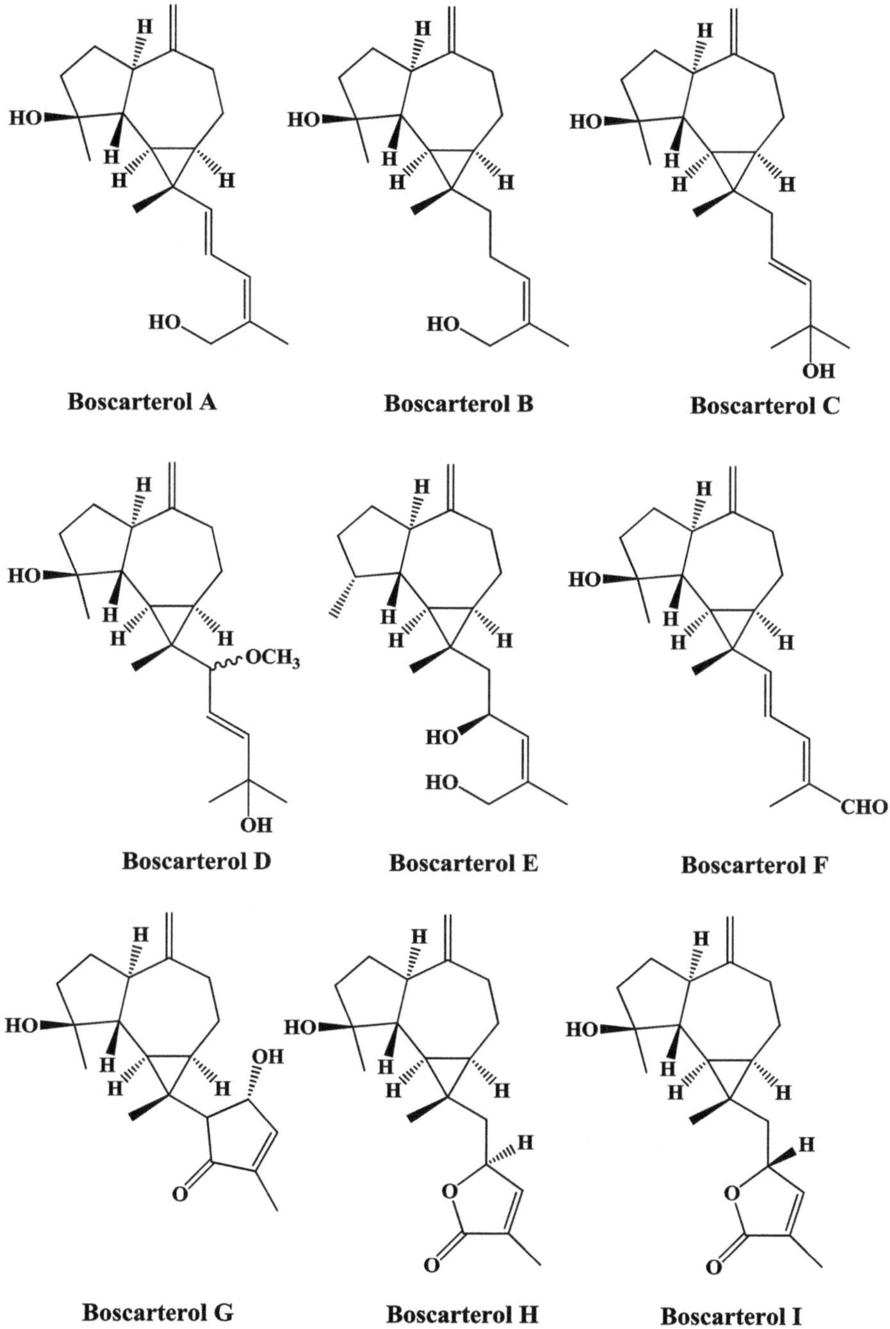

Fig. 28 Nine new boscarterols from *Boswellia carteri* (Wang et al. 2013)

where some export of spices occurs, rehabilitation of these ancient spice crops could provide a new source of income for the local population. Currently, replanting projects in the wild are under way in Ethiopia and Somalia. However, local scientists in Ethiopia report that the future of these projects is far from being assured. Today spices are such a minute item of export from Oman, Yemen, Ethiopia, and Somalia that they are not listed in databases of exports (Index Mundi, the FAO, and USDA). New initiatives and resources from the developed countries are required to reawaken interest in the neglected treasures of the ancient spices, which could have potential for new drugs. This review suggests that frankincense and myrrh are good candidates to start this effort.

Is it possible to revive the growth of the extinct apharsemon in the Dead Sea Basin in Israel? Most researchers with some familiarity with this highly reputed ancient spice would not consider this likely. This review has presented many reasons for this pessimistic view, including the vague identification of this plant, *Commiphora gileadensis* or *C. opobalsamum* by Forsskål and Linnaeus, as well as our lack of any remaining plants or even a residue of the plants that grew in the Dead Sea Basin. Pessimists claim that the glory of *C. gileadensis* belongs only to the past. However, some active researchers, including the present authors, believe that this plant still has a future due to its special medical characteristics established over a period longer than 1,000 years by the best physicians of many cultures.

A common exercise in modern biotechnology is the derivation of new, previously unknown medications from wild plants gathered in remote places, such as the Amazon. However, a more promising approach might be to trace medicinal plants of antiquity that is well adapted to the environment and known by tradition to produce medications. Such a project could lead to the restoration of the production of the ancient apharsemon in the Dead Sea Basin. Strengthened by this thesis, these researchers and several farmers in the Dead Sea Basin and elsewhere have managed, with the help of colleagues in other parts of the world, to raise many thousands plants of *Commphora gileadensis* opobalsamum, which are the closest candidates available for the ancient apharsemon. The exact identity of these plants is under studies. These plants exude the exclusively fragrant liquid resin that resembles what has been described for the ancient apharsemon. Furthermore, these plants grow well in the Dead Sea Basin, both in Ein Gedi and Almog near Jericho. The new discovery is that the resin from these plants has exhibited impressive cytotoxic activity against several cancer cell lines of lymphoma and carcinoma both of human and of mouse. Furthermore the major component – 20 % – of the essential oil of the apharsemon is β-caryophyllene and this compound was found at the first time to the best of our knowledge, to induce apoptosis in cell culture of the cancer cells studied. These cancer cells died because of this induction of apoptosis. Normal fibroblast cells were not affected by this beta caryophyllene. This compound is allowed by the Food and Drug Administartion of the USA and other countries as a food additive and is present in many foods of our diet. A request for a patent on our finding is pending. All these data would facilitate greatly our research to advance the development of apharsemon and the β-caryophyllene in the struggle against cancer. This research is just beginning, but the enthusiasm of the researchers is high. Other reports about anticancer activity by the resin of *C. gileadensis* were cited in this review.

Acknowledgments It is so pleasant to acknowledge the excellent help given by Ms. Carole Borowitz along the writing of this manuscript and to Ms. Idit Sofer for her special highly professional help toward the comprehensive collection of the literature along the last 1,500 years on the taxonomical identification of the apharsemon plant.

The senior author acknowledges the collaboration of the apharsemon research team and to the apharsemon growers in Israel whose enthusiasm and pioneering spirit is so commendable.

References

Abbas FA, Al-Massarany SM, Khan S, Al-Howiriny TA, Mossa JS, Abourashed EA (2007) Phytochemical and biological studies on Saudi *Commiphora opobalsamum L.* Nat Prod Res 21(5):383–391

Abdul-Ghani AS, Amin R (1997) Effect of aqueous extract of *Commiphora opobalsamum* on blood pressure and heart rate in rats. J Ethnopharmacol 57(3):219–222

AboMadyan AA, Morsy TA, Motawea SM (2004) Efficacy of myrrh in the treatment of schistosomiasis in Esbet El-Bakly. Egypt J Egypt Soc Parasitol 34(2):423–426

Adams F (trans and ed) (1844) The seven books of Paul of Aegina, vol VII. Sydenham Society, London, p 22

Adelakun EA, Finbar EA, Agina SE, Makinde AA (2001) Antimicrobial activity of *Boswellia dalzielii* stem bark. Fitoterapia 72(7):822–824

Ahmed F, Ali M, Singh O (2006) New compounds from *Commiphora myrrha* (Nees) Engl. Pharmazie 61(8):728–731

Al-Abdalall AHA (2013) Antibacterial properties and phytochemical analysis of aqueous extract of oleo-gum resins of *Commiphora myrrha* and *Commiphora molmol*. Can J Pure Appl Sci 7(2):2315–2323

Al Hubaishi A, Muller-Hohenstein K (1984) An introduction to the vegetation of Yemen. Ecological basis, floristic composition, human influence. GTZ, Eschborn

Alcorn J (1984) Huastec ethnobotany. University of Texas Press, Austin

Aldred C (1961) The Egyptians. Thames and Hudson, London

Aldredge LM (trans and ed) (1996) Benvenutus Grassus. The wonderful art of the eye. Michgan State University Press, E. Lansing

Alemika TE, Onawunmi GO, Olugbade TA (2004) Isolation and characterization of incensole from *Boswellia dalzielii* stem bark. J Pharm Bioresour 1(1):7–11

Al-Harrasi A, Al-Saidi S (2008) Phytochemical analysis of the essential oil from botanically certified oleogum resin of *Boswellia sacra* (Omani Luban). Molecules 13:2181–2189

Al-Harrasi A, Ali L, Ur Rehman N, Hussain J, Hussain H, Al-Rawahi A, Rizvi TS (2013) 11α-Ethoxy-β-boswellic acid and nizwanone, a new boswellic acid derivative and a new triterpene, respectively, from *Boswellia sacra*. Chem Biodivers 10:1501–1506

Al-Howiriny T, Al-Sohaibani M, Al-Said M, Al-Yahya M, El-Tahir K, Rafatullah S (2004) Hepatoprotective properties of *Commiphora opobalsamum* (Balessan). Drug Exp Clin Res 30(5–6):213–220

Al-Howiriny T, Al-Sohaibani M, Al-Said M, Al-Yahya M, El-Tahir K, Rafatullah S (2005) Effect of *Commiphora opobalsamum* (L) Engl. (Balessan) on experimental gastric ulcers and secretion in rats. J Ethnopharmacol 98(3):287–294

Al-Massarany SM, Abbas FA, Demirci B, Baser KHC, Khan SI, Al-Rehaily AJ, Mossa JS, Abourashed EA (2007) The chemical composition and biological evaluation of the essential oil of *Commiphora opobalsamum* L. J herb Spice Med Plant 13(4):111–121

Al-Saidi S, Rameshkumar KB, Hisham A, Sivakumar N, Al-Kindy S (2012) Composition and antibacterial activity of the essential oils of four commercial grades of Omani Luban, the oleo-gum resin of *Boswellia sacra* Flueck. Chem Biodivers 9:615–624

Amar Z (1998) The ash and the red material from Qumran. Dead Sea Discov 5(1):1–15

Amar Z (2002) The book of incense (in Hebrew). Erez, Tel Aviv. American Scientist. 2010. http://heritage-key.com
Amiel E, Dudai N, Rabinsky T, Rachmilevitch S, Ofir R, Ben-Yehoshua S (2011) Compositions comprising Beta–Caryophyllene and methods of utilizing the same. Patent Pending
Amiel E, Ofir R, Dudai N, Soloway E, Rabinsky T, Rachmilevitch S (2012) β-Caryophyllene, a compound isolated from the biblical balm of Gilead (*Commiphora gileadensis*), is a selective apoptosis inducer for tumor cell Lines. Evid Based Complement Altern Med 2012:872394
Ammon HP (2006) Boswellic acids in chronic inflammatory diseases. Planta Med 72(12):1100–1116
Ammon HPT, Safayhi H (2005) US Patent Application No. 20050209169
Anawati G (1987) La traite d'Averroes sur la theriaque et ses antecedents Grecs et Arabes. Quaderni di Studi Arabi 5–6:36–38. Venice
Anurekha J, Gupta VB (2006) Chemistry and pharmacological profile of Guggul: A review. Indian J Tradit Knowl 5(4):478–483
Arnott GAW (1839) Descriptions of some new or rare Indian plants. Ann Mag Nat Hist 3:85–92
Ashry KM, El-Sayed YS, Khamiss RM, El-Ashmawy IM (2010) Oxidative stress and immunotoxic effects of lead and their amelioration with myrrh (*Commiphora molmol*) emulsion. Food Chem Toxicol 48:236–241
Aurrens C, Laperdrix C (2008) Cosmetic composition for skin firmness and drainage. FR 2906718 A1 20080411
Ayalon E (1995) The iron age II pottery assemblage from Horvat Teiman (Kuntillet Ajrud). J Inst Archaeol Tel Aviv Univ 22:141–205
Baack LJ (2013) A naturalist of the northern enlightment: Peter Forsskål after 250 years. Arch Nat Hist 40(1):1–19
Babbitt FC (trans) (1928) Plutarch. Moralia: Isis and Osiris. Harvard University Press, Cambridge, MA
Badria F, Abou-Mohamed G, El-Mowal-fy A, Masoud A, Salama O (2001) Mirazid: a new schistosomicidal drug. Pharm Biol 39(2):127–131
Balfour B (1888) Botany of Socotra, vol 31, Transactions of the Royal Society of Edinburgh. R. Grant, Edinburgh
Barag D, Porat Y, Netzer E (1981) The synagogue at Ein Gedi. In: Levine LI (ed) Ancient synagogues revealed. The Israel Exploration Society, Jerusalem, pp 116–119
Bardel S (1935) Les dermatoses par bois toxiques. Thesis, Paris. Cited by Hausen BM (1970) in Untersuchungen über Gesundheitsschädigende Hölzer. Thesis, University of Hamburg
Basar S, Koch A, Konig WA (2001) A verticillane-type diterpene from *Boswellia carterii* essential oil. Flavour Fragr J 16(5):315–318
Baser KHC, Demirci B, Dekebo A, Dagne E (2003) Essential oils of some *Boswellia* spp., myrrh and opopanax. Flavour Fragr J 18(2):153–156
Bell CR (1967) Plant variation and classification. Wadsworth, Belmont
Bentham G, Hooker JD (1862–1883) Genera plantarum: ad exemplaria imprimis in Herberiis Kewensibus servata definite, vol 3. Londini: A. Black
Bentley R, Trimen H (1880) Medicinal plants. Being descriptions with original figures of the principal plants employed in medicine and an account of the characters, properties, and uses of their parts and products of medicinal value. J. & A. Churchill, London. (plate 59)
Ben-Yehoshua S, Mercier J (2005) UV irradiation, biological agents, and natural compounds for controlling post-harvest decay in fresh fruits and vegetables. In: Ben-Yehoshua S (ed) Environmentally friendly technologies for agricultural produce quality. CRC Press, Boca Raton, pp 265–301
Ben-Yehoshua S, Rosen B (2009) The secret of Ein Gedi. Cathedra 132:77–100. (In Hebrew)
Ben-Yehoshua S, Borowitz C, Hanuš LO (2012) Frankincense, Myrrh, and Balm of Gilead: ancient spices of Southern Arabia and Judea. In: Janick J (ed) Horticult Rev 39:1–76. Wiley-Blackwell, Hoboken
Ben-Yehoshua S, Ofir R, Rachmilevitc S, Amiel E, Soloway, Dudai N (2013) Renewal of persimmon production, *Commiphora Gileadensis* L. opobalsamum, in Israel: anticarcinogenic activity. Judea and Samaria Research Studies, Ariel University, Samaria and Jordan Rift Research Center, Ariel 22:377–389. (In Hebrew)

Bergstrom T, Persson A, Thulin M, Warfa AM (1982) Domestication of the frankincense trees. Unpublished travel report, Quoted in SEPASAL
Bhushan S, Kakkar V, Pal HC, Guru SK, Kumar A, Mondhe DM, Sharma PR, Taneja SC, Kaur IP, Singh J, Saxena AK (2013) Enhanced anticancer potential of encapsulated solid lipid nanoparticles of TPD: A novel triterpenediol from *Boswellia serrata*. Mol Pharm 10:225–235
Bible – Holy Bible: Genesis to Malachy. British and Foreign Bible Society, London 1911
Birdwood GCM (1862) Catalogue of the economic products of the Presidency of Bombay, Division I: Raw products. SOC Press, Bombay
Birdwood GCM (1870) On the genus *Boswellia*. Trans Linn Soc 27:143
Birkett MA, Abassi SA, Krober T, Chamberlain K, Hooper AM, Guerin PM, Petterson J, Pickett JA, Slade R, Wadhams LJ (2008) Antiectoparasitic activity of the gum resin, gum haggar, from the east African plant *Commiphora holtziana*. Phytochemistry 69(8):1710–1715
Bishnoi M, Patil CS, Kumar A, Kulkarni SK (2005) Protective effects of Nimesulide (COX inhibitor), AKBA (5-LOX inhibitor) andtheir combination in aging associated abnormalities in mice. Method Find Exp Clin Pharmacol 27(7):465–470
Bombardelli E, Spelta M (1991) Phospholipids-polyphenol complexes: a new concept in skin care ingredients. Cosmet Toilet 106:69–76
Bonwicke J (trans) (1725) P. Pomet 1694. A complete history of drugs. Printed for J. & J. Bonwicke, S. Birt, W. Parker, C. Hitch, and E. Wicksteed, London
Borrelli F, Capasso F, Capasso R, Ascione V, Aviello G, Longo R, Izzo A (2006) Effect of *Boswellia serrata* on intestinal motility in rodents: inhibition of diarrhea without constipation. Br J Pharmacol 148(4):553–560
Bostock L (trans) (1855) Pliny the elder. The natural histories. Bohn, London. http://penelope.uchicago.edu
Bowen MR (1989) Forestry research in Somalia. Mogadishu: National Range Agency, Ministry of Livestock, Forestry and Range, British Forestry Project Somalia, research section, working paper 16
Bressan D (2013) In Search of the lost land of gold (and mummified baboons too). American Scientist, 30 Aug 2013
Brieskorn CH, Noble P (1982) Constituents of the essential oil of myrrh. II: Sesquiterpenes and furanosesquiterpenes. Planta Med 44(2):87–89
Brieskorn CH, Noble P (1983) Two furanoeudesmanes from the essential oil of myrrh. Phytochemistry 22(1):187–189
Brun JP (2000) The production of perfumes in antiquity: The case of Delos and Paestum. Am J Archaeol 104(2):277–308
Brunet F (trans) (1933–1957) Alexander of Tralles. Oeuvres medicales d'Alexandre de Tralles 4:270–283. Geuthner, Paris
Budge EAW (trans and ed) (1913) Syrian anatomy, pathology and therapeutics. Oxford University Press, London
Cahen C (1947) Une traite d'armurerie compose pour Saladin. Bull d'Etudes Orientales (Damascus) 12:163
Carter HJ (1851) Comparative geography of the south-east coast of Arabia. J Bombay Branch R Asiat Soc 3:224–317
Cartwright J (1760) An essay upon the virtues of Balm of Gilead. London: printed for G. Kearsly, at the Golden Lion in Ludgate-Street, MDCCLX with a copper-plate of the balsam-tree, guarded by a Janissary
Catholic Encyclopedia. www.newadvent.org
Church AJ, Broadribb WJ (trans) (1876) Tacitus, Annals
Clarkson J, Chase MW, Harley MM (2002) Phylogenetic relationships in *Burseraceae* based on plastid rps16 intron sequences. Kew Bull 57:183–193
Costeo I, Mongio I (eds) (1608) Gerard of Cremona. Avicennae arabum medicorum principis, canon medicinae. Venetis, Apud Juntas, Venice
Cotton HM (2001) Ein Gedi between the two revolts. Scripta Clasicca Israelica 20:139–154
Cotton HM, Eck W (1997) Ein staatsmonopoland und seine folgen: Plinius Naturalis Historia 12, 123 und der preis fur Apharsemon. Rheinisches Museum fur Philologi 140:153–161

Cuaz-Perolin C, Billiet L, Bauge E, Copin C, Scott-Algara D, Genze F, Buchele B, Syrovets T, Simmet T, Rouis M (2008) Antiinflammatory and antiatherogenic effects of the NF-KB inhibitor acetyl-11-keto-beta-Boswellic acid in LPS-challenged ApoE-mice. Arterioscler Thromb Vasc Biol 28(2):272–277

Cui J, Huang L, Zhae A, Lew JL, Yu J, Sahoo S, Meinke PT, Royo I, Pelaez F, Wright SD (2003) Guggulesterone is a farnesoid X receptor antagonist in coactivator association assays but acts to enhance transcription of bile salt export pump. J Biol Chem 278(12):10214–10220

Dalby A (2002) Dangerous tastes: The story of spices. University of California Press, Berkeley

Darshan S, Doreswamy R (2004) Patented anti-inflammatory plant drug development from traditional medicine. Phytother Res 18(5):343–357

Dayagi-Mendels M (1989) Perfumes and cosmetics in the ancient world. Israel Museum, Jerusalem

De Candolle A, de Candolle C (1883) Monographiae phanerogamarum, vol 4. G. Masson, Paris

de Rapper S, Van Vuuren SF, Kamatou GPP, Viljoen AM, Dagne E (2012) The additive and synergistic antimicrobial effects of select frankincense and myrrh oils – a combination from the pharaonic pharmacopoeia. Lett Appl Microbiol 54:352–358

DeFenoyl R, Sauneron S (trans and eds) (1979) Alpin Prosper. Histoire naturelle de l'Egypte par Prosper Alpin, p 1581–1584. Collection des Voyageurs enEgypte. Archeolog Caire, Cairo

Deng R, Yang D, Radke A, Yang J, Yan B (2007) The hypolipidemic agent guggulsterone regulates the expression of human bile salt export pump: dominance of transactivation over farsenoid X receptor-mediated antagonism. J Pharmacol Exp Ther 320(3):1153–1162

Dictionnaire des sciences naturelles (1825) Par plusieurs professeurs du Jardin du Roi, et des principales Ecoles de Paris, vol 36. F.G. Levrault, Paris, p 224–225

Dolara P, Luceri C, Ghelardini C, Montserrat C, Aiolli S, Luceri F, Lodovici M, Menichetti S, Romanelli MN (1996) Analgesic effects of myrrh. Nature 379:29. Scientific Correspondence

Dolara P, Corte B, Ghelardini C, Pugliese AM, Cerbai E, Menichetti S, Lo Nostro A (2000) Local anesthetic, anti-bacterial and anti-fungal properties of sesquiterpenes from myrrh. Planta Med 66(4):356–358

Dominy N, Leigh Moritz G, Ikram S (2010) Baboons, stable isotopes, and the location of Punt. Reported at the 61st annual meeting of the American Research Center in Egypt, California

Duke JA (2008) Handbook of medicinal plants of the Bible. CRC Press, Boca Raton

Duncan A (1804) Edinburgh new dispensatory. Bell and Bradfute, Edinburgh

El Ashry ES, Rashed N, Salama OM, Saleh A (2003) Components, therapeutic value and uses of myrrh. Pharmazie 58(3):163–168

El-Mekkawy S, Meselhy MR, Nkobole N, Lall N (2013) Three new α-glucosidase inhibitors from guggul, the oleogum resin of *Commiphora wightii*. Nat Prod Res 27(2):146–154

El-Shahat M, El-Abd S, Alkafafy M, El-Khatib G (2012) Potential chemoprevention of diethylnitrosamine-induced hepatocarcinogenesis in rats: Myrrh (*Commiphora molmol*) vs. turmeric (*Curcuma longa*). Acta Histochem 114:421–428

El-Sherbiny GM, El-Sherbiny ET (2011) The effect of *Commiphora molmol* (myrrh) in treatment of *Trichomoniasis vaginalis* infection. Iran Red Crescent Med J 13(7):480–486

Erickson-Gini T (2007) The transformation of Nabatean society: acculturation or self- organization? In: Saidel BA, van der Steen EJ (eds) On the fringe of society: archaeological and enthnoarchaeological perspectives on pastoral and agricultural societies. BAR international series 1657, Oxford, p 45–55

Estrada AC, Syrovets T, Pitterle K, Lunov O, Büchele B, Schimana-Pfeifer J, Schmidt T, Morad SAF, Simmet T (2010) Tirucallic acids are novel pleckstrin homology domain-dependent Akt inhibitors inducing apoptosis in prostate cancer cells. Mol Pharmacol 77:378–387

Etman M, Amin M, Nada AH, Shams-Eldin M, Salama O (2011) Emulsions and rectal formulations containing myrrh essential oil for better patient compliance. Drug Discov Ther 5(3):150–156

Fabre A (2003) Use of ancient texts in modern therapeutic research. Revue d'Histoire Pharmacie 51(338):239–250. (In French)

Fattorusso E, Santacroce C, Xaasan CF (1983) 4(23)-Dihydroroburic acid from the resin (incense) of *Boswellia carterii*. Phytochemistry 22(12):2868–2869

Faulkner RO (1969) The ancient Egyptian pyramid texts. Oxford University Press, Oxford
Feliks Y (1968) Plant world of the Bible. Massada, Tel Aviv. (In Hebrew)
Feliks Y (1981) Nature and man in the Bible: chapters in biblical ecology. Soncino Press, London/New York
Fenwick A, Webster JP (2006) Schistosomiasis: Challenges for control, treatment and drug resistance. Curr Opin Infect Dis 19(6):577–582
Finkelstein I, Silberman NA (2006) David and Solomon. Simon & Schuster, New York
Fischer M, Gichon M, Tal O (2000) 'En Boqeq. Excavations in an oasis on the Dead Sea shore, vol 2, The officina: An early Roman building on the Dead Sea shore. Phillip von Zabern, Mainz
Forsskål P (1775) Amyris Opobalsamum – Flora Aegyptiaco-Arabica, sive, Descriptiones plantarum. Ex Officina Mölleri, Hauniae, pp 79–80. Class: VIII – Octandria
Frank MB, Yang Q, Osban J, Azzarello JT, Saban MR, Saban R, Ashley RA, Welter JC, Fung K-M, Lin H-K (2009) Frankincense oil derived from *Boswellia carteri* induces tumor cell specific cytotoxicity. BMC Complement Altern Med 9(6):1–11
Freese BL (1996) Medicinal myrrh. Archaeology (Publication of the Archaeology Institute of America) 49(3) May/June newsbrief
Gachathi FN (1997) Recent advances on classification and status of the main gum-resin producing species in the family *Burseraceae*. In: Mugah JO et al (eds) Conservation, management and utilization of plant gums, resins and essential oils. Proceedings of a regional conference for Africa held in Nairobi
Gayathri B, Manjula N, Vinaykumar KS, Lakshmi BS, Balakrishnan A (2007) Pure compound from *Boswellia serrata* extract exhibits anti-inflammatory property in human peripheral blood mononuclear cells. Int Immunopharmacol 7(4):473–482
Gebrehiwot K, Muys B, Haile M, Mitloehner R (2002) *Boswellia papyrifera* (Del.) Hoechst: a tropical key species in northern Ethiopia. In: Proceedings of the conference on international agricultural research for development, Deutscher Tropentag, Kassel-Witzenhausen
Genders R (1972) A history of scent. Hamish Hamilton, London
Ghazanfar SA (1994) Handbook of Arabian medicinal plants. CRC Press, Boca Raton
Greenhill T (1705) Nekrokedeia or the art of embalming. London: Printed for the author
Griffith RE (1847) Medical botany. Lea and Blanchard, Philadelphia
Groom N (1981) Frankincense and myrrh: A study of the Arabian incense trade. Longman, London/New York
Gunther RT (trans) (1933) Dioscorides. Materia Medica. Hafner, Chicago
Gupta I, Parihar A, Malhotra P, Gupta S, Ludtke R, Safayhi H, Ammon HP (2001) Effects of gum resin of *Boswellia serrata* in patients with chronic colitis. Planta Med 67(5):391–395
Gurib-Fakim A (2006) Medicinal plants: Traditions of yesterday and drugs of tomorrow. Mol Asp Med 27(1):1–93
Habeeb SM, El-Namaky AH, Salama MA (2009) Efficiency of *Allium cepa* and *Commiphora molmol* as a larvicidal agent against fourth stage larvae of Culex pipiens (Diptera: Culicidae). Am-Eurasia J Agric Environ Sci 5(2):196–203
Haefeli-Till D (ed) (1977) Constantine the African. Der liber de oculis des Constantine Africanus. Zircher Medizingeschichtliche, Zurich, Jarman
Haffor A-SA (2010) Effect of myrrh (*Commiphora molmol*) on leukocyte levels before and during healing from gastric ulcer or skin injury. J Immunotoxicol 7(1):68–75
Hamed MA, Hetta MH (2005) Efficacy of citrus reticulata and Mirazid in treatment of schistosoma mansoni. Mem Inst Oswaldo Cruz 100(7):771–778
Hamm S, Lesellier E, Bleton J, Tchapla A (2003) Optimization of headspace solid phase microextraction for gas chromatography/mass spectrometry analysis of widely different volatility and polarity terpenoids in olibanum. J Chromatogr 1018(1):73–83
Hamm S, Bleton J, Connan J, Tchapla A (2005) A chemical investigation by headspace SPME and GC-MS of volatile and semi-volatile terpenes in various olibanum samples. Phytochemistry 66:1499–1514
Hanuš L (2012) Unpublished results (fresh aromatic resin from a tree from eastern part of Yemen)

Hanuš LO, Řezanka T, Dembitsky VM, Moussaieff A (2005) *Commiphora* chemistry. Biomed Pap Med Fac Univ Palacky Olomouc Czech Repub 149(1):3–27
Hanuš LO, Moussaieff A, Řezanka T, Abu-Lafi S, Dembitsky VM (2007) Phytochemical analysis and comparison for differentiation of *Boswellia carterii* and *Boswellia serrata*. Nat Prod Commun 2(2):139–142
Hassler C (ed) (1843) Fabri. Evagatorium in Terrae Sanctae, Arabiae et Egypti peregri-nationem. Sumptibus Societatis Litterariae, Stuttgart
Hepper FN (1969) Arabian and African frankincense trees. J Egypt Archaeol 55:66–72
Hepper FN (1992) Illustrated encyclopedia of Bible plants. Baker Encyclopedia, Leicester
Hepper FN, Friis I (1994) The plants of Pehr Forsskal's Flora Aegyptiaco-Arabica. Royal Botanical Gardens, Kew
Hestrin R, Dayagi-Mendels M (1979) Inscribed seals. First temple period: Hebrew, Ammonite, Moabite, Phoenician and Aramaic. The Israel Museum and the Israel Department of Antiquities and Museums, Jerusalem
Hill AF (1952) Economic botany. McGraw-Hill, New York
Hirschfeld Y (2007) Ein Gedi Excavations II 1996–2002 Final Report. Israel Exploration Society, Jerusalem
Hooper D (1937) Useful plants and drugs of Iran and Iraq. Field Museum Natural History, Chicago
Hort AF (trans) (1916) Theophrastus: enquiry into plants, Loeb Classical Library. Harvard University Press, Cambridge, MA
Howes FN (1946) Vegetable gums and resins. Chronica Botanica Co., Waltham
Hwa JY (2007) US Patent No. 7223423
Ibn al-Baytar (1874) Kitab al-jami li-mufradat al-adwiya wa'l-aghdiya. Cairo: Ibn al-Bayṭār
Iluz D, Hoffman M, Gilboa-Garber N, Amar Z (2010) Medicinal properties of *Commiphora Gileadensis*. Afr J Pharm Pharmacol 4(8):516–520
Jacquin NJ (1797) Plantarum rariorum horti caesarei Schoenbrunnensis descriptiones et icones/ Opera et sumptibus Nicolai Josephi Jacquin. Viennae: Apud C. F. Wappler; Londini: Apud B. et J. White; Lugduni Batavorum: Apud S. et J. Luchtmans, 1797–1804
Jahier H, Noureddine A (eds) (1956) Ibn Sina Avicenne. Poeme de la medecine. Eddes "Belles Lettres", Paris
Jingai T, Shangwei S (1996) Studies on the constituents of essential oil of imported Myrrh and gum opoponax. Zhongguo Zhongyao Zazhi 21:235–237
Johnson DJ (1987) Nabatean trade: intensification and culture change. Ph.D. Dissertation, University of Michigan, Ann Arbor
Jones HL (trans) (1924) 1. Strabo geography. www.perseus.tufts.edu
Jones VM (1995) Qumran excavations, Cave of the column complex & environs. Vendyl Jones Research Institute, Arlington
Jones WHS, Omerod HA (trans) (1918) Pausanias. Description of Greece, Loeb Classical Library. Harvard University Press, Cambridge
Kai J, Kong F, Shen T, Wang X, Xu A, Yuan H, Zhang X (2008) Separation and identification of myrrh sesquiterpenoids and their anti-proliferation effect on tumor cells. Shandong Daxue Xuebao Yixueban 46(4):344–348
Keay J (2006) The Spice Route - a History. University of California Press, Berkeley
Kimura I, Yoshikawa M, Kobayashi S, Sugihara Y, Suzuki M, Oominami H, Murakami T, Matsuda H, Doiphode VV (2001) New triterpenes, myrrhanol A and myrrhanone A, from guggul-gum resins, and their potent anti-inflammatory effect on adjuvant-induced air-pouch granuloma of mice. Bioorg Med Chem Lett 11(8):985–989
Kitchen KA (1993) The land of Punt. In: Shaw T et al (eds) The archeology of Africa: Food, Metals, Towns. Routledge, London
Klein E, Obermann H (1978) (S)-1-isopropyl-4,8,12-trimethylcyclotetradeca-3E,7E,11E-trien-1-ol, a new cembrenol from the ethereal oil from olibanum. Tetrahedron Lett 4:349–352
Krieglstein CF, Anthoni C, Rijcken E, Laukotter M, Spiegel HU, Boden S, Schwei-zer S, Safayhi H, Senninger N, Schurmann G (2001) Acetyl-11-keto-Boswellic acid, a constituent of a herbal medicine from *B. serrata* resin, attenuates experimental ileitis. Int J Colorectal Dis 16(2):88–95

Kühn CG (1964) Claudii Galeni Opera Omnia, Leipzig 1821–1833; repr. Hildesheim, 1964
Langkaevel B (ed) (1868) Simeonis Seth. Syntagma per elementorum ordinem, de alimentorum facultate. Leipzig
Lemaire A (2010) Solomon & Sheba, Inc. Biblical Archeolo Rev 36:01, 54–58, 78
Lemenih M, Feleke S, Tadesse W (2007) Constraints to smallholders production of frankincense in Meteme district, north-western Ethiopia. J Arid Environ 71(4):393–403
Lev E (2003) Medicinal substances in Jerusalem from early times to the present day. Archaeo Press, Oxford
Levey M (trans and ed) (1966a) al-Kindi. The medical formulary or aqrabadhin of al-Kindi. University of Wisconsin, Madison
Levey M (trans and ed) (1966b) Ibn Wahshiyya. Medieval Arabic toxicology: the book of poisons of Ibn Wahshiyya and its relation to early Indian and Greek texts. Trans Am Philos Soc 56(7):1–30
Levey M (1973) Early Arabic pharmacology. Brill, Leiden
Levine LI (ed) (1981) Ancient synagogues revealed. The Israel Exploration Society, Jerusalem
Li F-S, Yan D-L, Liu R-R, Xu K-P, Tan G-S (2010) Chemical constituents of *Boswellia carterii* (Frankincense). Zhongguo Tianran Yaowu 8(1):25–27
Linnaeus C (1764) Species plantarum. Stockholm: Laurentii Salvii
Linnaeus C (1767) Systema naturae. *Mantissa plantarum. Impensis direct Laurentii Salvii.* Holmiae, p 65
Lotfy M, Badra G, Burham W, Alenzi FQ (2006) Combined use of honey, bee propolis and myrrh in healing a deep, infected wound in a patient with diabetes mellitus. Br J Biomed Sci 63(4):171–173
Low I (1924) Die Flora den Juden. Lowit, Vienna
Lui JJ, Nilsson A, Oredsson S, Badmaev V, Zhao WZ, Duan RD (2002) Boswellic acids trigger apoptosis via a pathway dependent on caspase-8 activation but independent on Fas/Fas ligand interaction in colon cancer HT-29 cells. Carcinogenesis 23(12):2087–2093
Majno G (1975) The healing hand: Man and wound in antiquity. Harvard University Press, Cambridge
Malandin G (ed) (1986) Platearius. 1232. Le Livre des simple medicines d'apres le manuscrip francais. Bibliotheque Nationale, Paris
Martial. Epigrams. Bohn (1897) (trans). http://ccel.org
Mazar B, Dothan T, Dunayevski I (1966) Ein Gedi. The first and second seasons of excavations, 1961–1962. Israel Exploration Society J. Atiquot 5, Jerusalem
Mazar B, Dotan T, Dunayevski A (1973) Ein Gedi: archeological excavations during the years 1961–1962. Israel Exploration Society J. Atiquot 5, Jerusalem
Melville A (trans) (1998) Ovid – metamorphosis. Oxford University Press, Oxford
Meyerhoff M (trans and ed) (1928) Hunayn ibn Ishaq. The book often treatises on the eye ascribed to Hunainibn Is-Haq (809–877 AD), The earliest existing systematic textbook on ophthalmology. Government Press, Cairo
Meyerhoff M, Sobhy G (trans and eds) (1932–1949) al-Ghafiqi. The abridged version of simple drugs of Ahmad ibn Muhammad al Ghafiqi Abu'l Farag (Barhebraeus) Al-Ettemad, Cairo
Miller JI (1998) The spice trade of the Roman Empire, 29 BC to AD 641. Oxford University Press, Oxford
Miller AG, Cope TA (eds) (1996) Flora of the Arabian peninsula and Socotra. Edinburgh University Press, Edinburgh
Miller AG, Morris M (1988) Plants of Dhofar: traditional, economic and medicinal uses. Oman, Office of the advisor for conservation of the environment. Diwanofthe Royal Court Sultanate of Oman, p 360
Milwright M (2003) The apharsemon of Matariyya: An exploration of a medieval panacea. SOAS Bull 66(2):193–209
Moldenke HN, Moldenke AL (1952) Plants of the Bible. Chronica Botanica Co., Waltham
Montagu LMW (1717) Complete letters. In: Halsband R (ed) Oxford University Press (1965). Letter to Lady Rich of 17.6.1717. Cited by Woods B and Calnan CD (1976) Toxic woods. Br J Dermatol 95(Suppl 13):1–97
Monti D, Manitto P, Tagliapietra S, Dada G, Speranza G (1986) The absolute stereochemistry of two furanogermacranes of myrrh as determined by the circular dichroism exciton chirality method. Gazz Chim Ital 116:303–306

Morikawa T, Oominami H, Matsuda H, Yoshikawa M (2010) Four new ursane-type triterpenes, olibanumols K, L, M, and N, from traditional Egyptian medicine olibanum, the gum-resin of *Boswellia carterii*. Chem Pharm Bull 58(11):1541–1544

Morikawa T, Oominami H, Matsuda H, Yoshikawa M (2011) New terpenoids, olibanumols D–G, from traditional Egyptian medicine olibanum, the gum-resin of *Boswellia carterii*. J Nat Med 65:129–134

Morris M (1989) Trees of knowledge (2): Frankincense. PDO News 1:16–19

Morteza-Semnani K, Saeedi M (2003) Constituents of the essential oil of *Commiphora myrrha* (Nees) Engl. var. molmol. J Essent Oil Res 15(1):50–51

Moussaieff A, Shohami E, Kashman Y, Fride E, Schmitz ML, Renner F, Fiebich BL, Munoz E, Ben-Neriah Y, Mechoulam R (2007) Incensole acetate, a novel anti- inflammatory compound isolated from *Boswellia* resin, inhibits nuclear factor-kappa B activation. Mol Pharmacol 72(6):1657–1664

Moussaieff A, Rimmerman N, Bregman T, Straiker A, Felder CC, Shoham S, Kashman Y, Huang SM, Lee H, Shohami E, Mackie K, Caterina MJ, Walker JM, Fride E, Mechoulam R (2008) Incensol acetate, an incense compound, elicits psychoactivity by activating TRPV3 channels in the brain. FASEB J 22:3024–3034

Mukhtar A, Al-Daghri N, Alokail MS, Hussain T (2013) Potential changes in rat spermatogenesis and sperm parameters after inhalation of *Boswellia papyrifera* and *Boswellia carterii* incense. Int J Environ Res Public Health 10:830–844

Muntner S (ed) (1971) Asaph Harofeh. Encyclopedia Judaica. Jerusalem: Sefer Harefuoth

Needham J, Lu GD (1974) Science and civilization in China, vol 5. Tao Hongjing: Materia Medica Mingyi Bielu (Miscellaneous records of physicians). Cambridge University Press, Cambridge

Neufeld E (1971) Hygiene conditions in ancient Israel (Iron Age). Biblic Archaeol 34(2):41–66

Niebuhr C (1792) Travels through Arabia and other countries in the East, vol 2. R. Morison and Son, Edinburgh. (Original titles: vol 1. Reisebeschreibung nach Arabien und andern umliegenden Ländern. vol 2. Beschreibung von Arabien)

Nielsen H (1974) Ancient ophthalmological agents. Acta Historica Scientiarium Natur-alium et Medicinalium 31:117

Nielsen K (1986) Incense in ancient Israel. Brill, Leiden

Oldfather CH (trans) (1935) Diodorus Siculus. Bibliotheca Historica, Loeb Classical Library. Harvard University, Cambridge

Orwa C, Mutua A, Kindt R, Jamnadass R, Simons A (2009) Agroforestree database: a tree reference and selection guide. Version 4.0. http://www.worldagroforestry.org/sites/treedbs/treedatabases.asp

Osbaldeston TA (ed) (2000) The herbal of Dioscorides. Book One: aromatics. www.Cancer-lynx.com

Owsley E, Chiang JY (2003) Guggulsterone antagonizes farnesoid X receptor induction of bile salt export pump but activates pregnane X receptor to inhibit cholesterol 7alpha-hydroxylase gene. Biochem Biophys Res Commun 304(1):191–195

Pagel JL (ed) (1893) Die Areolae des Johannes de Sancto Amando. Berlin 68:63–66

Papyrus Harris. British Museum, London, Serial 9999

Patrich J, Arubas B (1990) A juglet with apharsemon oil (?) from a cave near Qumran. Eretz Isr 21:321–332. (In Hebrew)

Pliny the Elder (1855) The natural history. Taylor and Francis, London

Poeckel D, Werz O (2006) Boswellic acids: biological actions and molecular targets. Curr Med Chem 13(28):3359–3369

Postlethwayt M (trans) (1766) Savary desBruslons. The universal dictionary of trade and commerce. Printed for H. Woodfall, London

Provan GJ, Gray AI, Waterman PG (1987) Chemistry of the *Burseraceae*. Part 5. Sesquiterpenes from the myrrh-type resins of some Kenyan *Commiphora* species. Flavour Fragr J 2:109–113

Qataya S (ed) (1981) IbnRidwan. Kitabal-kifayafi'l-tibb. Maktabatal-Wataniyya, Baghdad

Rahman MM, Garvey M, Piddock LJV, Gibbons S (2008) Antibacterial terpenes from the oleo-resin of *Commiphora molmol* (Engl.). Phytother Res 22:1356–1360

Ramzy F, Soheir M, Samia W (2010) Further assessment of Mirazid as antischistosomal drug in experimental schistosomiasis hematobium. Pharm Biol 48(7):775–779

Rao R, Khan Z, Shah A (2001) Toxicity studies in mice of *Commiphora momol* oleo gum resin. J Ethnopharmacol 76(2):151–154

Rawlinson G (trans and ed) (1859) The histories of Herodotus. D. Appleton & Company, New York

Rawlinson G (1897) D. Appleton & Company. The story of the nations: Egypt. Putnam, London

Rees A (1995) Frankincense and myrrh. New Plantsman 2(1):55–59

Remington JP, Wood HC (eds) (1918) The dispensatory of the United States of America, 20th edn. Lippincott. Online at: www.swsbm.com

Rosen B, Ben-Yehoshua S (2007) The agriculture of Roman-Byzantine Ein-Gedi and the enigmatic secret of the village 20: 626–641. In: Hirschfeld Y (ed) Ein Gedi excavations – final report 1996–2002. Israel Exploration Society, Jerusalem

Rosengarten F Jr (1970) The book of spices. Livingston, Wynnewood

Rosner F (trans) (1984) Maimonides. Treatise on poisons, hemorrhoids and cohabitation: Maimonides' medical writing. Maimonides Research Institue, Haifa

Rubin S (1974) Medieval English medicine. Barnes & Noble, New York

Said H (trans) (1973) Al-Biruni's book of pharmacy and medicine. Hamdad Foundation, Karachi

Santoro S, Superchi S, Messina F, Santoro E, Rosati O, Santi C, Marcotullio MC (2013) Agarsenone, a cadinane sesquiterpenoid from *Commiphora erythraea*. J Nat Prod 76:1254–1259

Sayyid A (ed) (1985) al-Umari. Masalik al-absar. Institut Francais d'Archeologie Orientale, Cairo

Schoff WH (1922) Aloes. J Orient Soc 42:171–185

Scott KD, McIntyre CL, Playford J (2000) Molecular analyses suggest a need for a significant rearrangement of Rutaceae subfamilies and a minor reassessment of species relationships within Flindersia. Plant Syst Evol 223:15–27

Sengupta K, Alluri KV, Satish AR, Mishra S, Golakoti T, Sarma KVS, Dey D, Raychaudhuri SP (2007) A double blind, randomized, placebo controlled study of the efficacy and safety of 5-Loxin® for treatment of osteoarthritis of the knee. Arthritis Res Ther 10. Online/content/10/4/R85. SEPASAL Online Database. Survey of economic plants for arid and semi-arid lands. Royal Botanic Gardens, Kew, London. www.kew.org/ceb/sepasal

Shah BA, Qazi GN, Taneja SC (2009) Boswellic acids: A group of medicinally important compounds. Nat Prod Rep 26(1):72–89

Sharma A, Mann AS, Gajbhiye V, Kharya MD (2007) Phytochemical profile of *Boswellia serrata*: An overview. Pharmacogn Rev 1(1):137–142

Shen T, Wan W, Yuan H, Kong F, Guo H, Fan P, Lou H (2007) Secondary metabolites from *Commiphora opobalsamum* and their antiproliferative effect on human prostate cancer cells. Phytochemistry 68(9):1331–1337

Shen T, Yuan HQ, Wan WZ, Wang XL, Wang XN, Ji M, Lou HX (2008a) Cycloartane-type triterpenoids from the resinous exudates of *Commiphora opobalsamum*. J Nat Prod 71(1):81–86

Shen T, Wan W, Wang XN, Sun LM, Yuan HQ, Wang XL, Ji M, Lou HX (2008b) Sesquiterpenoids from the resinous exudates of *Commiphora opobalsamum* (*Burseraceae*). Helvetica Chimica Acta 91(5):881–887

Shen T, Wan W-Z, Wang X-N, Yuan H-Q, Ji M, Lou H-X (2009) A triterpenoid and sesquiterpenoids from the resinous exudates of *Commiphora myrrha*. Helv Chim Acta 92:645–652

Shen T, Li G-H, Wang X-N, Lou H-X (2012) The genus Commiphora: A review of its traditional uses, phytochemistry and pharmacology. J Ethnopharmacol 142:319–330

Shoemaker M, Hamilton B, Dairkee SH, Cohen I, Campbell MJ (2005) In vitro anti-cancer activity of twelve Chinese medicinal herbs. Phytother Res 19(7):649–651

Shuaib M, Ali M, Naquvi KJ (2013) Tetraterpenyl esters from the oleo-resin of *Commiphora myrrha* (Nees) Engl. Pharma Chemica 5(2):133–138

Shulan S, Wang T, Duan J-A, Zhou W, Hua Y-Q, Tang Y-P, Yu L, Qian D-W (2011) Anti-inflammatory and analgesic activity of different extracts of *Commiphora myrrha*. J Ethnopharmacol 134:251–258

Shu-Lan S, Jin-Ao D, Yu-Ping T, Xu Z, Li Y, Feng-Rong J, Wei Z, Dan L, An-Wei D (2009) Isolation and biological activities of neomyrrhaol and other terpenes from the resin of *Commiphora myrrha*. Planta Med 75(4):351–355

Siddiqui M (ed) (1928) al-Tabari. Paradise of wisdom. Kunstdruckerei "Sonne", Berlin
Simmons AJ, Collins RV, Sayed DN, Setzer W (2013) Cytotoxicity of *Commiphora myrrha* on MCF-7 cells. Abstracts of papers, 245th ACS national meeting & exposition, New Orleans, LA, USA, April 7–11, 2013, BIOT-339
Singer-Avitz L (1996) Household activities during the iron period at Tel Beer Sheva and activities. Erez Yisrael 25:166–174. (In Hebrew)
Singer-Avitz L (1999) Beer Sheva – a gateway community in southern Arabia for longdistance trade in the 8th century BCE. Tel Aviv 26:1–75
Singh SK, Bhusari S, Singh R, Saxena A, Mondhe D, Qazi GN (2007) Effect of AKBA on metastatic growth factor responsible for angiogenesis. Vasc Pharmacol 46(5):333–337
Singh S, Khajuria A, Taneja SC, Johri RK, Singh J, Qazi GN (2008) Boswellic acids: A leukotriene inhibitor also effective through topical application in inflammatory disorders. Phytomedicine 15(6–7):400–407
Southgate VR, Rollinson D, Tchuem Tchuente LA, Hagan P (2005) Towards control of schistosomiasis in sub-Saharan Africa. J Helminthol 79:181–185
Spencer WG (trans) (1938) Celsus. De medicina, Loeb Classical Library. Harvard University Press, Cambridge, MA
Stannard J (1966) Dioscoridies and renaissance Materia Medica, Analecta Medico Historica I: Materia Medica in the XVI century. Pergamon, Oxford
Starkey J, Pitt M (trans) (1678) Charas. The royal pharmacopoea: galenical and chymical. Printed for Starkey and Pitt, London
Stephenson J (1831) Medical botany, vol IV. John Churchill, London
Stuart M (1979) Reference section. In: Stuart M (ed) The Encyclopedia of Herbs and Herbalism. Orbis Publishing, London, pp 141–283
Su S, Wang T, Duan J-A, Zhou W, Hua Y-Q, Tang Y-P, Yu L, Qian D-W (2011) Anti-inflammatory and analgesic activity of different extracts of *Commiphora myrrha*. J Ethnopharmacol 134:251–258
Su S, Hua Y, Wang Y, Gu W, Zhou W, Duan J-A, Jiang H, Chen T, Tang Y (2012) Evaluation of the anti-inflammatory and analgesic properties of individual and combined extracts from *Commiphora myrrha*, and *Boswellia carterii*. J Ethnopharmacol 139:649–656
Suhail MM, Wu W, Cao A, Mondalek FG, Fung K-M, Shih P-T, Fang Y-T, Woolley C, Young G, Lin H-K (2011) *Boswellia sacra* essential oil induces tumor cell-specific apoptosis and suppresses tumor aggressiveness in cultured human breast cancer cells. BMC Complement Altern Med 11:129
Szapary PO, Wolfe MJ, Bloedon LT, Cucchiara AJ, Dermarderosian AH, Cirigliano MD, Rader DJ (2003) Guggulipid for the treatment ofhypercholesterolemia: A randomized controlled trial. JAMA 290(6):765–772
Talmud. Tractate Critot 6:71. Babylonian Talmud
Talmud. Tractate Sanhedrin 43a, attributed to Rabbi Hiyya ben Ashi
Talmud. Tractate Shabbath 25b, 26a [Laws of the Sabbath and festivals]. Jerusalem Talmud
Talmud. Tractate Yoma 41:74. Jerusalem Talmud
Teketay G (2003) Frankincense and myrrh resources of Ethiopia: Medicinal and industrial use. Ethiop J Sci 26(2):161–172
Temkin O (1973) Galenism: rise and decline of a medical philosophy. Cornell University Press, Ithaca
Theophrastus Eresius (2009) Historia plantarum. In: Wimmer F (ed) Kessinger Publishing, LLC
Thesiger W (1959) Arabian sands. Penguin, London
Thieret JW (1996) Frankincense and myrrh. Lloydiana 1(4):6–9
Thomas B (1932) Arabia Felix. Scribner, New York
Thulin M, Warfa AM (1987) The frankincense trees of n. Somalia and s. Kew Bull 42(3):487–500
Tipton DA, Hamman NR, Dabbous MK (2006) Effect of myrrh oil on IL-1beta stimulation of NF-kappaB activation and PGE(2) production in human gingival fibro- blasts and epithelial cells. Toxicol In Vitro 20(2):248–255
Tucker AO (1986) Frankincense and myrrh. Econ Bot 40(4):425–433
Uhlworm O, Behrens WJ (eds) (1883) Botanisches Zentralblatt 15:135–136

Uphof JCT (1968) Dictionary of economic plants. Verlag Cramer, Wurzburg
Urizar NL, Moore DD (2003) Gugulipid: A natural cholesterol lowering agent. Annu Rev Nutr 23:303–313
Van Beek GW (1958) Frankincense and myrrh in ancient south Arabia. J Am Orient Soc 78(3):141–152
van Beek GW, Jamme A (1958) An inscribed southern Arabian clay stamp from Bethel. BASOR 151:9–16. Wageningen University, Forest Ecology and Management Group FRAME project 2006–2010: Myrrh and gum arabic sustainable use of dry woodlandresourcesinEthiopia. Wageningen
Van Vuuren SF, Kamatou GPP, Viljoen AM (2010) Volatile composition and antimicrobial activity of twenty commercial frankincense essential oil samples. S Afr J Bot 76:686–691
Wang F, Li Z-L, Cui H-H, Hua H-M, Jing Y-K, Liang S-W (2011) Two new triterpenoids from the resin of *Boswellia carterii*. J Asian Nat Prod Res 13(3):193–197
Wang, Yan-gai, Jin Ren, Ai-guo Wang, Jian-bo Yang, Teng-fei Ji, Qin-Ge Ma, Jin Tian, Ya-lun Su (2013) Hepatoprotective Prenylaromadendrane-Type Diterpenes from the Gum Resin of *Boswellia carterii*. J Nat Prod 76:2074–2079
Watson G (1966) Theriac and Mithridatum: A study of therapeutics. Welcome History of Medicine Library, London
Weeks A, Simpson BB (2007) Molecular phylogenetic analysis of *Commiphora* (*Burseraceae*) yields insight on the evolution and historical biogeography of an "impossible" genus. Mol Phylogen Evol 42:62–79
Whiston W (trans) (1737) The complete works of Josephus Flavius. Mazin, London
Willdenow CL (1799) An accurate description of the Balm of Gilead; its growth, and remarkable properties. Med Phys J (Lond) 1:33–39
Wolf CU (1971) The Onomasticon of Eusebius Pamphili: Compared with the version of Jerome and annotated. Catholic University of America Press, Washington, DC
Wood JRI (1997) A handbook of the Yemen flora. Royal Botanic Gardens, Kew
Wood C (trans and ed) (1936) AliibnIsaal-Kahhal. Memorandum book of a tenth-century oculist for the use of modern ophthalmologists. Northwestern University Press, Chicago
Woods B, Calnan CD (1976) Toxic woods. Br J Dermatol 95(Suppl 13):1–97
Woodville W (1810) Medical botany, vol 3, 2nd edn. W. Phillips, London
Woolley CL, Suhail MM, Smith BL, Boren KE, Taylor LC, Schreuder MF, Chai JK, Casabianca H, Haq S, Lin H-K, Al-Shahri AA, Al-Hatmi S, Young DG (2012) Chemical differentiation of *Boswellia sacra* and *Boswellia carterii* essential oils by gas chromatography and chiral gas chromatography–mass spectrometry. J Chromatogr A 1261:158–163
Wreszinski W (1912) Der papyrus Ebers, Altagyptische Medizin. British Museum Publications 10059, London
Xiao N, Suhail MM, Yang Q, Cao A, Fung K-M, Postier RG, Woolley C, Young G, Zhang J, Lin H-K (2012) Frankincense essential oil prepared from hydrodistillation of Boswellia sacra gum resins induces human pancreatic cancer cell death in cultures and in a xenograft murine model. BMC Complement Altern Med 12:253
Xu J, Guo Y, Zhao P, Xie C, Jin D-Q, Hou W, Zhang T (2011a) Neuroprotective cadinane sesquiterpenes from the resinous exudates of *Commiphora myrrha*. Fitoterapia 82:1198–1201
Xu J, Guo Y, Li Y, Zhao P, Liu C, Ma Y, Gao J, Hou W, Zhang T (2011b) Sesquiterpenoids from the resinous exudates of *Commiphora myrrha* and their neuroprotective effects. Planta Med 77(18):2023–2028
Xu J, Guo Y, Zhao P, Guo P, Ma Y, Xie C, Jin D-Q, Gui L (2012) Four new sesquiterpenes from *Commiphora myrrha* and their neuroprotective effects. Fitoterapia 83:801–805
Yang J-L, Shi Y-P (2012) Cycloartane-type triterpenoids and sesquiterpenoids from the resinous exudates of *Commiphora opobalsamum*. Phytochemistry 76:124–132
Yoshikawa M, Morikawa T, Oominami H, Matsuda H (2009) Absolute stereostructures of olibanumols A, B, C, H, I, and J from olibanum, gum-resin of *Boswellia carterii*, and inhibitors of nitric oxide production in lipopolysaccharide-activated mouse peritoneal macrophages. Chem Pharm Bull 57(9):957–964

Yuan HQ, Kong F, Wang XI, Young CY, Hu XY, Lou HX (2008) Inhibitory effect of acetyl-11-keto-beta-boswellic acid on androgen receptor by interference of Sp1 binding activity in prostate cancer cells. Biochem Pharmacol 75(11):2112–2121
Yunnan Baiyao Company. Manufacturers of Yunnan Baiyao Yunnan, China
Zhu N, Sheng S, Sang S, Rosen RT, Ho C-T (2003) Isolation and characterization of several aromatic sesquiterpenes from *Commiphora myrrha*. Flavour Fragr J 18(4):282–285
Ziegler GM (1932) The diuturnal use of perfumes and cosmetics. Sci Mon 34(3):222–237
Zohary M (1982) Plants of the Bible. Cambridge University Press, Cambridge

Sarcopoterium spinosum

Zalmen Henkin, Tovit Rosenzweig, and Zohara Yaniv

Abstract *Sarcopoterium spinosum* (L.) Spach (previously identified as *Poterium spinosum* L.) is a spiny rosaceaeous dwarf shrub, 30–60 cm in height, with branches ending in dichotomous and leafless thorns. It appears over a wide range of climatic conditions from the semi-arid ecotone between the Mediterranean and Irano-Turanian vegetation zones to the sub-humid Mediterranean regions. The combination of the plant's clonal and sexual reproduction contributes to its long term survival and dominance. The whole bush is used as fuel, for making fences and sheep pens, for making brooms and as stuffing for mattresses. One of the most quoted medicinal indications is the use of its root bark as a popular cure against diabetes and for dissolving kidney stones. Traditionally it is used as a tranquilizer and a potion prepared from its leaves is said to dispel fears. Green leaves, salted and spiced and soaked in saliva are used to cure eye complaints, and also as a lucky charm. Five pentacyclic triterpenoids were identified in *Sarcopoterium spinosum*, Tormentic acid is suggested to be the major constituent extract, and to mediate its anti-proliferative activity on several cancer cell lines.

Keywords Clonal and sexual reproduction • Diabetes • Irano-Turonian steppe • Kidney stones • Mediterranean • Pentacyclic triterpenoids • *Sarcopoterium spinosum* • Successional processes • Thorny shrub • Traditional medicine • Tranquilizer

Taxonomic Characteristics and Morphological Description

Sarcopoterium spinosum (L.) Spach (previously identified as *Poterium spinosum* L.) is a spiny rosaceaeous dwarf shrub, 30–60 cm in height, with branches ending in dichotomous and leafless thorns (Litav and Orshan 1971) (Figs. 1 and 2).

Z. Henkin (✉)
Neve Yaar; ARO, The Volcani Center, Bet-Dagan, Israel
e-mail: henkinz@volcani.agri.gov.il

T. Rosenzweig
Ariel University Center, Tel-Aviv, Israel

Z. Yaniv
Department of Plant Introduction and Seed Research, Institute of Plant Sciences, Agricultural Research Organization, Bet-Dagan, Israel
e-mail: zoharayb@netvision.net.il

Z. Yaniv, N. Dudai (eds.), *Medicinal and Aromatic Plants of the Middle-East*, Medicinal and Aromatic Plants of the World 2, DOI 10.1007/978-94-017-9276-9_7

Fig. 1 *Sarcopoterium spinosum in* the wild. Judea Mt (Taken by the author Z.Y.)

Fig. 2 *Sarcopoterium spinosum*. Close look (Picture taken by the author)

It is called in Hebrew: Sira; in Arabic: Bilan or Natch; and its common name is: Prickly Shrubby Burnet. It is found in the Eastern Mediterranean and revealed in early geobotanical, eco-physiology, life cycle and successional aspect studies (Eig 1946; Litav 1967; Litav et al. 1963; Zohary 1973).

The combination of the plant's clonal and sexual reproduction contributes to its long term survival and dominance (Reisman-Berman and Henkin 2007). Adventitious roots, formed at prostrate branch nodes, are also common. Germination is epigeous and occurs inside the intact fruit while each fruit with its several seeds usually produces one seedling (Litav and Orshan 1971). The first year seedlings growing under mature plants develop long and thin taproots reaching at least 40 cm of depth. It was shown that seedlings that survived the first year had a very good chance of surviving the next few years (Henkin and Seligman 2007). Those that survived beyond 5 years began producing clonal structures and expanded laterally, peaking at about 15 years. Grazing increased the thorniness of the shrubs, evidently as a defense response to defoliation (Ronel et al. 2007).

Crude Drug

Minced roots are boiled in water for 30 min, and then simmered over a small flame for 10–12 additional hours (Yaniv Bachrach 2007).

In Arab villages the whole bush is used as fuel, for making fences and sheep pens, for making brooms from its branches and as stuffing for mattresses. In addition the branches are used to cover tender young plants to protect them from birds and animals (Dafni et al. 1984).

Geographical Distribution

The dwarf-shrub *Sarcopoterium spinosum* is thought to have originated from the semi-arid transition zone between the Mediterranean and Irano-Turonian steppe vegetation zones and then expanded westward (Litav and Orshan 1971; Zohary 1973). It invaded abandoned agricultural land in the more humid true-Mediterranean zone, especially in hilly, rocky regions (Litav and Orshan 1971; Baruch 1986, 1990).

Today it appears over a wide range of climatic conditions from the semi-arid ecotone between the Mediterranean and Irano-Turanian vegetation zones (Zohary 1973), where average annual rainfall is in the 200–300 mm range, to the sub-humid Mediterranean regions with more than 800 mm average annual rainfall. *Sarcopoterium spinosum* dominates many of the hilly parts of the eastern Mediterranean region. It is common in Greece and Turkey, where it appears also in some locations along the Black Sea. Its northern limit is the north Dalmatian cost. According to Proctor (1968) and Martinoli (1969), in the West Mediterranean basin the species appears in locations on the Balkan coast, Italy, and Tunisia (Djerba). In addition, a small population is also recognizable in Malta. The species reaches its western limit of distribution in Italy, with the populations in Sardinia (Gargano et al. 2007). In Greece, the species has received considerable research attention as a major component of the widespread fire-induced dwarf-shrub communities (phrygana). There, considerable variation has been identified among a wide range of phrygana communities (Tsiourlis et al. 2007).

Sarcopoterium spinosum appears in a wide range of habitats and on soils overlying different substrates, including soft chalk, hard limestone and sand-stone (Litav and Orshan 1971). Accordingly, the species appears on a wide range of soils in both xeric and mesic habitats, from close to sea level and up to 1,300 m a.s.l. In the Middle East it dominates the batha vegetation of the Mediterranean coast as well as the semi-steppic plant aggregates of the hinterlands (Zohary 1973). In these areas the species shows a great ability to persist and expand even amid extensive disturbance (Litav and Orshan 1971; Seligman and Henkin 2000, 2002). Along the Eastern Mediterranean coast, as described for Greece (Lavrentiades 1969), the main vegetation type hosting *Sarcopoterium spinosum* is the phrygana. In Italy the plant appears in a few regions (Pignatti 1982), inhabiting a narrow range of habitats in isolated and generally small areas close to the shoreline.

In Israel *Sarcopoterium spinosum* appears in many soils and throughout a wide climatic range. It tends to dominate on hard limestone overlain with terra rossa soil and on soft chalk overlain with light-colored rendzina in the sub-humid Mediterranean zone and on various substrates in the semi-arid region. It is also common on the low calcareous sandstone ('kurkar') hillocks in the coastal plain. It commonly invades abandoned cultivated areas, even though the survival rate of seedlings of this species is often very low, especially where there is vigorous herbaceous vegetation. Competition with herbaceous vegetation over water has been demonstrated under controlled experimental conditions (Litav et al. 1963). In addition, it has been observed that there is little encroachment of *Sarcopoterium spinosum* into well vegetated areas. However, in more favorable habitats, such as rendzina soils on soft chalk, it can form almost continuous stands, expand, coalesce and cover over tens to hundreds of square meters. After fires there is massive seedling establishment even where herbaceous vegetation is quite vigorous.

Ecological and Climatic Requirements

Sarcopoterium spinosum dwarf-shrub communities naturally dominate relatively poor sites including phosphorus-poor habitats in many landscapes in Eastern Mediterranean countries (Eig 1946; Zohary 1973; Litav and Orshan 1971; Reisman-Berman 2004), especially where long periods of exploitation have eliminated most of the tree species and their seed sources (Lev-Yadun 1997). Following the removal of utilization pressure, the cover and biomass of the shrub and woody vegetation increase (Perevolotsky and Seligman 1998). Successional processes and natural disturbances such as fire cause the area to be dominated by thorny shrub thickets for many decades (Henkin et al. 1999, 2007).

When the landscape becomes dominated by spiny dwarf shrub thickets, its biodiversity, forage and amenity values become low. Survival of *Sarcopoterium spinosum* seedlings is highly sensitive to soil moisture stress, especially during the first summer, so often it is absent where vigorous herbaceous growth has exhausted the available soil moisture by the beginning of the summer (Henkin et al. 1998). Where it has become established, especially in karstic habitats or on chalky substrates (Litav 1967) that have available moisture well below the rooting depth of the (mainly annual) herbaceous vegetation, vegetative propagation of *Sarcopoterium spinosum* often occurs (Seligman and Henkin 2000; Reisman-Berman 2004).

Where resources are irregularly distributed and establishment sites are fully occupied by perennial species, localized phalanx-type clonal proliferation can reduce the risk of genet extinction. On favorable sites this strategy confers upon *Sarcopoterium spinosum* genets the tenacity required to hold off replacement by other species and to dominate large areas of the landscape for much longer than the average life span of individual ramets (Seligman and Henkin 2002). *Sarcopoterium spinosum* growing on chalk substrate also demonstrated exceptionally high resilience in the case of disturbance or attempted eradication (Perevolotsky et al. 2001).

For thousands of years (Baruch 1986, 1990) these dwarf-shrub communities have been subject to many disturbances, mainly fire and grazing, but also to periodic cultivation and harvesting for fuel or construction material (e.g. for 'thicket fences' around holding paddocks in traditional villages). Recovery after disturbance is often rapid. Regeneration can occur from seedlings and from surviving root-crown meristems (Litav and Orshan 1971), and *Sarcopoterium spinosum* persists for decades in many habitats, sometimes as uninterrupted, continuous stands. Yet plants have rarely been found to reach the age of 19 years (Litav and Orshan 1971). *Sarcopoterium spinosum*, with its well-defined dimorphic growth habit – large mesophyllic leaves during the active growth period in winter and spring, and small sclerophyllous leaves during the summer – is an indigenous dwarf shrub that is relatively dormant in the late summer and winter.

The strongly peaked age structure, together with the appearance of clonal regeneration on the older plants, suggests that subsequent maintenance of the stand does not depend on recurrent seedling recruitment. This can explain the persistence of *Sarcopoterium spinosum* in stands of herbacaeous vegetation despite the extreme sensitivity of the seedlings to competition under such conditions (Seligman and Henkin 2000).

Sarcopoterium spinosum often dominates a landscape to the virtual exclusion of other species, but it is also used in landscape gardening, especially along highways. Depending on the context, the shrub is a desirable, undesirable or neutral element in the landscape. Efficient management of the shrub will require further elucidation of its response to diverse ecological conditions.

Traditional Use and Common Knowledge

Many authors consider the "**thorns**" mentioned in the Bible to be *S. spinosum*. "And ***thorns*** shall come up her palaces, nettles and brambles in the fortress thereof…" (Isaiah, 34, 13); and "For as the crackling of ***thorns*** under a pot, so is the laughter of the fools; this also is vanity" (Ecclesiastes, 7, 6). Also: "Therefore I will hedge up her way with ***thorns***; and I will build a wall against her, so that she cannot find her paths"(Hosea, 2:6).

The plant is mentioned in the Mishnah as a means of curing inflammation: "With the ***thorns*** they would remove the affected parts and separate them from the healthy flesh" (Mishnah, Keritot, 6, 2).

Dioscorides recommended a potion prepared from the leaves of S. *spinosum* for the treatment of dysentery and earache (Günter 1968).

Israeli Arabs traditionally use the plant as a tranquilizer; a potion prepared from the leaves is said to dispel fears. Green leaves, salted and spiced and soaked in saliva, are used to cure eye complaints, and also as a lucky charm. The plant roots are widely used as a folk remedy for diabetes and for dissolving kidney stones (Yaniv et al. 1987).

The following is a list of the medicinal uses of *Sacropoterium spinosum* as reported by the informants (interviewed healers). The preparation process is described.

1. To treat diabetes:
   - Boil the minced roots in water for 30 min. Then simmer over a small flame for 10–12 additional hours. Drink the resulting concentrate first thing in the morning on an empty stomach, and later several times during the day.
2. To relieve toothache:
   - Boil the roots in water and gargle with the liquid.
   - Boil Sira roots together with an equal quantity of raspberry roots, cloves and a little salt, until the decoction turns red. Then use it as a gargle.
3. To relieve stomach aches and headaches and to help digestion:
   - Boil the roots in water and drink the decoction.
4. To rectify the blood circulation:
   - Cook the seeds in water and drink the decoction.
5. To cure external infections and to relieve pain:
   - Boil the root in water, and spread in a dressing for the affected area.
6. For relieving kidney constipation:
   - The patient sits in a tub filled with water boiled with fruits of *S. spinosum.*
7. For curing temporary paralysis.
   - The roots are boiled and the decoction is given a few times a day.

As seen above, *S. spinosum* is used as a traditional medicine to treat a number of ailments including diabetes (Yaniv Bachrach 2007).

Recent research has found insulin-like effects on metabolic pathways in classic insulin-responsive tissues (Rosenzweig et al. 2007).

Major Chemical Constituents and Bioactive Compounds

Chemical constituents of *Sarcopoterium spinosum* and their bioactivity have not been fully identified yet. Most of the analysis was performed by Reher who obtained several compounds from the aerial and underground parts of the herb. Five pentacyclic triterpenoids (Reher et al. 1991a, b; Reher and Buděšínský 1992) were identified by comparing their NMR and mass spectra with those of chemically identified samples, among them tormentic and ursolic acids (Reher and Buděšínský 1992). Tormentic acid is suggested to be the major constituent of *S. spinosum* extract, and to mediate its anti-proliferative activity on several cancer cell lines (Loizzo et al. 2013). In addition, there are some indications of the beneficial effect of tormentic acid and tormentic acid glucoside isolated from *S. spinosum* in reducing blood glucose in alloxan-induced diabetic mice (Reher et al. 1991a, b). A hypoglycemic effect of tormentic acid was also demonstrated by Ivorra et al., showing induction of insulin secretion by the compound (Ivorra et al. 1988a, b). However, while *in-vitro* studies demonstrated that *S. spinosum* extract increases insulin secretion, *in-vivo* studies performed on diabetic mice show that the hypoglycemic effect of the extract originated from its effect on the target tissue of insulin, rather than the induction of insulin secretion (Smirin et al. 2010). Thus, more studies must be performed in order to clarify whether tormentic acid improves insulin sensitivity in addition to its effect on insulin secretion, or other compounds in *S. spinosum* extract act in concert with tormentic acid to carry out its anti-diabetic activity.

The plant sterol β-sitosterol has been identified in *S. Spinosum* root extract as well, but its contribution to the bioactivity of *S. spinosum* has not been tested. β-sitosterols are well known for their beneficial effect of reducing blood cholesterol. In addition, there is some evidence of potential anti-diabetic properties of the compound, as was demonstrated in β-sitosterol isolated from *Solanum surattense* (Gupta et al. 2011) and β-sitosterol 3-beta-glucoside isolated from *Centaurea seridis*, both showing an anti-hyperglycemic effect in diabetic rats, presumably via a mechanism of increasing insulin secretion (Ivorra et al. 1988a, b).

Catechines and epicatechines were also identified in the active extract. In addition, the existence of a dimeric form of catechines had been suggested, based on the retention times of intense peaks obtained by LC/MS (Smirin et al. 2010). The role of these compounds in mediating the activity of *S. spinosum* has not been elucidated yet. Proanthocyanidines were also isolated from the root bark of *S. spinosum*; however, there are no clear indications of their involvement in the bioactivity of the herb (Reher et al. 1991a, b).

Modern Medicine Based on Traditional Medicine Uses (Biological Activities)

According to ethnopharmacological studies *S. Spinosum* extract is used for the treatment of several disorders. The primary use, mentioned in most surveys, is of an aqueous extract prepared from the root bark for the treatment of diabetes (Hamdan and Afifi 2004; Said et al. 2002; Friedman et al. 1986; Al-Qura'n 2009; Yaniv et al. 1987; Steimetz 1965; Yaniv Bachrach 2007). Several studies investigated the anti-diabetic function of *S. Spinosum*. Reduced blood glucose was demonstrated in normal rabbits (Mishkinsky et al. 1966) and in diabetic rodents, including the KK-Ay mice, a spontaneously developing diabetes model, as a result of genetic defect (Smirin et al. 2010), and in alloxan-treated rats, a chemically induced diabetic model (Mishkinsky et al. 1966).

Antidiabetic activity might be mediated by several mechanisms; inhibiting intestinal digestion and absorption of carbohydrates, inducing insulin secretion by the pancreas or enhancing glucose disposal from the blood by target tissues such as muscle, adipose or liver tissues, either by improving insulin sensitivity or mimicking its action. Kasabri et al. (2011) showed that the acute effect of *S. Spinosum* on blood glucose is mediated through the inhibition of starch digestion, as was demonstrated both *in vitro* and *in-vivo*, showing reduced glycemia following an oral starch tolerance test, and no effect in an oral glucose tolerance test (OGTT). Both tests were performed 30 min after the administration of the extract. Unlike these results, improved glucose tolerance was observed in mice given the extract for longer time (24 h or several weeks) before the performance of an intra-peritoneal glucose tolerance test (IPGTT) (Smirin et al. 2010). These results suggest that other mechanisms, not involved in digestion or absorption of carbohydrates, are involved in the anti-diabetic function of *S. spinosum*, which required at least 24 h in order to achieve the hypoglycemic activity. It was found that *S. spinosum* extract leads to the induction of insulin secretion as well as insulin-like effects; increased glucose uptake by skeletal muscle cells, adipocytes and hepatocytes, increased GSK phosphorylation and reduced lipolysis. However, long-term treatment with the extract was followed by the correction of hyperinsulinemia, suggesting that the most prominent mechanisms of action are those of the target tissues of insulin, mediated by improved insulin sensitivity or mimicking insulin action, rather than increasing insulin secretion (Smirin et al. 2010).

Secondary theraputic applications of *S. spinosum* mentioned in ethnopharmacological surveys are for pain relief, mainly toothache (Yaniv Bachrach 2007), disorders of the digestive system (Friedman et al. 1986; Al-Qura'n 2009; Ali-Shtayeh et al. 2000) asthma (Friedman et al. 1986), renal calculi (Al-Qura'n 2009), poisoning (Friedman et al. 1986) and cancer (Durodola 1975). The anti-proliferative properties of *S. spinosum* extract were examined on several cancer cell lines, showing cytotoxic effect, which are partially attributed to the action of tormentic acid (Loizzo et al. 2013). This supports traditional knowledge regarding the anti-cancer properties of the extract. The other mentioned medicinal uses of *S. spinosum* have not been validated using scientific tools.

Despite the well-documented traditional medicinal use of the plant, *S. Spinosum* is not recognized by the U.S. pharmacopiea (herbal medicines compendium), European pharmacopiea or the British pharmacopoeia (2012). Accordingly, the availability of commercialized *S. spinosum* extract is very limited.

Conclusion

Sarcopoterium spinosum is one of the most common plants in the flora of Israel. The plant was extensively used in the Holy Land for practical purposes, such as making brooms, building fences and domestic heating. However, recent traditional medicine of the Holy Land emphasizes the medicinal importance of *S. spinosum* as evident from our ethno botanical survey (Yaniv et al. 1987).

One of the most quoted medicinal indications is the use of the root bark as a popular cure against diabetes. There is now supportive evidence that the roots can be used as a source of remedy for diabetes. Efforts should be made to develop practical applications of root extract for necessary uses.

References

Ali-Shtayeh MS, Yaniv Z, Mahajna J (2000) Ethnobotanical survey in the Palestinian area: a classification of the healing potential of medicinal plants. J Ethnopharmacol 73(1–2):221–232

Al-Qura'n S (2009) Ethnopharmacological survey of wild medicinal plants in Showbak, Jordan. J Ethnopharmacol 123(1):45–50

Baruch U (1986) The late Holocene vegetational history of Lake Kinneret (Sea of Galilee), Israel. Paléorient 12:37–48

Baruch U (1990) Palynological evidence of human impact on the vegetation as recorded in Late Holocene lake sediments in Israel. In: Bottema S, Entjes-Nieborg G, van Zeist W (eds) Man's role in the shaping of the eastern Mediterranean landscape. A.A. Balkema, Rotterdam, pp 283–293

Dafni A, Yaniv Z, Palevitch D (1984) Ethnobotanical survey of medicinal plants in northern Israel. J Ethnopharmacol 10(3):295–310

Durodola JI (1975) Antitumour effects against sarcoma 180 ascites of fractions of *Annona senegalensis*. Planta Med 28(1):32–36

Eig A (1946) Synopsis of the phytogeographical units of Palestine. Palest J Bot (Jerus Ser) 3:183–246

Friedman J, Yaniv Z, Dafni A, Palevitch D (1986) A preliminary classification of the healing potential of medicinal plants, based on a rational analysis of an ethnopharmacological field survey among Bedouins in the Negev desert, Israel. J Ethnopharmacol 16(2–3):275–287

Gargano D, Fenu G, Medagli P, Sciandrello S, Bernardoa L (2007) The status of *Sarcopoterium spinosum* (Rosaceae) at the western periphery of its range: ecological constraints lead to conservation concerns. Isr J Plant Sci 55:1–13

Gunther RT (1968) Greek herbal of Dioscorides, illustrated by a Byzantine. 512. In: English by I. Goodyear, 1655 (Facsimile of the 1934 edition). Hafner Publisher, London/New York

Gupta R, Sharma AK, Dobhai MP, Sharma MC, Gupta RS (2011) Antidiabetic and antioxidant potential of beta-sitosterol in streptozotocin-induced experimental hyperglycemia. J Diabetes 3(1):29–37

Hamdan II, Afifi FU (2004) Studies on the in vitro and in vivo hypoglycemic activities of some medicinal plants used in treatment of diabetes in Jordanian traditional medicine. J Ethnopharmacol 93(1):117–121

Henkin Z, Seligman NG (2007) Survival of *Sarcopoteriun spinosum* seedlings growing on Terra rossa soil. Isr J Plant Sci 55:45–51

Henkin Z, Seligman NG, Kafkafi U, Prinz D (1998) End-of-season water depletion in relation to growth of herbaceous vegetation in a sub-humid Mediterranean dwarf-shrub community on two contrasting soils. Plant Soil 202:317–326

Henkin Z, Seligman NG, Noy-Meir I, Kafkafi U (1999) Secondary succession after fire in a Mediterranean dwarf-shrub community. J Veg Sci 10:503–514

Henkin Z, Seligman NG, Noy-Meir I (2007) Successional transitions and management of a phosphorus-limited shrub land ecosystem. Rangel Ecolo Manag 60:453–463

Ivorra MD, Paya M, Villar A (1988a) Hypoglycemic and insulin release effects of tormentic acid: a new hypoglycemic natural product. Planta Med 54(4):282–285

Ivorra MD, D'Ocin MP, Paya M, Villar A (1988b) Antihyperglycemic and insulin-releasing effects of beta-sitosterol 3-beta-D-glucoside and its aglycone, beta-sitosterol. Arch Int Pharmacodyn Ther 296:224–231

Kasabri V, Afifi FU, Hamdan I (2011) In vitro and in vivo acute antihyperglycemic effects of five selected indigenous plants from Jordan used in traditional medicine. J Ethnopharmacol 133(2):888–896

Lavrentiades GJ (1969) Studies on the flora and vegetation of the Ormos Archangelou in Rhodos Island. Vegetatio 19(1–6):308–329

Lev-Yadun S (1997) Flora and climate in southern Samaria: past and present. In: Finkelstein I, Lederman Z, Bunimovitz S (eds) Highlands of many cultures, Monograph series. The Institute of Archaeology, Tel Aviv University, Tel Aviv, pp 85–102

Litav M (1967) Micro-environmental factors and species interrelationships in three batha associations in the foothill region of the Judean hills. Isr J Bot 16:79–99

Litav M, Orshan G (1971) Biological flora of Israel. 1. *Sarcopoterium spinosum* (L.) sp. Isr J Bot 20:48–64

Litav M, Kupernik G, Orshan G (1963) The role of competition as a factor in determining the distribution of dwarf shrub communities in the Mediterranean territory of Israel. J Ecol 51:467–480

Loizzo MR, Bonesi M, Passalacqua NG, Saab A, Menichini F, Tundis R (2013) Antiproliferative activities on renal, prostate and melanoma cancer cell lines of *Sarcopoterium spinosum* aerial parts and its major constituent tormentic acid. Anticancer Agent Med Chem 13(5):768–776

Martinoli G (1969) *Poterium spinosum* L. elemento del Mediterraneo Orientale a Calamosca (Capo S. Elia, Sardegna). Giorn Bot Ital 103:325–340

Mishkinsky J, Menczel E, Sulman FG (1966) Hypoglycaemic effect of *Poterium spinosum* L. (Rosaceae). Arch Int Pharmacodyn Ther 161(2):306–313

Perevolotsky A, Seligman NG (1998) Grazing in Mediterranean ecosystems: inversion of a paradigm. Bioscience 48:1007–1017

Perevolotsky A, Ne'eman G, Jonathan R, Henkin Z (2001) Resilience of the dwarf shrub prickly burnet to management in East Mediterranean rangelands. J Range Manag 54:561–566

Pignatti S (1982) Flora d'Italia, vol 1. Edagricole, Bologna

Proctor MCF (1968) Sarcopoterium Spach. In: Tutin TG, Heywood VH, Burges NA, Moore DM, Valentine DH, Walters SM, Webb DA (eds) Flora Europaea, vol 2. Cambridge University Press, Cambridge, p 34

Reher G, Buděšínský M (1992) Triterpenoids from plants of the sanguisorbeae. Phytochemistry 31(11):3909–3914

Reher G, Reznicek G, Baumann A (1991a) Triterpenoids from *Sarcopoterium spinosum* and *Sanguisorba minor*. Planta Med 57(5):506

Reher G, Slijepcevic M, Kraus L (1991b) Hypoglycemic activity of triterpenes and tannins from [the root bark of *Sarcopoterium spinosum* and two *Sanguisorba* species [*S. minor* and *S. officinalis*]. Planta Med 57(8):A57–A58

Reisman-Berman O (2004) Mechanisms controlling spatio-temporal dynamics of shrubland patchiness: case study of *Sarcopoterium spinosum* (L.) Spach. Ph.D. thesis, Ben-Gurion University of the Negev, Beer-Sheva, Israel

Reisman-Berman O, Henkin Z (2007) *Sarcopoterium spinosum*: revisiting shrub development and its relationship to space occupation with time. Isr J Plant Sci 55:53–61

Ronel M, Malkiel H, Lev-Yadun S (2007) Quantitative characterization of the thorn system of the common shrubs *Sarcopoterium spinosum* and *Calicotome villosa*. Isr J Plant Sci 55:63–72

Rosenzweig T, Abitbol G, Talerc D (2007) Evaluating the anti-diabetic effects of *Sarcopoterium spinosum* extracts in vitro. Isr J Plant Sci 55:103–109

Said O, Khalil K, Fulder S, Azaizeh H (2002) Ethnopharmacological survey of medicinal herbs in Israel, the Golan Heights and the West Bank region. J Ethnopharmacol 83(3):251–265

Seligman NG, Henkin Z (2000) Regeneration of a dominant Mediterranean dwarf-shrub after fire. J Veg Sci 11:893–902

Seligman NG, Henkin Z (2002) Persistence in *Sarcopoterium spinosum* dwarf-shrub communities. Plant Ecol 164:95–107

Smirin P, Taler D, Abitbol GFT, Brutman-Barazani T, Kerem Z, Sampson SR, Rosenzweig T (2010) *Sarcopoterium spinosum* extract as an antidiabetic agent: in vitro and in vivo study. J Ethnopharmacol 129(1):10–17

Steinmetz EF (1965) A sensational drug, Peterii Spinosi Cortex radicis, (Poterium spinosum root-bark): new agent against diabetes. Publisher, Amsterdam

Tsiourlis G, Konstantinidis P, Xofis P (2007) Taxonomy and ecology of phryganic communities with *Sarcopoterium spinosum* (L.) Spach of the Aegean (Greece). Isr J Plant Sci 55:15–34

Yaniv Bachrach Z (2007) Ethnobotanical studies of *Sarcopoterium spinosum* in Israel. Isr J Plant Sci 55:111–114

Yaniv Z, Dafni A, Friedman J, Palevitch D (1987) Plants used for the treatment of diabetes in Israel. J Ethnopharmacol 19(2):145–151

Zohary M (1973) Geobotanical foundations of the Middle East. Gustav Fischer Verlag/Swets and Zeitlinger, Stuttgart/Amsterdam

Traditional Uses of *Pistacia lentiscus* in Veterinary and Human Medicine

S. Landau, H. Muklada, A. Markovics, and H. Azaizeh

Abstract *Pistacia lentiscus* L. is an evergreen shrub or tree of the Anacardiaceae family that is ubiquitous in Mediterranean areas under 1,100 m above sea level. The essential oil obtained from the gum/resin is commonly called mastic oil, whereas the oil obtained from the leaves is termed lentisk oil. Mastic, foliage, and fruit essential oils have served in traditional human medicine for millennia. In human medicine, aqueous extracts of leaves and young twigs are prepared as infusions or decoctions. Leaves and young twigs are used "as is" in veterinary medicine. Extracts exhibit a high concentration of phenolics and terpenes of varying composition. In addition to the traditional use of mastic resin in human medicine against stomach ulcers, evidences for anti-bacterial, antifungal, antioxidant, hepato-protective, and anti-carcinogenic properties of products sourcing from *P. lentiscus*, some of them controversial, are accumulating. The anthelmintic and anticoccidial properties of *P. lentiscus* foliage have been recently demonstrated in small ruminants. Metabolomics, encompassing new analytical procedures will probably help in elucidating the rationale of traditional medicine in using *P. lentiscus* for therapies or prevention. It can be expected that new cultivars will be cultivated on the basis of bio-activity.

Keywords Mastic • Lentisk • Folk medicine • Ethno-veterinary • Herbal medicine • Organic agriculture

S. Landau (✉) • H. Muklada
Department of Agronomy and Natural Resources, Institute of Plant Sciences,
Agricultural Research Organization, the Volcani Center, P.O. Box 6, Bet Dagan 50250, Israel
e-mail: vclandau@volcani.agri.gov.il

A. Markovics
Department of Parasitology, Kimron Veterinary Institute, Bet Dagan 50250, Israel

H. Azaizeh
The Regional Research and Development Center, The Galilee Society,
Shefa-Amr 20200, Israel

Z. Yaniv, N. Dudai (eds.), *Medicinal and Aromatic Plants of the Middle-East*,
Medicinal and Aromatic Plants of the World 2, DOI 10.1007/978-94-017-9276-9_8

Taxonomy Characteristics

Pistacia lentiscus L. – also termed lentisk or mastic tree is one of the 11 species in the genus *Pistacia* L. of the Anacardiacae (Zohary 1952). *P. lentiscus* is an evergreen 1–5 m high shrub or tree, growing in semi-arid areas of the Mediterranean area from Morocco and the Iberian peninsula in the west through southern France and Turkey to Iraq and Iran in the east. It is also native to all Mediterranean islands.

The genus *Pistacia* is characterized by its dioecious reproductive system (male and female plants) and by its homeochlamydic perianth of flowers ("naked" flowers; Mabberley 1997). Based on morphological characters, Zohary (1952) divided the genus *Pistacia* into four sections: *Lentiscella* (New World species), *Lentiscus* (evergreens), *Butmela* (*P. atlantica* Desf.), and *Terebinthus* (*P. chinensis* Bunge, *P. khinjuk* Stocks, *P. Palaestina* Boiss., and *P. vera* L.). Kafkas and Perl-Treves (2002) divided the genus into single-trunk trees (*P. atlantica*, *P. eurycarpa, P. integerrima, P. khinjuk*, and *P. vera*) and shrubs and small trees (*P. lentiscus*, *P. mexicana*, *P. palaestina*, *P. texana*, and *P. terebinthus*) and claimed that this morphological classification is supported by DNA analysis (RAPD). Parfitt and Badenes (1997) suggested dividing the *Pistacia* genus into two sections, i.e., *Lentiscus* and *Terebinthus*, based on plastid restriction site analysis and morphological characters: the *Lentiscus* section includes the evergreen species with paripinnate (no terminal leaflet) leaves and smaller seeds, while the *Terebinthus* section consists of deciduous trees with imparipinnate leaves and larger seeds. However hybridization is common between *Pistacia* species: in particular, what was once classified as *Pistacia saporta* (Zohary 1952) is now (Yi et al. 2008) considered as a hybrid of *P. lentiscus* (of the *lentiscus* section) and *P. palaestina* (of the *terebinthus* section). Indeed, in the most recent phylogeny analysis of *Pistacia*, based on nuclear ribosomal internal transcribed spacers (ITS) and nuclear DNA subsets, the species of section *lentiscus* were not represented as a monophyletic group, but instead, as a few parallel clades with the clade of section *terebinthus* nested within it (Yi et al. 2008).

A new insight into the taxonomy of Pistacia was offered by Inbar (2008): *Pistacia* spp. are the obligate host for specialized gall-forming aphids (Homoptera: Fordinae). As the galling process involves intimate crosstalk between the plant and insect genotypes, gall-forming insects can serve as a tool in plant systematics. *P. lentiscus*, *P. aethiopica* and two New World species emerged in a clear distinct position from all other species in the aphid-based association. Two groups emerged from the preferences of aphid genera: the "khinjuk" group included *P. khinjuk*, *P. palaestina*, *P. terebinthus*, *P. chinensis* and *P. integerrima*; and the "Vera" group, encompassing *P. vera*, *P. mutica*, and *P. atlantica*. Unfortunately, as *P. lentiscus* hosted only exclusive aphid fauna, no phylogenetic information was provided by aphids.

Natural History, Morphology, and Requirements

P. lentiscus is a thermophilous species, growing in warm regions at low altitudes and in sunny sheltered spots at medium altitudes (<1,100 m a.s.l.), featuring resilience to both heavy frosts (Palacio et al. 2005) and prolonged drought (Filella et al. 1998). *P. lentiscus* thrives on various types of soils. It is able to tolerate and accumulate salt, which probably explains its frequency in Mediterranean coastal regions (Barazani and Golan-Goldhirsch 2009). It is frequently found in communities with carob (*Ceratonia siliqua*, see Israel, coastal plain, Fig. 1), *Phillyrea* spp., and varieties of oak (*Quercus* spp.) and buckthorn (*Rhamnus* spp.) trees. Under anthropogenic pressure, encompassing logging, grazing (Fig. 1a), and fires, or even with minimal disturbance like in Nature reserves, it is found in the form of round brush islets less than 3 m in height and 10 m in diameter but when pruned, it can develop some large trunks and numerous thicker and longer branches and it often becomes a tree of up to 7 m, as found in the mastic grooves on the Chios island (Greece; Fig. 1e).

All *P. lentiscus* trees share alternate, leathery, winged leaf rachis and fasciculate inflorescence, but morphology can vary: oblong, lanceolate or elliptical leaflets (5 or 6 pairs; Fig. 1b), fleshy or dry mesocarp, bony or leathery endocarp. In particular, trees from sub-cultivars of the *chia* variety cultivated on the Greek island of Chios are classified according to leaflet morphology (Hagidimitriou 2009). The genetic variability of morphological attributes in *P. lentiscus* was studied by Barazani et al. (2003), working on accessions of *P. lentiscus* from Israel, Cyprus, Spain, and Tunisia, and growing under similar conditions in the same location in Israel. High polymorphism was found in morphological attributes (Leaf length, number of leaflets, leaflet texture and color, trunk diameter, canopy area) with no significant differences for geographical location or gender.

Flowering takes place between Mid-March and the end of April. Male inflorescences (Fig. 1c) aggregate 8–10 flowers. The female flower has one seminal primorde. Flowers on male trees are deep red, and on female trees, yellow. Fruits are ripe occurs 150–230 days after the onset of flowering (Spain; Jordano 1989). The fruit is a drupe, first red (Fig. 1b) and then black when ripe, about 4 mm in diameter. *P. lentiscus* seeds are dispersed by birds.

Even though most drugs from *P. lentiscus* source from trees that grow wild, the species is cultivated for mastic – resin from *P. lentiscus* var. *chia* – in 21 villages of southern Chios (Hagidimitriou 2009). The island enjoys 520 mm of rain falling from October to May. Summer is warm; winter is pleasant with no frost. Average humidity ranges between 75 % in winter to 60 % in summer. Trees are not irrigated (Table 1).

Chalky rocks of the Chios island, which does not hold surface moisture seem to be particularly well appropriate for *P. lentiscus* (Hagidimitriou 2009) but the plant also thrives in Mediterranean coastal sandy soils (Landau et al. 2002).

Fig. 1 *Pistacia lentiscus* thicket under grazing (Israel, coastal plain; (**a**), young twigs (**b**), fruit (**c**), male inflorescence (**d**), and in tree-like thicket (Chios, Greece)

Table 1 Climate data for Chios, Greece (Source: www.weather-to-travel.com)

Month	Jan	Feb	Mar	Apr	May	Jun	Jul	Aug	Sep	Oct	Nov	Dec	Year
Average high °C	11	11	13	17	22	26	28	28	25	20	16	13	19.2
Average low °C	5	5	6	9	13	17	19	19	16	12	9	6	11.3
Precipitation mm	100	78	61	44	24	4	1	0	8	23	55	122	520

Table 2 Some historical mentions of *P. lentiscus* in *Materia medica* and prescription descriptions

Drug	Period AD	Document	References
Oil	First century	Pliny the Elder, Natural History	Holland(1601)
Oil and resin	First century	Dioscorides, *de Materia Medica*	Leonti et al. (2009)
Resin	Fifth century	Talmud, tractate Shabbat, 8:7	Shifman et al. (2002)
Resin	Tenth–nineteenth century	Genizah, Ben Ezra synagogue, Cairo	Lev and Amar (2000)
Resin	Fifteenth–sixteenth century	Chilandar Medical Codex, Serbia	Jaric et al. (2011)

The Crude Drug Used

The essential oil obtained from the gum/resin is commonly called mastic oil, whereas the oil obtained from the leaves is termed lentisk oil. The resin (Lev and Amar 2000, Israel; Leonti et al. 2009, Sardinia and Italy; Hagidimitriou 2009, Greece), foliage essential oil (Leonti et al. 2009, Sardinia and Italy), and fruit oil (Charef et al. 2008, Algeria), as well as galls (Inbar 2008), all have served in traditional human medicine for millennia (Table 2). In human medicine, aqueous extracts of leaves and young twigs are prepared as infusions (Said et al. 2002, Israel) or decoctions (Guarrera et al. 2005, Spain).

Leaves and young twigs are used “as is” in veterinary medicine (Muklada et al. 2013, Israel; Bullitta et al. 2007, Sardinia; Pieroni et al. 2006, Cyprus).

Collection Practice

Mastic

Mastic is produced from the resin secreted after incision made in the bark of trunks and branches of *P. lentiscus*. The harvest takes place between mid-July and mid-October. Each tree is harvested twice. Before harvesting, the area around each tree is cleaned and flattened. The trunks are well cleaned, scratched and wiped. Then the producers make an incision in a depth of 2–4 cm by using a crescent shaped hammer. Each tree is incised 20–100 times yearly. The sap is allowed to drip onto the ground below that is covered with white clay. Originally liquid, it hardens when the weather turns cold into drops of hard, brittle, translucent resin. After collection, the mastic is washed manually with cold water and soap and dried in shaded areas. Male trees produce more than females. They start to produce when aged 5 years and reach their higher production at 15 (Hagidimitriou 2009). A new collection technique involves a stimulating agent (ethephon) that stimulates resin excretion after incision of the tree. Resin is then collected in a liquid form.

Foliage

P. lentiscus is an evergreen and foliage is available all year long. There is no clearcut report of seasonal variation of polyphenol contents but the highest yield of essential oil was obtained when foliage was harvested in March-June in three locations in Morocco (Zrira et al. 2003) and in Greece (Gardeli et al. 2008). In Sardinia and Sicily, leaves and young twigs from *Pistacia lentiscus* are also collected in the spring, at blooming time. Distillation is carried out immediately after harvesting, yielding about 0.2 % of essential oil in 4 h (Sardinia and Sicily; Leonti et al. 2009)

Major Chemical Constituents and Bioactive Compounds

P. lentiscus seed oils (Charef et al. 2008) are characterized by a high content of unsaturated fatty acids (approximately 80 %) with a dominance of C 18:1 (>50 %) and C 18:2 (>20 %), and more than 8 mg/g of tocopherols (Dhifi et al. 2013), inferring anti-oxidative activity. However, the two major classes of bio-actives source in *Pistacia lentiscus* are phenolics and terpenes. Even though all parts of the plants contain both phenolics and terpenes, the phenolics concentrate mainly in crude, whole leaves and twigs, whereas terpenes are the major bio-active in resin or essential oils obtained by hydro-distillation of chopped leaves or resin.

Phenolics

The foliage of *P. lentiscus* is rich in tannins (proanthocyanidins), but concentration depends on the methodology and standards selected. Depending on standards, phenolics represented 7–12 % of water extract and 5–7 % of ethanol 70 % extracts of foliage in Israel (Azaizeh et al. 2012). Estimations of condensed (Decandia et al. 2000) and polyethylene-glycol-binding tannins (Landau et al. 2004) are approximately 20 % of dry matter.

Table 3 Comparative composition of phenolics (mg/g, DM basis), as determined by HPLC in Italy (Romani et al. 2002), Israel (Azaizeh et al. 2012), and Algeria (Rodriguez-Perez et al. 2013) in *P. lentiscus* leaves

Compound name	Israel	Italy	Algeria
3,4,5 Tri-O-galloyquinic acid	15.9	10	
3,5 Di-O-galloyquinic acid	10.8	27	
5-O-galloyl quinic acid		9.6	
Beta-glucogallin			0.3
Beta-glucogallin isomer			2.2
Catechin	5.7	1.8	3.4
Chlorogenic acid	17.5		
Cyanidin 3-O-glucoside		0.4	
Delphidin 3-O-glucoside		0.8	
Gallic acid	1.22	3.7	0.3
Kaempferol glycoside			0.1
Luteolin			0.1
Monogalloyl glucose		2.8	
Myricetin 3-O rhamnoside		6.8	
Myricetin-3-O-rhamnoside1			6.2
Myricetin-3-O-rhamnoside2			1.0
Myricetin-3-O-rutinoside	6.8	4.5	
Myricetin-galactoside			1.7
Myricetin-rhamnopyranoside1			0.3
Myricetin-rhamnopyranoside2			0.1
Myricetin-xyloside			0.5
Pistafolin A			0.1
Quercetin 3-O-rhamnoside		3.7	
Quercetin arabinopyranoside			0.4
Quercetin glycoside			0.1
Rutin (Quercetin-3-O-rutinoside)	13.6		

The first identification of polyphenols in *P. lentiscus* foliage in Italy was pioneered in Italy by Romani et al. (2002), based on HPLC, HPLC-diode array detection (DAD), HPLC-MS, together with ^{1}H- and ^{3}C NMR: they established that *P. lentiscus* polyphenols belong to three distinct classes, i.e., galloyl derivatives (hydrolizable tannins), flavonol glucosides, and glucosides of myricetin, cyanidin and quercetin (anthocyanins). Further studies in Israel (Azaizeh et al. 2012) and Algeria (Rodriguez-Perez et al. 2013) confirmed this classification. It seems that some compounds are ubiquitous in *P. lentiscus* foliage, e.g., catechin and gallic acid, but variety exists in others (Table 3).

In Italy and Israel, galloyl derivatives (5.3 % of leaf dry weight) detected by HPLC/DAD formed more 70 and 64 % of polyphenols, respectively, in the foliage of *P. lentiscus*. In Italy, myricetin derivatives represented 20 % of polyphenols, with quercetin, delphinidin, and cyanidin glucosides making for the rest (Table 3); and in

Israel flavonol glucosides represented 29 % of polyphenols (Azaizeh et al. 2012). In a recent metabolomic study with methanolic extracts of *P. lentiscus* foliage in Algeria (Rodriguez-Perez et al. 2013) where HPLC, mass spectrometry (MS) and electrospray ionization (ESI) tandem mass spectrometry were combined (HPLC-ESI-TOF-MS), beta-glucogallin was found instead of galloyl derivatives of quinic acid. The carbohydrate moieties in anthocyanins showed great variety between the three countries.

In addition to geographical variety, Romani et al. (2002) suggest that light density affects polyphenol concentration. Gardeli et al. (2008) found significant, but small (<2 %) seasonal effects on total phenolics (ng gallic acid/g plant) in methanolic extracts of *P. lentiscus* foliage.

The concentration of phenolics in resin is extremely low: they consist of tyrosol, p-hydroxybenzoic acid, p-hydroxyphenylacetic acid, and vanillic acid, with 375, 136, 107, 41 ng/g resin, respectively, with traces of gallic and trans-cinnamic acid (Kaliora et al. 2005).

Terpenes

Variety is the rule for terpenes in *P. lentiscus* foliage. In Spain, total terpene in the foliage of *P. lentiscus* was positively correlated with soil moisture, but negatively with air temperature (Llusia et al. 2006). Analyses of essential oil distilled from the leaves of *P. lentiscus* reveal great intra-specific biochemical diversity, encompassing monoterpenes, monoterpene esters, monoterpene alcohols, and sesquiterpenes (Ben Douissa et al. 2005). Barazani et al. (2003) reported that the proportions of α-pinene, limonene, sabinene, β-caryophyllene, and germacrene D are greatly affected by the gender of plant and geographic origin (Table 4). In Corsica (Castola et al. 2000), cluster analysis and discriminant analysis of 105 plants allowed to classify leaf essential oils into three groups, based on the content of terpinen-4-ol, α-pinene, limonene and myrcene: group I was featured high terpinen-4-ol (25 %, SD 8 %) or α-pinene (32 %, SD 8 %); group II, high limonene (47 %, SD 5 %), with moderate terpinen-4-ol (11 %, SD 4 %) and α-pinene (5.2 %, SD 1.5 %); and group III had very high concentration of myrcene (77 %, SD 11 %). In Morocco, terpene composition was affected by season of harvest and location (Zrira et al. 2003) and in Greece, by phenological stage (Gardeli et al. 2008). Koutsoudaki et al. (2005) compared terpene composition in mastic gum and essential oil sourcing from plants in the same location: the oil contained 50 % and threefold more α-pinene and β-myrcene, respectively. Also, the gum contained 22 % of (Z,Z, farnesol), whereas only traces were found in the oil.

Many analyses of resin focus on tetra- and penta-cyclic triterpenes because much of medical activity is attributed to them. The major tri-terpenes in *P. lentiscus* var. Chia resin are isomasticadienonic acid, masticadienonic acid, and 28-norolean-17-en-3-one (24, 9.3, and 19 % of triterpenic fraction, respectively, but 11 and 20 different triterpenes were identified in the acidic and neutral fractions, respectively (Assimopoulou and Papageorgiou 2005).

Table 4 Terpenes in *P. lentiscus* leaf essential oil (After Barazani et al. 2003)

	Essential oil content (%)							
	Israel		Cyprus		Spain			
	Male	Female	Male	Female	Male 1	Female 1	Male 2	Female 2
α-thujene	0	0	0.3	0	11.1	0	0	0
α-pinene	13.8	16.2	24.1	49.0	22.7	6.5	5.1	19.4
Camphene	0.5	1.0	0	1.2	2.8	0	0	0.9
Sabinene	0.7	0.7	1.2	12.2	24.3	10.2	11.0	9.6
β-pinene	1.9	3.9	7.4	8.6	4.4	0.9	0	7.3
Myrcene	1.9	1.1	1.6	1.0	1.1	1.5	1.3	0.3
α –phellandrene	1.1	0.4	0	0.8	4.0	1.1	0	0
Limonene	45.5	29.5	40.6	6.1	3.3	29.7	28.7	1.6
β-phellandrene	3.3	4.2	1.2	6.4	3.0	1.2	0.9	1.5
β-Ocimene Z	0	0.2	0	0	0	0	0	0
β-Ocimene E	0	0.6	0	0	0	1.4	0	1.7
γ-terpinene	0	0	0	0.2	0.2	0	0	0
2-Undecanone	1.1	0.7	0.3	0	0	0.5	0	0.6
β-Elemene	1.1	0	2.9	0	0	0.8	0.8	0
Caryophyllene	5.4	13.2	0	3.7	10.7	20.7	22.4	6.0
α –humulene	1.3	2.4	1.2	0.7	1.8	2.9	2.2	3.0
γ-muurolene	0	0	0.4	0	0	1.0	1.1	2.2
Germacrene D	22.3	26.0	18.8	10.0	10.2	20.3	23.7	41.1
α -muurolene	0	0	0	0	0	0	0.5	0.8
δ -Cadinene	0	0	0	0	0.3	1.1	1.3	2.1
Total (μg/g)	974	545	1,181	1,084	793	762	904	749

To summarize, the intra-specific variability of terpenes in the foliage, gum resin, and essential oils sourcing from *P. lentiscus* is very high, which is promising if selection of plants for specific bioactives is anticipated.

Traditional Use and Common Knowledge

The first century Greek physician and botanist, Dioscorides, first wrote about the medicinal properties of mastic products in his classic treatise *De Materia Medica* (see Leonti et al. 2009). Mastic has been used as a medicine since antiquity and is still used in traditional folk human and veterinary medicine around the Mediterranean basin. Relative to humans, mastic gum and leaf infusions were deemed good for prevention of digestive problems, useful for bronchitis, teeth sanitation and against jaundice, bed-wetting, and headaches caused by colds. Veterinary uses encompass treatment of ecto-parasites, wound healing (external application of leaf and fruit oil), bloating and belly-aches (leaves, ingested per os) and even disinfection of water wells by using foliage brooms (Table 5).

Table 5 Traditional uses of *P. lentiscus* in human and veterinary medicine

	Way of administration	Ailment	Animal	Location	Reference
Leaves, decoction	Mouth washes	Tooth ache	Human	Maratea (Italy)	Guarrera et al. (2005)
Leaf, infusion	Per os	Jaundice, respiratory problems		Israel, Palestine	Said et al. (2002)
Leaves	Eaten as is	Bed wetting			
	External	watering eyes			
Leaf aqueous extract	External	Mycotic scalp infection		Palestine	Ali-Shtayeh and Abu-Ghdeib (1999)
Branches broom		Disinfection of water wells and pools		Israel	Muklada et al. (2013)
Whole leaves and twigs	Eaten as is	Diarrhea, belly-ache	Goat kids	Israel	Muklada et al. (2013)
	Masticated with saliva and spit on	Ophtalmitis	Cattle	Cyprus	Pieroni et al. (2006)
Fruit and twig decoction	Per os	Constipation	All livestock	Sardinia	Bullitta et al. (2007)
		Bloat	Cattle	" "	" "
Fruit oils, powdered stems	External application, application on incised veins	Wounds	All livestock		Bullitta et al. (2007)
		Scabies	All livestock	" "	" "

Modern Medicine Based on Traditional Medicine Uses

Evidences for anti-bacterial, antifungal, anthelmintic and anticoccidial, antioxidant, hepato-protective, and anti-carcinogenic properties of products sourcing from *P. lentiscus*, some of them controversial, are accumulating (Table 6).

Antibacterial

The antibacterial effects under research concern the generally pathogenic *Escherichia coli, Staphylococcus aureus*, and *Bacillus subtilis*; *Helicobacter pylori* associated with gastric ulcers; and *Streptococcus mutans* that is involved in dental plaque accumulation. Analysis of the activity of crude mastic oil and 12 of its components, some of them at trace levels, against *Escherichia coli, Staphylococcus aureus*, and *Bacillus subtilis* showed synergism of different fractions. When a

Table 6 Scientific evidence for the traditional uses of *P. lentiscus* products

Product	Property	Method	Disease	Activity	Authors
Mastic oil	Antibacterial	Disk diffusion	*Escherichia coli, Staphilococcus Aureus, Bacillus subtilis*	Medium activity for mastic oil, verbenone, α – terpineol, linalool	Koutsoudaki et al. (2005)
Mastic gum in ethanol				No activity	
Mastic gum in acetone	Antibacterial (anticariogenic)	Disk diffusion	*Streptococcus mutans*	As effective as vancomycin	Aksoy et al. (2006)
		In vivo		Reduces *S. mutans* in saliva	
Mastic gum	Antibacterial	Clinical	*Helicobacter pylori*	Bactericidal in part of the patients	Dabos et al. (2010)
Mastic gum	Antibacterial	Clinical	*Helicobacter pylori*	Reduced colonization, no effect on gastritis	Paraschos et al. (2007)
Leaf aqueous extract	Antifungal	In vitro, mycelial inhibition	*Microsporum canis, Trychophyton mentagrophytes, Trichophyton violaceum*	>94 % inhibition	Ali-Shtayeh and Abu-Ghdeib (1999)
Leaf methanolic extract	Antioxidant	FRAP assay		High free radical-scavenging	Gardeli et al. (2008)
Leaf aqueous extract	Antioxidant	In vitro, rat liver cells	Iron-induced lipid peroxidation	75 % maximum inhibition	Ljubuncic et al. (2005a)
Leaf aqueous extract	Antioxidant	In vivo	Iron-induced lipid peroxidation in, TAA-treated rats	Aggravation of TAA liver fibrosis, hepatotoxic	Ljubuncic et al. (2005b)
Leaf aqueous extract	Hepatoprotective	In vivo, rats	CCl_4 infection	Reduced liver enzyme, Non – boiled > boiled	Janakat and Al-Merie (2002)
Mastic oil	Tumor cell-growth suppression and angiogenesis	In vitro, K562 human leukemia cells		Inhibited Growth of leukemia cells, attenuated angiogenesis	Loutrari et al. (2006)
Mastic gum	Hepatoprotective	In vivo, humans, 5 g/day, 18 months		Decrease in serum total cholesterol, LDL, total cholesterol/HDL ratio, liver enzymes	Triantafyllou et al. (2007)

correlation was established between disk diffusion and control antibiotic concentration (gentamycin and tetracyclin against. *E. coli*; gentamycin and vancomycin against *S. aureus* and *B. subtilis*), it revealed high efficacy of mastic oil, and some minor components, *i.e.*, verbenone, α–terpineol, and linalool against *S. aureus* and *B. subtilis*, and medium activity against *E. coli* (Koutsoudaki et al. 2005). All bacteria were resistant to α–pinene. Interestingly, even though verbenone, α–terpineol, and linalool represent only 0.07, 0.01, and 0.5 % of mastic oil, respectively, they still were as anti-bacterially potent as the whole oil.

As to the activity of mastic gum and its extracts against *Helicobacter pylori*, results are controversial: Dabos et al. (2010) reported partial success in eradication of *H. pylori*, as evidenced by urea breath test, in four groups of patients who already harboured the bacteria. Mastic gum tended to eradicate *H. pylori* in two groups: 4/13 (350 mg/day, 14 days; $P<0.08$), 5/13 (1,050 mg/day, 14 days, $P<0.06$), but failed in one, and eradication was more complete in a group treated with pantoprazole, amoxillin and clarithromycin. The conclusion by authors that mastic gum has bactericidal activity is supported by Paraschos et al. (2007) who showed that a poly-β-myrcene-free mastic extract, and in particular, its acidic fraction (minimum bactericidal concentration [MBC], 0.139 mg/ml) containing isomasticadienolic acid (MBC, 0.202 mg/ml), reduces *H. pylori* colonization. However, it is agreed upon that mastic gum has some bactericidal activity but does not systematically eradicate *H. pylori* in mice (Loughlin et al. 2003). However, effect on gastritis alleviation was minor, if any. This is in agreement with Al-Said et al. (1986) who claimed that mastic gum at 500 mg/kg/day mildly reduced the damage caused by pyloric ligation, aspirin, and phenylbutazone on gastric mucosa in rats.

Aksoy et al. (2006) examined if the popular use of mastic gum in mouth and tooth health was scientifically based and compared the inhibitory effect of chewed mastic gum with that of a placebo gum. Significantly fewer concentrations of *S. mutans* (ARCC 27351) were found in the saliva of volunteers after 15, 45, 75, 105, and 135 min after chewing mastic gum was initiated.

Anthelmintic and Anticoccidial (Table 7)

In the Samaria Hills of Israel, shepherds will tether diarrheic young goats close to a *P. lentiscus* bush (Muklada et al. 2013). Indeed, consuming *P. lentiscus* foliage was associated with decreased nematode fecal egg count (FEC) values in goat kids (Landau et al. 2010) and lambs (Manolaraki et al. 2009), as well as decreased coccidial oocyst fecal counts in kids (Markovics et al. 2012). Azaizeh et al. (2012) showed that aqueous, and even more, ethanol 70 % extracts of *P. lentiscus* foliage impair larval exsheathment processes, and acetone extracts of the plant inhibit larval migration (Manolaraki et al. 2009), hence impairing development of parasitic nematodes to adulthood. When tannins were specifically bound to by using Polyethylene-glycol (MW 4,000; Landau et al. 2010) *in vivo*, or polyvinyl polypyrrolidone (PVPP) *in vitro* (Manolaraki et al. 2009), intermediate values were found for the

Table 7 Scientific evidence for efficacy of *P. lentiscus* in veterinary treatments

	Procedure	In	Use	Organism	Country	Reference
Whole leaves and twigs	Eaten, as is	Vivo	Anthelmintic	Goat kids	Israel	Landau et al. (2010)
Fruit powder	Eaten, as is	Vivo	Anthelmintic	Lambs	Crete	Manolaraki et al. (2009)
Whole leaves and twigs	Eaten, as is	Vivo	Coccidiostat	Goat kids	Israel	Markovics et al. (2012)
Leaf ethanol 70 % extract	Larval ex- sheathment inhibition	Vitro	Anthelmintic	Parasitic nematodes	Israel	Azaizeh et al. (2012)
Leaf acetone: water extract	Larval migration inhibition	Vitro	Anthelmintic	Parasitic nematodes	Crete	Manolaraki et al. (2009)

inhibition rates of migration and ex-sheathment, respectively, an evidence that tannins are not the only anthelmintic moiety in the foliage of *P. lentiscus*. In contrast, tannins contributed all the coccidiostatic activity (Markovics et al. 2012).

Anti-ectoparasital

Anti-ectoparasital even though anecdotal reports hint at anti-scabies activity (Bullitta et al. 2007), we found no published scientific verification of this property.

Antifungal

Aqueous extract of *P. lentiscus* leaves had 94.2, 94.0, 100, and 96.1 % of the antifungal potency of griseofulvin against dermatophytes *Microsporum canis, Trychophyton mentagrophytes, Trichophyton violaceum* isolated from tinea capitis cases (Ali-Shtayeh and Abu-Ghdeib 1999).

Antioxidant and Hepato-Protective

The traditional use of mastic gum against "jaundice" triggered research into antioxidant and hepatoprotective properties of *P. lentiscus* as liver diseases are conditions in which oxidative stress is important. The inhibition of iron (Fe^{2+})-induced lipid peroxidation in rat liver homogenates by water extracts of *P. lentiscus* and their effect on rat pheochromocytoma and human hepatoblastoma cells was studied by Ljubuncic et al. (2005a): 50 % of peroxidation (IC50) was suppressed by as little as 10 μg/ml of the extract and the maximal inhibition reached 90 % of inhibition by

Trolox (Vit.E equivalent). Antioxidant capacity, as determined by DPPH (2,2-diphenyl-1-picrylhdrazyl) and FRAP (ferric reducing-antioxidant power) assays was highest for methanol extract of leaves harvested in the spring (Gardeli et al. 2008). The IC50 was extremely low (5 μg/ml). The ability to scavenge free radicals so efficiently was attributed to flavonoid and galloyl derivatives in leaves (Gardeli et al. 2008) and to di-gallic acid in fruit (Bhouri et al. 2010, 2012): digallic acid showed an important free radical scavenging activity towards the 2,2′-azino-bis(3-ethylbenzothiazoline-6-sulphonic acid) – ABTS(+) – radical (99 %) and protection against lipid peroxidation (68 %) and inhibited the xanthine oxidase, involved in the generation of free radicals, and the lipid peroxidation induced by H_2O_2 in the K562 cell line. In rats intoxicated with CCl_4, aqueous extracts of *P. lentiscus* leaves reduced three liver enzymes associated with cell integrity (ALP, ALT, AST) and bilirubin in blood, consequent with the use of these extracts against jaundice in Jordanian folk medicine (Janakat and Al-Merie 2002). In contrast, when hepatic injury was caused in rats by thioacetamide (TAA), followed by lipid-peroxidation-induced cirrhosis and fibrosis, treatment with aqueous extracts of *P. lentiscus* leaves (gavage 15, 75 or 200 mg/kg in a 1 ml bolus) did not reduce liver enzymes and was associated with more acute liver inflammation and fibrosis (Ljubuncic et al. 2005b). A long-term study in Greece showed that high ingestion of mastic gum power (5 g/day, 18 months) resulted in decreased levels of total serum cholesterol, LDL, total cholesterol/HDL ratio, lipoprotein (a), apolipoprotein A-1, apolipoprotein B, SGOT, SGPT and gamma-GT, suggestive of hepato-protective effects (Triantafyllou et al. 2007).

Apoptosis and Angiogenesis

Recent evidence point at potential effects of *P. lentiscus* products on cancer development. Digallic acid from *P. lentiscus* fruit possessed pro-apoptotic effects, as shown by provoking DNA fragmentation of K562 human keukemia cells (Bhouri et al. 2010). A pro-apoptotic effect of mastic oil on K562 cells was also reported by Loutrari et al. (2006). Mastic oil also inhibited the release of vascular endothelial growth factor (VEGF) from K562 and B16 mouse melanoma cells. Moreover, mastic oil caused a concentration-dependent inhibition of endothelial cell (EC) proliferation without affecting cell survival and a significant decrease of microvessel formation both in vitro and in vivo. Authors suggest that mastic oil, through its multiple effects on malignant cells and ECs, may be a useful natural dietary supplement for cancer prevention.

Toxicity

There are conflicting safety reports regarding the use of *P. lentiscus* products, based on *in vitro* and *in vivo* evidence. Concentrations of 75 mg/ml of mastic gum were not toxic to Hep-2 cells (Aksoy et al. 2006); working with the same type of cultured cells, Ljubuncic et al. (2005a) confirmed non-toxicity for aqueous extracts of *P. lentiscus* foliage. However, the same authors, working with TAA-treated rats, reported that not only gavage with the extract did not alleviate the TAA-induced condition, it had hepato-toxic effects in the controls not treated with TAA (Ljubuncic et al. 2005b): In healthy rats, long-term administration of the extract induced hepatic fibrosis and an inflammatory response, mild cholestasis and depletion of reduced glutathione associated with an increase in its oxidized form. The explanation for this discrepancy is not yet clear.

Because of its richness in tannins, foraging on *P. lentiscus* is associated with increased defecation of dietary and microbial N in goats (Silanikove et al. 1996a), resulting in marked reduction of blood (Hanovice-Ziony et al. 2010) and milk (Decandia et al. 2000) urea concentrations. However, it is not toxic, with no effect on serum liver enzymes (Silanikove et al. 1996b). Goats, which thrive on shrublands rich in *P. lentiscus* (Landau et al. 2002) have probably acquired liver detoxication mechanisms, but N depletion can certainly a problem in other species.

Perspective

Metabolomics, encompassing new analytical procedures (Rodríguez-Pérez et al. 2013) will probably help in elucidating the rationale of traditional medicine in using *P. lentiscus* for therapies or prevention. It can be expected that new cultivars will be cultivated on the basis of bio-activity.

References

Aksoy A, Duran N, Koksal F (2006) In vitro and in vivo antimicrobial effects of mastic chewing gum against *Streptococcus mutans* and *mutans streptococci*. Arch Oral Biol 51:476–481

Ali-Shtayeh MS, Abu-Ghdeib SI (1999) Antifungal activity of plant extracts against dermatophytes. Mycoses 42:665–672

Al-Said MS, Ageel AM, Parmar NS, Tariq M (1986) Evaluation of mastic, a crude drug obtained from *Pistacia lentiscus* for gastric and duodenal anti-ulcer activity. J Ethnopharmacol 15:271–278

Assimopoulou AN, Papageorgiou VP (2005) GC-MS analysis of penta- and tetra-cyclic triterpenes from resins of Pistacia species Part I *Pistacia lentiscus* var Chia. Biomed Chromatogr 19:285–311

Azaizeh H, Halahleh F, Abbas N, Markovics A, Muklada H, Ungar ED, Landau SY (2012) Polyphenols from *Pistacia lentiscus* and *Phillyrea latifolia* impair the exsheathment of gastro-intestinal nematode larvae. Vet Parasitol 191:44–50

Barazani O, Golan-Goldhirsch A (2009) Salt-driven interactions between *Pistacia lentiscus* and *Salsola inermis*. Environ Sci Pollut Res 16:855–861

Barazani O, Dudai N, Golan-Goldhirsh A (2003) Comparison of Mediterranean *Pistacia lentiscus* genotypes by random amplified polymorphic DNA, chemical, and morphological analyses. J Chem Ecol 29:1939–1952

Ben Douissa F, Hayder N, Chekir-Ghedira L, Hammami M, Ghedira K, Mariotte AM, Dijoux-Franca MG (2005) New study of the essential oil from leaves of *Pistacia lentiscus* L. (Anacardiaceae) from Tunisia. Flavour Fragr J 20:410–414

Bhouri W, Derbel S, Skandrani I, Boubaker J, Bouhlel I, Sghaier MB, Kilani S, Mariotte AM, Dijoux-Franca MG, Ghedira K, Chekir-Ghedira L (2010) Study of genotoxic, antigenotoxic and antioxidant activities of the digallic acid isolated from *Pistacia lentiscus* fruits. Toxicol In Vitro 24:509–515

Bhouri W, Skandrani I, Ben Sghair M, Dijoux- Franca MG, Ghedira K, Chekir Ghedira L (2012) Digallic acid from *Pistacia lentiscus* fruits induces apoptosis and enhances antioxidant activities. Phytother Res 26:387–391

Bullitta S, Piluzza G, Viegi L (2007) Plant resources used for traditional ethnoveterinary phytotherapy in Sardinia. Genet Resour Crop Evol 54:1447–1464

Castola V, Bighelli A, Casanova G (2000) Intraspecific chemical variability of the essential oil of *Pistacia lentiscus* L. from Corsica. Biochem Syst Ecol 28:79–88

Charef M, Yousfi M, Saidi M, Stocker P (2008) Determination of the fatty acid composition of Acorn (*Quercus*), *Pistacia lentiscus* seeds growing in Algeria. J Am Oil Chem Soc 85:921–924

Dabos KJ, Sfika E, Vlatta LJ, Giannikopoulos G (2010) The effect of mastic gum on *Helicobacter pylori*: a randomized pilot study. Phytomed 17:296–299

Decandia M, Sitzia M, Cabiddu A, Kababya D, Molle G (2000) The use of polyethylene glycol to reduce the anti-nutritional effects of tannins in goats fed woody species. Small Rumin Res 38:157–164

Dhifi W, Jelali N, Chaabani E, Beji M, Fatnassi S, Omri S, Mnif W (2013) Chemical composition of lentisk (*Pistacia lentiscus* L.) seed oil. Afr J Agric Res 8:1395–1400

Filella I, Llusia J, Pinol J, Penuelas J (1998) Leaf gas exchange and fluorescence of *Phillyrea latifolia, Pistacia lentiscus* and *Quercus ilex* saplings in severe drought and high temperature conditions. Environ Exp Biol 39:213–220

Gardeli C, Papageorgiou V, Mallouchos A, Theodosis K, Komaitis M (2008) Essential oil composition of *Pistacia lentiscus* L. and *Myrtus communis* L.: evaluation of antioxidant capacity of methanolic extracts. Food Chem 107:1120–1130

Guarrera PG, Salerno G, Caneva G (2005) Folk phytotherapeutical plants from Maratea area (Basilicata, Italy). J Ethnopharmacol 99:367–378

Hagidimitriou M (2009) The resin of chios mastic tree: an old product with modern use. wwwunipait/arbor/varie/Presentation_MHagidimitriou_2pdf. Accessed 27 July 2013

Hanovice-Ziony M, Gollop N, Landau SY, Ungar ED, Muklada H, Glasser TA, Perevolotsky A, Walker JW (2010) No major role for binding by salivary proteins as a defense against dietary tannins in Mediterranean goats. J Ecol Chem 36:736–742

Holland P (1601) Pliny the elder, natural history, vol VI, books 20–23. Harvard University Press, 1999,532 pages

Inbar M (2008) Systematics of Pistacia: insights from specialist parasitic aphids. Taxon 57:238–242

Janakat S, Al-Merie H (2002) Evaluation of hepato-protective effect of *Pistacia lentiscus*, *phillyrea latifolia*, and *Nicotiana glauca*. J Ethnopharmacol 83:135–138

Jaric S, Mitrovic M, Djurdjevic L, Galic G, Pavlovic D, Pavlovic P (2011) Phytotherapy in medieval Serbian medicine according to the pharmacological manuscripts of the Chilandar Medical Codex (15–16th centuries). J Ethnopharmacol 137:601–619

Jordano P (1989) Pre-dispersal biology of *Pistacia lentiscus* (Anacardiaceae): cumulative effects on seed removal by birds. Oikos 55:375–386

Kafkas S, Perl-Treeves R (2002) Interspecific relationships in Pistacia based on RAPD fingerprinting. HortSci 37:168–171

Kaliora AC, Mylona A, Chiou A, Petsios DG, Andrikopoulos NK (2005) Detection and identification of simple phenolics in *Pistacia lentiscus* resin. J Liq Chromatogr Relat Technol 27:289–300

Koutsoudaki C, Krsek M, Rodger A (2005) Chemical composition and antibacterial activity of the essential oil and the gum of *Pistacia lentiscus* Var "chia". J Agric Food Chem 53:7681–7685

Landau SY, Perevolotsky A, Kababya D, Silanikove N, Nitzan R, Baram H, Provenza FD (2002) Polyethylene-glycol increases the intake of tannin-rich Mediterranean browse by ranging goats. J Range Manag 55:598–603

Landau S, Dvash L, Decandia M, Cabiddu A, Shapiro F, Molle G, Silanikove N (2004) Determination of poly(ethylene glycol)-binding to browse foliage, as an assay of tannin, by near-infrared reflectance spectroscopy. J Agric Food Chem 52:638–642

Landau S, Azaizeh H, Muklada H, Glasser T, Ungar ED, Baram H, Abbas N, Markovics A (2010) Anthelmintic activity of *Pistacia lentiscus* foliage in two Middle Eastern breeds of goats differing in their propensity to consume tannin-rich browse. Vet Parasitol 173:280–286

Leonti M, Casu L, Sanna F, Bonsignore L (2009) A comparison of medicinal plant use in Sardinia and Sicily – de Materia Medica revisited? J Ethnopharmacol 121:255–267

Lev E, Amar Z (2000) Ethnopharmacological survey of traditional drugs sold in Israel at the end of the 20th century. J Ethnopharmacol 72:191–205

Ljubuncic P, Azaizeh H, Portnaya I, Cogan U, Said O, Abu-Saleh K, Bomzon A (2005a) Antioxidant activity and cytotoxicity of eight plants used in traditional Arab medicine in Israel. J Ethnopharmacol 99:43–47

Ljubuncic P, Sing H, Cogan U, Azaizeh H, Bomzon A (2005b) The effects of aqueous extracts prepared from the leaves of *Pistacia lentiscus* in experimental liver disease. J Ethnopharmacol 100:198–204

Llusià J, Peñuelas J, Alessio GA, Estiarte M (2006) Seasonal contrasting changes of foliar concentrations of terpenes and other volatile organic compound in four dominant species of a Mediterranean shrubland submitted to a field experimental drought and warming. Physiol Plant 127:632–649

Loughlin MF, Ala'Aldeen DA, Jenks PJ (2003) Monotherapy with mastic does not eradicate *Helicobacter pylori* infection from mice. J Antimicrob Chemother 51:367–371

Loutrari H, Magkouta S, Pyriochou A, Koika V, Kolisis FN, Papapetropoulos A, Roussos C (2006) Mastic oil from *Pistacia lentiscus* var chia inhibits growth and survival of human K562 leukemia cells and attenuates angiogenesis. Nutr Cancer 55:86–93

Mabberley D (1997) The plant book. Cambridge University Press, Cambridge

Manolaraki F, Sotiraki S, Stefanakis A, Skampardonis V, Volanis M, Hoste H (2009) Anthelmintic activity of some Mediterranean browse plants against parasitic nematodes. Parasitol 7:1–12

Markovics A, Cohen I, Muklada H, Glasser T, Dvash L, Ungar ED, Azaizeh H, Landau SY (2012) Consumption of *Pistacia lentiscus* foliage alleviates coccidiosis in young goats. Vet Parasitol 186:165–169

Muklada H, Azaizeh H, Abu-Rabia A, Kaadan S, Landau SY (2013) *Pistacia lentiscus*. In: Muklada H (ed) Veterinary uses of Israel plants. Agricultural Research Organization and the Galilee Society Publication, Shefa-Amr, pp 8–9 (Hebrew)

Palacio S, Milla R, Montserrat-Marti G (2005) A phenological hypothesis on the thermophilous distribution of *Pistacia lentiscus* L. Flora 200:527–534

Paraschos S, Magiatis P, Mitakou S (2007) In vitro and in vivo activities of Chios mastic gum extracts and constituents against *Helicobacter pylori*. Antimicrob Agents Chemother 51:551–559

Parfitt DE, Badenes ML (1997) Phylogeny of the genus Pistacia as determined from analysis of the chloroplast genome. Proc Nat Acad Sci USA 94:7987–7992

Pieroni A, Giusati ME, de Pasquale C, Lenzarini C, Censorii E, Gonzales-Tejero MR, Sanchez-Rojas CP, Ramiro-Gutierrez JM, Skoula M, Johnson C, Sarpaki A, Della A, Paraskeva-Hadjichambi D, Hadjichambis A, Hmamouchi M, El-Johri S, El-Demerdash M, El-Zayat M, Al-Shahaby O, Houmani Z, Scherazed M (2006) Circum Mediterraneum cultural heritage and medicinal plant uses in traditional animal healthcare: a field survey in eight selected areas within the RUBIA project. J Ethnobiol Ethnomed 24:2–16

Rodríguez-Pérez C, Quirantes-Piné R, Amessis-Ouchemouk NH, Madani K, Segura-Carretero A, Fernández-Gutierrez A (2013) A metabolite-profiling approach allows the identification of new compounds from *Pistacia lentiscus* leaves. J Pharm Biomed Anal 77:167–174

Romani A, Pinelli P, Galardi C, Mulinacci N, Tattini A (2002) Identification and quantification of galloyl derivatives, flavonoid glycosides and anthocyanins in leaves of *Pistacia lentiscus*. Phytochem Anal 13:79–86

Said O, Khalil K, Fulder S, Azaizeh H (2002) Ethno-pharmacological survey of medicinal herbs in Israel, the Golan Heights and the West Bank region. J Ethnopharmacol 83:251–265

Shifman A, Orenbuch S, Rosenberg M (2002) Bad breath – a major disability according to the Talmud. Isr Med Assoc J 4:843–847

Silanikove N, Gilboa N, Nir I, Perevolotsky A, Nitsan Z (1996a) Effect of a daily supplementation of polyethylene glycol on intake and digestion of tannin-containing leaves (*Quercus calliprinos, Pistacia lentiscus*, and *Ceratonia siliqua*) by goats. J Agric Food Chem 44:199–205

Silanikove N, Gilboa N, Perevolotsky A, Nitsan Z (1996b) Goats fed tannin-containing leaves do not exhibit toxic syndromes. Small Rumin Res 21:195–201

Triantafyllou A, Chaviaras N, Sergentanis TN, Protopapa E, Tsaknis J (2007) Chios mastic gum modulates serum biochemical parameters in a human population. J Ethnopharmacol 111:43–49

Yi T, Wen J, Golan-Goldhirsch A, Parfitt DE (2008) Phylogenetics and reticulate evolution in Pistacia (Anacardiaceae). Am J Bot 95:241–251

Zohary D (1952) A monographical study of the genus Pistacia Palestine. J Bot (Jerus Ser) 5:187–228

Zrira S, Elamrani A, Benjilali B (2003) Chemical composition of the essential oil of *Pistacia lentiscus* L. from Morocco – a seasonal variation. Flavour Fragr J 18:475–480

Farming Amphetamines: Khat (*Catha edulis* Forsk.) a Traditional Plant with Mild Stimulating Psychoactive and Medicinal Properties

Shimon Ben-Shabat, Pierre Goloubinoff, Nativ Dudai, and Efraim Lewinsohn

Abstract Khat (*Catha edulis* Forsk., Celastraceae) is a flowering perennial shrub with a long history of use and cultivation in East Africa and the Arabian Peninsula. Young khat leaves are traditionally chewed in social gatherings to attain special states of mind, aimed especially at awakeness and enhanced mind focus. Since khat chewing contains amphetamine-like molecules and reportedly causes addiction among users it is banned in most countries, but it is part of social life and legal in some countries. The main pharmacoactive compounds present in khat leaves are the phenylpropylamino alkaloids (S)-cathinone and (S)-cathine.

L-Phenylalanine serves as a key biosynthetic precursor of phenylpropylalkaloids. Phenylalanine is converted by a series of not yet fully characterized reactions involving chain-shortening to benzaldehyde, then ligation to decarboxylated pyruvate, oxidation and incorporation of an amino group to yield (S)-cathinone, the most active compound accumulating in young leaves. (S)-Cathinone is subsequently reduced to (S)-cathine, the main compound accumulated in mature leaves, but pharmacologically less active than (S)-cathinone. The pharmacological prospects of khat uses and some personal experiences of one of the authors in khat chewing are described here.

Keywords Khat • *Catha edulis* • Celastraceae • Cathinone • Phenylpropylamino alkaloids • Amphetamines

---

S. Ben-Shabat (✉)
Department of Biochemistry and Pharmacology, Faculty of Health Sciences, Ben-Gurion University of the Negev, Beer-Sheva, Israel
e-mail: sbs@bgu.ac.il

P. Goloubinoff
Department of Plant Molecular Biology, Faculty of Biology and Medicine, University of Lausanne, CH-1015 Lausanne, Switzerland
e-mail: Pierre.Goloubinoff@unil.ch

N. Dudai • E. Lewinsohn
The Aromatic, Medicinal and Spice Crops Unit, Newe Ya'ar Research Center, Agricultural Research Organization, Ramat Yishay, Israel
e-mail: nativdud@gmail.com; twefraim@volcani.agri.gov.il

Z. Yaniv, N. Dudai (eds.), *Medicinal and Aromatic Plants of the Middle-East*, Medicinal and Aromatic Plants of the World 2, DOI 10.1007/978-94-017-9276-9_9

Introduction to the Plant, Its Uses and Botany

Khat (*Catha edulis* Forsk., Celastraceae) is a perennial leafy bush with little white flowers that can reach up to 4 m height. Khat is thought to have originated in Ethiopia, where it may still grow in the wild. It is increasingly cultivated in Ethiopia and in the surrounding countries of Somalia, Eritrea, Djibouti, Sudan, Kenya and in Yemen across the Red Sea. Since the mid twentieth century, khat is intensely cultivated and consumed mostly in Yemen, where it is generally planted in rows on narrow mountainous terraces above 1,000 m altitude. Emigrants from these countries have spread the habit of khat chewing to the Western world (Cox and Rampes 2003). The immigration of Yemenite Jews brought khat culture into Israel and also the plants. Khat cuttings were often personally carried by immigrants in their own suitcases as a valued commodity. Khat is increasingly being cultivated in Israel for local consumption mainly by the respected elders of the Yemenite Jewish community and by a growing number of young Israelis of different ethnicity and socioeconomic situation (Fig. 1).

Legends mention that both coffee and khat plants were first brought from Ethiopia to the Yemen by Muslim scholars to prevent sleeping during their night-long meditations and studies of scriptures. At present, young khat leaves are traditionally chewed in social gatherings to attain a mild euphorical state of mind and a stimulant effect among users. Traditionally, Jewish men in Yemen and in Israel chew Khat on Friday nights after religious prayers for overnight bible studies for the sake of concentration and alertness during the long scholar sessions. Despite its mild narcotic nature, but since the amount of leaves needed to produce an effect and that can be effectively chewed in one day is limited (Kalix 1990), khat is not a controlled plant under the Israeli law. Therefore growing, selling or using khat is a common practice and the plant has gained economic importance to meet the demand of traditional and young new chewers (Figs. 2 and 3).

Fig. 1 Geographical distribution of khat cultures (map from Google Earth). Khat is though to originate from Ethiopia where it also grows in the wild. It is cultivated in the neighboring African countries and across the Red Sea in Yemen, the 'Assir province of Saudi Arabia and in Israel

Fig. 2 Khat fields in the Mashabsha area, Hajja province Yemen (*Left*). Khat merchant in souk Shaffar, Tihama, Yemen (*Right*)

Fig. 3 Khat shrubs cultivated in the Newe Yaar research center in Israel. The young leaves (*left*, the part that is chewed) and flowers (*bottom*, used in celebrations as teas)

The Swedish botanist Peter Forsskal named the plant as *Catha edulis* in 1775. The species name – *edulis* means that the plant is edible. Before the nineteenth century, there was virtually no khat in Yemen, just like before the sixteenth century, there was no coffee in the Arabian Peninsula, both plants originating from eastern Africa. The Turks had a demand for coffee and thus large bushes of famous "Arabica beans" started to populate the Arabia Felix's terraces, sometimes under the shadow of fruit trees (Botta 1841) at middle altitude in the Yemeni mountains. Whereas Arabica coffee beans were exported through the famous Al Mocha port on the red sea (making "Mocha" a synonym of coffee) to the Turkish empire and from there, *via* Vienna, to whole of Europe, the sweet caffeine-less shells of the coffee beans ("guishr" in Arabic), remained in Arabia and became the national Yemenite drink that was consumed as a warm sweet infusion spiced with cinnamon and ginger. By mid twentieth century, coffee became massively produced in South America, and the famous Arabica coffee was gradually supplanted by khat, another plant originating from Ethiopia's mountains demanding similar growth conditions of temperature, humidity and altitudes between 1,200 and 2,000 m above the sea level. Yet, contrary to coffee, khat is consumed fresh where it is produced. Thus, the major water resources of this very arid country and the rare arable lands became dedicated to this non-nutritious self-consumed crop. Although it may provide consumers with most daily doses of fresh vitamins, the damage from khat use are more ecologic and economic than addictive and behavioral *per se*. Moreover, it is not unusual that a third of the daily income is spent on khat, literally taking nutritious food and clean water away from malnourished children and reducing their access to education and health.

The Yemenite Jews that emigrated to Palestine and later to Israel between 1914 and 1962 brought with them khat chewing habits, but this was done in general with great moderation: typically once a week, exclusively among men studying *Torah*, *Gemara* or *Zohar*. In contrast, the last wave of Jewish emigration from Yemen that started on December 12 1990 and was still trickling in 2013, of about 1,000 persons, brought modern habits of excessive daily khat consumption both among men and women, which met a rising fashion among second and third generation of Israeli Yemenites that rapidly spread to other communities, including Ashkenazis from wealthy neighborhoods such as north Tel Aviv (Tawil et al. 1998). Whereas the Israeli law remains thus far lenient with a moderate traditional consumption of khat leaves by the elders, it had to legislate recently in the particular case of the potent drug "hagigat", which is a dangerous, powerful concentrate of khat extract with many unrelated artificial additives (http://en.wikipedia.org/wiki/Cathinone).

It is however not clear why most western countries, including the USA have included raw khat fresh leaves into their list of forbidden drugs. In general, a rapid irreversible addiction, in a few takes, is a major criterion to include a new substance into this infamous list. Yet, whereas 2–3 takes of Crack cocaine suffice to establish a strong addiction, about 100-fold of daily khat chewing may cause a strong, albeit generally reversible addiction. Thus, when Arab Yemenites travel abroad, they generally report being merely "unhappy" and nervous during the first 3–4 days of khat withdrawal. When several weeks later, they return to Yemen they report being "happy" from being able to chew khat again. Possibly khat has been classified in the list of forbidden substances because suddenly deprived severely addicted khat users have been anecdotally reported in the media to be violent. Yet, to our knowledge, these are very exceptional cases, certainly compared to alcohol addicts that so often cause social violence, hooliganism and severe familial brutality. Thus, compared to addictions such as cigarette smoking, alcohol or even coffee drinking, it is not clear why, with the exception of Yemen, Africa's horn and Israel, khat is banned from most countries. In 1980, the world health organization classified khat as a drug of abuse that can produce moderate psychological dependence although khat is not considered to be seriously addictive. Still, in 1993, cathinone was placed along with several psychoactive substances in Schedule I of the Controlled Substances Act of the USA, while cathine is ruled as Schedule IV substance. Strictest control is imposed on Schedule I substances in the US. Griffiths et al. (2010) recently reviewed the complex control and monitoring issues in Europe of khat leaves *versus* the pure compounds derived from it.

Chemistry of Khat

Khat leaves contain a large number of specialized metabolites, such as, alkaloids, terpenoids, tannins, and other glycosides. More than 50 different structures have been detected within the alkaloids group. Among them the most important are the phenylalkylamines due to their demonstrated pharmacological activity in the central nervous system (Kalix 1990, 1996). Three phenylalkylamines, (−)-cathinone [(S)-α-aminopropiophenone], (+)-cathine [(1S)(2S)-norpseudoephedrine], and (−)-norephedrine [(1R)(2S)-norephedrine] well account for the CNS stimulant effects of khat leaves (Lee 1995). Cathinone is more potent as amphetamine-like stimulant than cathine and (−)-norephedrine (Geisshuler and Brenneisen 1987; Kalix 1990) (Fig. 4).

The alkaloid precursor 1-phenylpropane-1,2-dione was previously found in the volatile fraction emitted by young khat leaves (Krizevski 2007; Krizevski et al. 2008, 2010). In addition, traces of (*R*)-phenylacetylcarbinol (*R*-PAC) and (*S*)-phenylacetylcarbinol (*S*-PAC) that are known as building blocks for the chemical synthesis of ephedrine alkaloids and as products of bacterial acetolactate synthases were also detected in khat (Krizevski 2007; Krizevski et al. 2008, 2010). Other interesting compounds detected in young leaf extracts are probably a result of phenylalkyl alkaloids and acetaldehyde conjugation reactions. Two oxazolidine derivatives, 2,4-dimethyl-5-phenyloxazolidine and 4-methyl-2-(*trans*-1-pentenyl)-5-phenyloxazolidine (Fig. 4) found in khat could readily be derived by condensation of norephedrine and acetaldehyde or *trans*-2-hexenal, respectively (Krizevski et al. 2012a). It was suggested that oxazolidines derived from (−)-ephedrine may be considered as potentially useful prodrug candidates, therefore we believe that these new oxazolidine derivatives of norephedrine, found in young leaf extract of khat, could be prodrugs of norephedrine tending to

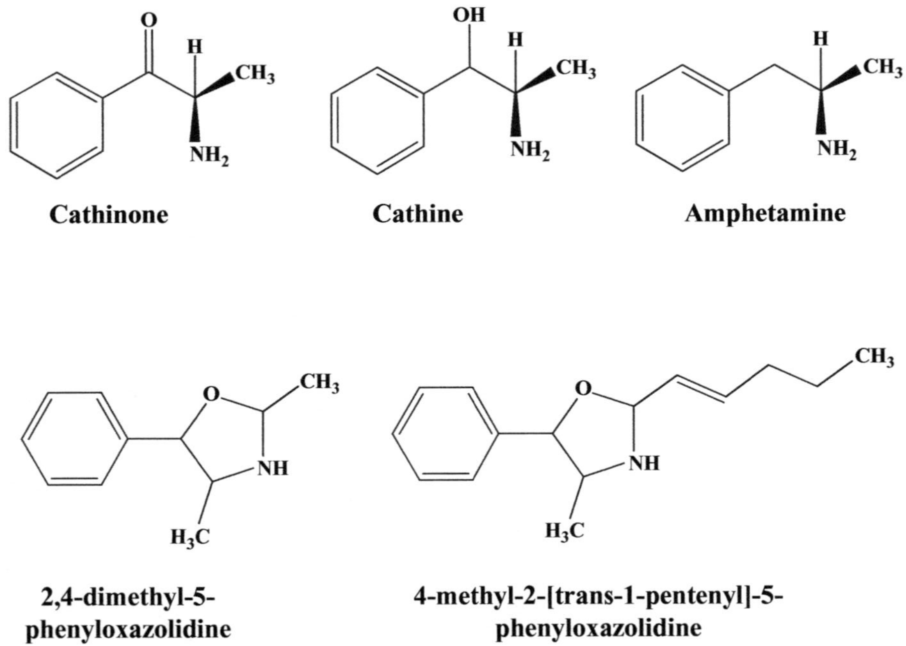

Fig. 4 The main amphetamine-like constituents of khat: Cathinone and cathine (norpseudoephedrine). The structure of amphetamine is given on *left* for comparison. Two oxazolidine derivatives, 2,4-dimethyl-5-phenyloxazolidine and 4-methyl-2-(trans-1-pentenyl)-5-phenyloxazolidine with unproven pharmacological activity have been found in khat leaves (Krizevski et al. 2012a)

decompose back to norephedrine during stress conditions (Bundgaard and Johansen 1982). Phenylalkylamines are also known to be unstable in the presence of oxygen and tend to decompose and react to generate dimeric derivatives (Szendrei 1980).

Catha edulis plants also possess substantial antioxidative capacity and activity. The antioxidative activity and the total phenols content in khat is remarkably very high, 1.5-fold as compared to *Rosmarinus officinalis* aerial parts. Thus, *Catha edulis* varieties featuring top antioxidative are promising candidates for future commercialization (Dudai et al. 2008).

Phenylalkylamine Biosynthesis in Khat

The biosynthetic pathways to phenylpropylamino alkaloids in plants are largely unknown, unlike the biosynthesis of the related phenylethylamino alkaloids (such as tyramine and dopamine) that are better studied and found to be biosynthesized via decarboxylation of aromatic amino acids (Facchini and De Luca 1995). (L)-Phenylalanine serves as the initial precursor of phenylpropylamino alkaloids, in khat and in *Ephedra sinica*, a Chinese medicinal plant that contains ephedrine and pseudoephedrine (Grue-Sorensen and Spenser 1994; Leete 1958). Pulse chase experiments in *Ephedra* suggested that phenylpropylamino alkaloids are biosynthesized through the condensation of pyruvate and benzoic acid to form the volatile compound 1-phenylpropyl-1,2-dione (PPD), which possibly undergoes transamination to release (S)-cathinone, then reduced to (1S)(2S)-norpseudoephedrine and (1R)(2S)-norephedrine (Fig. 5) (Grue-Sorensen and Spenser 1994). Thus, the direct alkaloid precursor of (S)-cathinone both in khat and in ephedra is apparently 1-phenylpropane-1,2-dione (Grue-Sorensen and Spenser 1994), a unique compound that was identified in the volatile fraction emitted from khat and ephedra plants (Abdulsalam et al. 2004; Krizevski 2007; Krizevski et al. 2012b). Furthermore, 1-phenylpropane-1,2-dione and (S)-cathinone were detected in young leaves, stems and inflorescences, but both compounds were absent in old leaves, stems and fruits (Krizevski et al. 2007; Krizevski 2007). Yet, we found experimentally that the actual substrate for cathinone biosynthesis is benzaldehyde and not benzoic acid in *Ephedra sinica* (Fig. 5) and preliminary evidence indicates that this step is also operational in khat leaves. Benzaldehyde carboxyligase reaction probably involves a thiamine diphosphate dependent enzyme of the acetolactate synthase family since it does not require the addition of FAD as a cofactor (Stanislav et al. 2003; Chipman et al. 2005). Further biochemical characterization of this novel enzymatic activity is underway.

(S)-Cathinone reductase enzymatic activity was found in cell-free protein extracts of *Catha edulis* and *Ephedra sinica* (Krizevski et al. 2012b). These enzymes convert (S)-cathinone into the diastereoisomers (1S)(2S)-norpseudoephedrine and (1R)

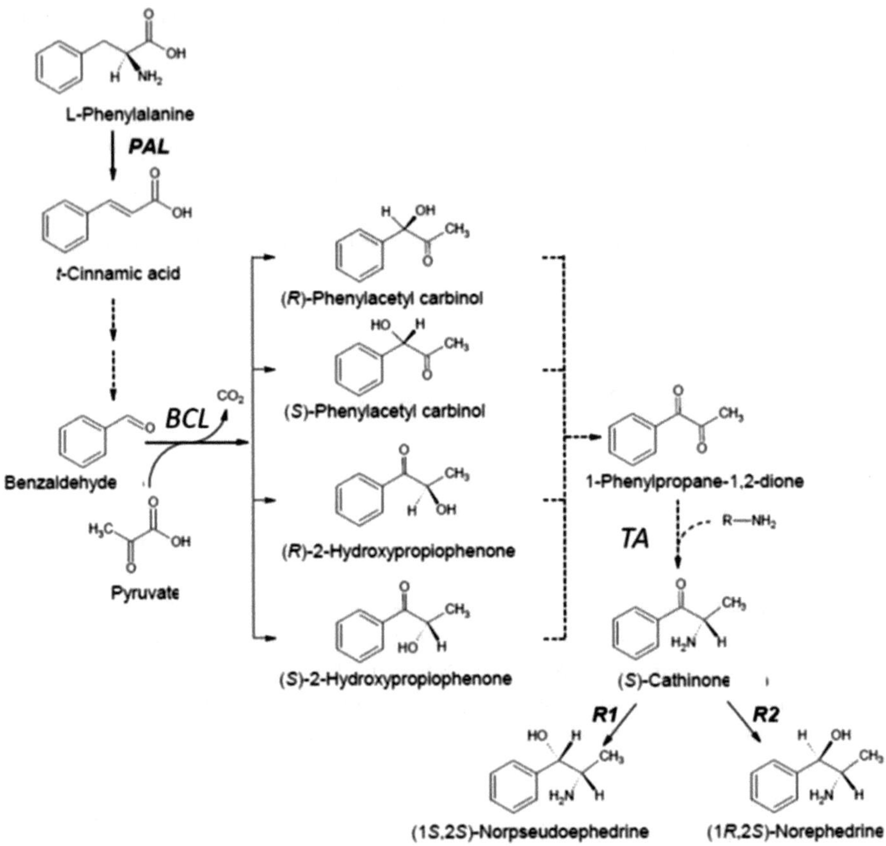

Fig. 5 Proposed biosynthetic pathways to phenylpropylamino alkaloids. The pathway is based on evidence obtained from studies on the pioneering work of Leete (1958) and khat and on studies on *Ephedra* spp. a plant that also accumulates phenylpropylalkaloids. Solid arrows represent biochemically resolved reactions including phenylalanine ammonia lyase (PAL); (*S*)-cathinone reductases (R1, R2) (Krizevski et al. 2010) and benzaldehyde carboxyligase BCL (Krizevski et al. 2012b) while *dashed arrows* represent biochemically unconfirmed reactions. The biosynthetic steps from *trans*-cinnamic acid to benzaldehyde and the amine donor for the biosynthesis of (*S*)-cathinone (TA) are still unknown (The figure was modified from Krizevski et al. 2012b)

(2S)-norephedrine in the presence of NADH in both plants. Further characterization of these enzymes is needed.

Recently, we prepared cDNA libraries from young khat leaves and generated sequence information of the transcriptome of young khat leaves (Hagel et al. 2011, 2012). Putative functions were assigned to > 98 % of the ESTs, providing a key resource for gene discovery. Candidates potentially involved at various stages of phenylpropylamino alkaloid biosynthesis from L-phenylalanine to (1S,2S)-cathine were identified.

Pharmacokinetics of Khat

When khat is chewed, absorption of cathinone is slow, with maximal plasma concentrations occurring at approximately 2 h (Widler et al. 1994). The terminal elimination half-life is approximately 4.3 h. Similar effects are achieved with orally administered pure cathinone (Toennes et al. 2003). Cathinone is the keto-analog of cathine and because it is more lipophilic it penetrates the blood-brain barrier more easily. Cathinone was only detectable in blood samples up to 10 h after ingestion (Widler et al. 1994). For cathine given in solution, the mean ± SD half-life was found to be 5.1 ± 2.0 h (Hoogkamer et al. 1994). The average amount of time that cathine spends in the system (mean residence time (MRT)) was longer than that of cathinone, which can be explained by its longer elimination half-life (Hoogkamer et al. 1994; Toennes et al. 2003). S-(−)-Cathinone has been shown to be metabolized stereoselectively to R,S-(−)-norephedrine (Brenneisen et al. 1986). It was found that only 7 % or less of the absorbed cathinone is recovered in urine and the mean ± SD amount of norephedrine excreted is much higher than the amount ingested (Toennes and Kauert 2002).

Mode of Action

Various studies have shown that khat has psychostimulating properties, which are comparable to those of amphetamine but less intense. Phenylpropylamino alkaloids mimic the action of adrenaline and cause the release of endogenous catecholamines from the post-ganglionic sympathetic fibers (Bruneton 1995; Kalix 1990, 1996). Khat has a moderate potential for psychic dependence and chronic use may lead to a certain degree of tolerance. (1*R*)(2*S*)-Ephedrine and (1*R*)(2*S*)-norephedrine are used for their vasoconstrictive properties, to treat acute asthma attacks as well as rhinitis, sinusitis and rhinopharyngitis (Bruneton 1995). (1*R*)(2*S*)-Ephedrine is also widely used during anesthesia; it is frequently prescribed for treatment of bronchial asthma and allergies and for countering an overdose of depressants (Lewis and Elvin-Lewis 1977; Nencini and Ahmed 1989). (1*S*)(2*S*)-Pseudoephedrine is used to treat nasal congestion (Bruneton 1995).

The presence of other substances with possible or proven pharmacological activity in khat was found in organic extract as well as in water extracts. Administration of the flavonoid fraction, isolated from khat, produced a significant anti-inflammatory activity and was comparable with the standard anti-inflammatory drug oxyphenbutazone (Al-Meshal et al. 1986). Cytotoxicity of Khat organic extract was found to be effective on cultured mammalian cells. Treatment of oral carcinoma (KB cells) with chloroform extract of khat leaves was found to be cytotoxic and it is apparent that the main cause of cytotoxicity of the khat extract may be the inhibition of de novo RNA synthesis (Al-Ahdal et al. 1988). No specific compound was detected from the above extracts.

The Traditional Uses of Khat in Yemen and in Israel

The concentration of the active principles in khat leaves is rather low (500–1,000 μg/g FW). Therefore, a typical khat chewing session takes a few hours to complete and involves the consumption of about 100–200 g fresh young leaves. The practice is not considered immoral among traditional users, but the bitter taste of the leaves and its slow effect deter its wider acceptance. Only young khat leaves are traditionally chewed, usually in social gatherings being perceived as a way to attain a mild euphorical state of mind and stimulant effect among users (Fig. 2). Often a tea made from fresh flowers is also used in celebrations and weddings. Although many reports on the physiological effects of khat are available in the literature, in this chapter, one of the authors (P.G.) will share his own experiences in khat chewing.

The Effects of Khat: A Personal Subjective Account

I am dyslexic and suffer from an attention deficit. This is a crippling condition for a university professor such as myself. I am often called to write scientific papers, review articles and prepare lectures. Optimally, fulfilling such elaborate intellectual tasks necessitates hours of uninterrupted and focused attention in complete isolation. However, even when hiding behind the locked door of my office during a silent weekend, I rapidly feel a growing sense of unease that reaches almost painful levels, and following 10–20 min of concentration, I need to take a break (make coffee, check and answer emails, check how my transgenic bacteria and plants grow, etc.) that very often carries me away irreversibly from my initial task.

Personally, khat intake (while being in Israel or Yemen where this is perfectly legal, of course!) decreases the threshold of this unease and painful attention deficit. Moreover, it puts me in a kind of semi-indifferent state of mind with regards to incoming external stimuli. Thus, while under the influence of mild khat intake (a "roubta" of about 50–80 young fresh leaves), I am not tempted to answer the phone. I don't even care if it keeps ringing! Same with incoming emails, or Facebook notifications. Even when members of my family physically force my door and barge into my workspace, no trivial remarks or questions from them will divert me from my tasks of reading or writing. Whereas this self-centered attitude may at first, rise a few eyebrows and generate even acerbic remarks from my colleagues and plain indignation from my family members, soon they appreciate this exceptional khat-

induced state of mine because, when I finally answer to their solicitations, I do so with much more patience and less aggressiveness than I usually do when I am being disturbed. In short, despite inducing in me a slightly autistic attitude, khat may paradoxically convert me into a nicer person to others. But this is of course transient and fades away. Within a day or two, I become as irascible as usual.

One may wonder about the quality of an academic work performed under the influence of khat. It is well know that drunk drivers are convinced that they drive better under the influence of alcohol, whereas physiological tests clearly demonstrate less attentiveness, less control and much slower reactions than when sober. When being driven between 3 and 6 pm in a Yemeni taxi or bus, by drivers with bulging cheeks dripping green saliva, I was initially quiet worried. But soon, I realized that khat caused them to drive excessively... Slowly! So slowly that the western man that I am got often upset. Looking outside, I realized that even on rare straight paved roads, all cars drove as sluggishly as us. Thus, even if the reaction time of my khat-soaked driver was half that of a normal driver, the risk of causing a severe frontal accident, which exponentially decreases with the speeds of the opposing cars, strongly reduced the risks of damages. This being said, Yemen is also blessed with curvy unpaved mountain roads and there are clear signs that cars and trucks sometimes took the short cuts into the abyss, possibly because their drivers were chewing khat and therefore did not care too much to turn their steering wheel when needed...

But writing a scientific article under khat influence is apparently less a risky exercise. Moreover, the quality of the production can be objectively judged by oneself once sober, and also by others, such as Editors and referees of peer-reviewed journals. In the last decade, I had three to four opportunities, while being in Israel, to write a scientific article or a review with a short submission deadline while chewing khat. When on the next day I reread my texts, I always had amazingly good surprises. It was as if someone wrote them for me, but exactly as I wanted them to be! It was as if I underwent partial anesthesia of my attention deficit. The scientific texts were in general precise, polished, and almost ready to submit.

Other Physiological and Physiological Effects of Khat

Because it reduces anxiety and it generally increases indifference for stimuli, including daily and exceptional worries of life, I found khat also acted on me as an anti-depressant.

Effects on sleep: *Personally and also according to many Khat users, Khat intake causes sleeplessness. I had about three strong Khat hangovers in my life (due socio-political circumstances in Yemen) and these resulted in complete sleepless nights. Yet, contrary to the "regular" sporadic sleepless nights where you suffer from the anxiety from being unable to fall asleep and turn crazy from roaming fruitless thoughts, the sleepless night under khat influence are very quite and meditative. You don't really sleep but you rest and remain calm. Moreover, you are not upset by the thoughts that may visit you. It is a kind of a nightlong meditation resulting in less tiredness on the next day, compared to normal insomnia. Although I would not recommend doing this too often, the negative consequences on the next day performances are less severe than following a "normal" sleepless night caused by a combination of a heavy dinner and the worries of late mid life.*

Traditionally, Jews and Muslim scholars were taking Khat to stay awaken whole nights and study sacred texts. Typically, on Shabbat eve, Jews used to sleep no more than 2–3 h and wake up at 3 am to study "parashot", "Zohar" and other sacred books. Khat was of help for them to fulfill these exhausting sleepless religious duties. However, khat intake on Shabbat may clash with another important religious precept, highly recommended to be carried out on the holly Shabbat: "grow and multiply" (Genesis 1:28).

Reported Effects on the Libido

Speaking from personal experience and half-whispered testimonies from close Yemeni male friends, I would certainly not recommend a middle aged man to take khat when intending to conclude thereafter a gallant encounter with a lady. Forty-eight hours of detoxification is absolutely necessary between the two activities. Within 1–2 h of the first khat leaf intake, men of all ages generally report strange sensations at the level of their prostate. It delays and diffuses ejaculation, which may be of benefit in particular cases of men suffering from premature discharges. Otherwise, contrary to the popular tales, khat is counter-beneficial to sexual activity.

The following story that happened to my father-in-law illustrates this point. He was an 11-year-old Jewish boy named Salem ben Yehia. He used to offer his services as a blacksmith on the market of a mountainous town named Ash-Shahel at about 1,500 m above the red sea level, in the Yemenite province of Hajja, where the most reputed potent "red" khat grows. Using a small portable charcoal stove and a bellows, Salem used to make a living out of soldering small broken metal objects brought to him by the peasants of the region and by selling a few homemade nails. One morning, a hesitant old man in his 50 approached him. His typical white gown and large straw hat clearly informed the boy that this client was a peasant from the Tihama, a nearby arid plain on the red sea shores where temperatures can reach 50 °C at mid day. Yet the man didn't seem to be typical client. He carried under his right arm a huge pack of local red khat carefully wrapped into banana leaves, to protect the precious content from heat and dehydration on the long way back to the scorching plain bellow. The dreadful climatic conditions led the Jews of Yemen to identify the Tihama plain as being the original "tehom" abyss of the bible (Genesis 1:2). The old parched-skinned Zaranik looked embarrassed. While changing weight from one foot to another, he started praising the Jewish people for being a holly people and knowing "secrets". Following a long silence, so atypical from a Yemenite men in a market, Salem replied:

- *Say no more, yah hadj, you have a "weakness" with your… wives.*
- *By Allah! You Jewish people really are magicians! Do you have a medicine, a Dawa for me?*

Salem dropped his tools and fetched for a small piece of paper and with a sharp charcoal wrote in Hebrew "refu'a shlema" (full healing), than ripped the message in small pieces, which he mixed in water in a mug. He gave the man to drink to the last drop and ordered him to swallow up to the last fragment of paper.

– *Come back in an hour and, God willing, I'll have a second dawa for you.*

By the end of the market, Salem presented his client with a compact loaf of date paste mixed with almonds, figs and a good quarter of pure garlic!

– *Take three thin slices per day. Pay me one ryal now and bring me a goat kid when, God willing, your young new wife will be pregnant.*
– *This is it? Asked the old man. Are you sure it will work?*
– *Yes, yes, ya Hadj, Allah willing. You said it yourself, we Jews are holly, we have secrets. This is a very potent medicine.*

Salem let the man go away, carrying under his arm red khat and in the other hand the precious smelly medicine. But after a few seconds, Salem ben Yehia called the man back.

– *I almost forgot. This is extremely important! This medicine is very potent but there is one counter-indication that may completely neutralize its effects: khat chewing. You must absolutely avoid khat during the whole treatment! Promise this to me!*

The man looked quiet disappointed but at last reluctantly promised:

– *Allah is my witness; khat not will go under my tongue before my new wife is pregnant!*

Two month later, a young man from the Tihama, who was the elder son of the old man, searched for Salem in the market. He had two live goat kids as thanking gift in return for Salem's very successful medical and psychological services.

Effects on Appetite

One interesting aspect is that khat clearly acts as an appetite suppressant. Chewing and swallowing young khat leaves while drinking abundant water during a few hours before lunch strongly reduces appetite. It allows to simply skip lunch and even dinner, while not becoming increasingly obsessed by hunger. During fasting, it allows you to keep performing optimally various intellectually demanding tasks, such as article writing and reading and even teaching biochemistry classes, without being sidetracked by increasing obsessing thoughts about food. Thus, pharmacological exploitation of khat as an appetite suppressant should be more investigated, alongside its remarkable ameliorating effects on attention deficits.

Addiction and Cheek Cancers

Sporadic khat intake does not develop dependence. Khat is more dangerous by its indirect effects, such as more intense cigarette smoking while chewing. Yet, per se, unlike smoking, there is no passive khat chewing. Regular khat intake may lead in the very long term to throat cancers due to repetitive microlesions to the inner cheek by abrasive microparticles of lignin (wood), exposure to chemicals such as astringent tannins, and the mechanic shearing of daily excessive dilatation of the cheek skin. Thus, the incidence of inner mouth cancers in Yemen is the highest in the world, matching only that of trumpetists (see http://en.wikipedia.org/wiki/Oral_cancer*).*

Personally, as a very sporadic khat user, my inner mouth cavity always feels the consequences of a short afternoon of chewing that may last the 2–3 following days: an all out mouth cavity inflammation, accompanied by muscle pains, as if my jaw muscles ran a chewing marathon! You better eat then only "'assit ou zom", a Yemenite recipe for kind of porridge with garlic and curcuma-flavored yoghourt, than thick crusted bread topped with "zahug" chili sauce!

Farming of Khat in Ethiopia, Yemen and Israel

Although khat grows wild in high altitudes in East Africa and the Arabian Peninsula, it is also cultivated in these areas (Krikorian 1984; Kalix and Braenden 1985). In Ethiopia khat is the second most important foreign exchange earner of the country after coffee (Ashenef and Engidawork 2013). According the IRIN humanitarian news and analysis, khat cultivation raised from about 80,000 ha in 1997 to 124,000 ha in 2005 in Yemen. Khat propagation is done usually by rooting cuts or splitting plants. However, in a selection study 200 seedlings from seeds that were collected from cultivated plants, were grown in the Newe Yaar Center and yielded a high variation in plant morphology and chemical composition (Dudai, unpublished data). The cultivated khat shrub has two forms, "red khat" and "green khat", also known as "white khat". The "red" type has dark red stems while the leaves get stronger coloration under low temperature during the winter. In Israel there is a distinct user's tendency to prefer red type over the green type, while most of the original introduced plants were green. In the selection process seeds of the green type plants yielded both red and green plants. This result suggests that at least a part of the mother green plant were heterozygotes. This raises the question if red khat has larger amounts of the psychostimulating principles than green khat. In these 200 progeny plants there was no correlation between the coloration and phenylpropylamino alkaloids concentration was noted (Dudai, unpublished data). From the experiments that were conducted, it became evident that there are significant variations in the phenylpropylamines composition and content during the course of year (Krizevski et al. 2007). During the spring season maximal levels of phenylpropylamines and 1-phenyl-1,2-propanedione were recorded, while the levels during the autumn season were minimal. In terms of composition, it was found that during the autumn and winter seasons, the highest (–)-cathinone percentage 60 % out of the

Fig. 6 Commercial selling of khat products near Bet Dagan, Israel. Young leaves in modern packages are available, some of them labeled "organically grown"

phenylpropylamino alkaloids fraction was detected. During the summer season the lowest (−)-cathinone percentage 33 % out of the phenylpropylamino alkaloids fraction was detected. It was found that there is a strong influence of genetic factors on traits associated with the variations in the phenylpropylamines composition and content during the course of year (Krizevski et al. 2007). These findings open up the possibility of using different methods to produce khat of superior quality that will compete with the khat that is currently produced in Ethiopia. In Israel, khat leaves are marketed fresh, and often sanitized and distributed in simple packages (Fig. 6).

In conclusion, khat is a plant with a long tradition of use in the Middle East and Eastern Africa. The mild stimulatory effect attained during traditional chewing probably dictates its controversial legal status in different countries. In any case, despite its intense bitter taste and the low levels of alkaloid content, khat chewing has gained acceptance beyond traditional use where legal and is experiencing increased cultivation in Israel. Khat carries with it an unexplored and underexploited pharmacological potential. More research is needed to evaluate the potential risks and benefits of this controversial plant.

Acknowledgments We thank Edward E. Farmer for critically reviewing this manuscript and Grant 31003A-140512/1 from the Swiss National Fund, for funding. We Agricultural Research Fund grant This work was partially supported by a grant from the Israel Science Foundation (Grant No. 814/06) and the Canada-Israel Binational Agricultural Research and Development Fund (Project CA-9117-09).

References

Abdulsalam AF, Jian-Kai Z, Xue-Feng Y (2004) Solid-phase microextraction for flavor analysis in Harari khat (*Catha edulis*) stimulant. J Zhejiang Univ (Sci) 5:428–431

Al-Ahdal MN, McGarry TJ, Hannan MA (1988) Cytotoxicity of khat (*Catha edulis*) extract on cultured mammalian cells: effects on macromolecule biosynthesis. Mutat Res 204:317–322

Al-Meshal IA, Tariq M, Parmar NS, Ageel AM (1986) Anti-inflammatory activity of the flavonoid fraction of khat (*Catha edulis* Forsk). Agents Actions 17(3–4):379–380

Ashenef A, Engidawork E (2013) Fluoride levels and its safety in tea (*Camellia sinensis*) and khat (*Catha edulis*) imported and produced in Ethiopia. Ethiop J Environ Stud Manag 6:149–158

Botta PE (1841) Relation d'un voyage dans l'Yemen entrepris en 1837 pour le Museum d'Histoire naturelle de Paris. Benjamain Duprat, Paris

Brenneisen R, Geisshüsler S, Schorno X (1986) Metabolism of cathinone to (−)-norephedrine and (−)-norpseudoephedrine. J Pharm Pharmacol 38:298–300

Bruneton J (1995) Pharmacognosy, phytochemistry, medicinal plants. Intercept, Andover

Bundgaard H, Johansen M (1982) Pro-drugs as drug delivery systems XX. Oxazolidines as potential pro-drug types for γ-aminoalcohols, aldehydes or ketones. Int J Pharm 10:165–175

Chipman DM, Duggleby RG, Tittmann K (2005) Mechanisms of acetohydroxyacid synthases. Curr Opin Chem Biol 9:475–481

Cox G, Rampes H (2003) Adverse effects of khat: a review. Adv Psychiatr Treat 9:456–463

Dudai N, Fischer R, Segev D, Chaimovitsh D, Rosenzweig N, Shimoni M (2008) Antioxidative activity of khat (*Catha edulis* Forsk.). Acta Hortic (ISHS) 778:85–92

Facchini PJ, De Luca V (1995) Expression in *Escherichia coli* and partial characterization of two tyrosine/dopa decarboxylases from opium poppy. Phytochemistry 38:1119–1126

Geisshuler S, Brenneisen R (1987) The content of psychostimulant phenylpropyl and phenylpentenyl khatamines in *Catha edulis* Forsk. of different origin. J Ethnopharmacol 19:269–277

Griffiths P, Lopez D, Sedefov R, Gallegos A, Hughes B, Noor A, Royuela L (2010) Khat use and monitoring drug use in Europe: the current situation and issues for the future. J Ethnopharmacol 132:578–583

Grue-Sorensen G, Spenser ID (1994) Biosynthetic route to the *Ephedra* alkaloids. J Am Chem Soc 116:6195–6200

Hagel JM, Krizevski R, Kilpatrick K, Sitrit Y, Marsolais F, Lewinsohn E, Facchini PJ (2011) Expressed sequence tag analysis of khat (*Catha edulis*) provides a putative molecular biochemical basis for the accumulation of phenylpropylamino alkaloids. Genet Mol Biol 34:640–646

Hagel JM, Krizevski R, Marsolais F, Lewinsohn E, Facchini PJ (2012) Biosynthesis of amphetamine analogues in plants. Trend Plant Sci 17:404–412

Hoogkamer JFW, Al-Hamdan Y, Kleinbloesem C et al (1994) Pharmacokinetics of D-norpseudoephedrine in healthy subjects employing GC-MS. Br J Clin Pharmacol 37:500

Kalix P (1990) Pharmacological properties of the stimulant khat. Pharmacol Ther 48:397–416

Kalix P (1996) *Catha edulis*, a plant that has amphetamine effects. Pharm World Sci 18:69–73

Kalix P, Braenden O (1985) Pharmacological aspects of the chewing of khat leaves. Pharmacol Rev 37:149–164

Krikorian AD (1984) Khat and its use: an historical perspective. J Ethnopharmacol 12:115–178

Krizevski R (2007) Phenylpropylamine alkaloids accumulation in khat (*Catha edulis* Forsk.). M.Sc. thesis, The Hebrew University of Jerusalem

Krizevski R, Dudai N, Bar E, Lewinsohn E (2007) Developmental patterns of phenylpropylamino alkaloids accumulation in khat (*Catha edulis*, Forsk., Celastraceae). J Ethnopharmacol 114:432–438

Krizevski R, Dudai N, Bar E, Dessowi I, Ravid U, Lewinsohn E (2008) Quantitative stereoisomeric determination of phenylpropylamino alkaloids in khat (*Catha edulis* Forsk.). Isr J Plant Sci 56:207–213

Krizevski R, Bar E, Shalit O, Sitrit Y, Ben-Shabat S, Lewinsohn E (2010) Composition and stereochemistry of ephedrine alkaloids accumulation in *Ephedra sinica* Stapf. Phytochemistry 71:895–903

Krizevski R, Bar E, Dudai N, Levy A, Lewinsohn E, Sitrit Y, Ben-Shabat S (2012a) Naturally occurring norephedrine oxazolidine derivatives in khat (*Catha edulis*). Planta Med 78:838–842

Krizevski R, Bar E, Shalit OR, Levy A, Hagel JM, Kilpatrick K, Marsolais F, Facchini PJ, Ben-Shabat S, Sitrit Y, Lewinsohn E (2012b) Benzaldehyde is a precursor of phenylpropylamino alkaloids as revealed by targeted metabolic profiling and comparative biochemical analyses in *Ephedra* spp. Phytochemistry 81:71–79

Lee M (1995) The identification of cathinone in khat (*Catha edulis*): a time study. J Forensic Sci 40:116–121

Leete E (1958) Biogenesis of d-norpseudoephedrine in *Catha edulis*. Chem Ind 1958:1088–1089

Lewis WH, Elvin-Lewis MPF (1977) Medical botany, plants affecting man's health. Wiley Interscience, New York

Nencini P, Ahmed AM (1989) Khat consumption: a pharmacological review. Drug Alcohol Depend 23:19–29

Stanislav E, Vyazmensky M, Geresh S, Barak Z, Chipman DM (2003) Acetohydroxyacid synthase: a new enzyme for chiral synthesis of *R*-phenylacetylcarbinol. Biotechnol Bioeng 83:833–840

Szendrei K (1980) La chimie du khat. Bull Stupef XXXII:5–35

Tawil H, Miodownik S, Goloubinoff P (1998) Operation Esther: opening the door for the last Jews of Yemen. Belkis Press, New York

Toennes SW, Kauert GF (2002) Study on the excretion and detection of cathinone, cathine and phenylpropanolamine in urine after khat chewing. Clin Chem 48:1715–1719

Toennes SW, Harder S, Schramm M, Niess C, Kauert GF (2003) Pharmacokinetics of cathinone, cathine and norephedrine after the chewing of khat leaves. Br J Clin Pharmacol 56:125–130

Widler P, Mathys K, Brenneisen R, Kalix P, Fisch HU (1994) Pharmacodynamics and pharmacokinetics of khat: a controlled study. Clin Pharmacol Ther 55:556–562

Citron Cultivation, Production and Uses in the Mediterranean Region

Joshua D. Klein

Abstract The citron (*Citrus medica* L.) is one of the three primordial citrus fruit (the others being pummelo and mandarine) from which most other citrus originated. It is an evergreen tree, ranging in height from 3 to 5 m, with fruit borne on thorny branches in 2–3 waves during the year. The tree is relatively short-lived (up to 15–18 years) and is sensitive to many insects and soil diseases, as well as to high and low temperatures. Fruit range in size from 200 to 800 g, although they can grow much larger. Fruit attain their size while the peel is still green, and then ripen to yellow or even orange. Although raw citron peel and extracts are highly regarded in Asia for their many medicinal uses (ranging from Alzheimer's to cancer to diabetes to ulcers to intestinal parasites) and as an insect repellent, the majority of fruit grown in the Mediterranean Basin are used for Jewish ritual during the autumn harvest festival of Sukkot (Tabernacles). Specific varieties of etrog (pl. etrogim), as such citrons are called, are grown under meticulous conditions of orchard care, harvest, and postharvest handling so that only fruit of the highest quality reach the religious market. Other uses for citrons in the Mediterranean region are as an ingredient in perfumes, as sweetened and/or brined peel for confectionery and baking, and as flavoring for candy and alcoholic beverages.

Keywords Etrog • Succade • Confectionery • Perfume • Flavoring • Medicine • Jewish ritual • Alcohol • Chilling-sensitive • Mal secco

J.D. Klein (✉)
Agricultural Research Organization-Volcani Center, Institute of Plant Sciences, Bet Dagan 50250, Israel
e-mail: vcjosh@agri.gov.il

Z. Yaniv, N. Dudai (eds.), *Medicinal and Aromatic Plants of the Middle-East*, Medicinal and Aromatic Plants of the World 2, DOI 10.1007/978-94-017-9276-9_10

General Aspects

Description of the Plant

The citron is an evergreen medium-sized tree, ranging in height up to 5 m. In commercial practice, the tree is usually pruned to 3 m maximum height, for ease in harvesting the fruit. Like many citrus, the branches of the citron can have numerous thorns, which range in length from 0.5 to 6 cm.

Citron leaves are ovate, ranging in length from 7 to 15 cm. They are thick and stiff with short, sometimes winged, petioles, and have a distinctive scent (as does the bark of the tree) that is reminiscent of the citron fruit itself. In fact, when the Jews of ancient Israel identified citron as the "Fruit of the goodly tree" indicated in Leviticus 23:40, they noted that it is one of the very few fruits whose aroma is matched by that of tree on which it grows (Babylonian Talmud Sukka 35a). Initially a dark purple in color, the leaves are medium green when mature.

The tree bears flowers in 2–3 waves, with major blooming taking place in early winter and in mid-summer, and a minor wave of flowering in late spring. Flowers are borne in clusters of 5–20 blooms, with a few individual inflorescences. Petals are intensely purple on the outside at initial bud stage, but the color fades to light purple, pink or white as the bud swells, depending on variety. At anthesis there are 4–5 petals, 1–2 cm long with white-to-cream coloration on the inside. Each cluster usually has some perfect flowers, but many are male. Flowers usually self-pollinate even before opening, whereupon a thick (5 × 10 mm), yellow-and-purple, and often persistent style surrounded by more than 30 stamens is revealed atop a green fruitlet (5 × 15 mm). Fruitlets often abscise before reaching 3 cm in length, especially at temperatures over 35°. Style abscission is more influenced by fruit variety than by environment.

The citron fruit has a fragrance highly reminiscent of lemons, of which it is an evolutionary ancestor (Barkley et al. 2006; Barrett and Rhodes 1976; Ollitrault et al. 2012; Pessina et al. 2011). Fruits are often oval in shape, but some varieties are pronouncedly pyriform, and there are occasionally nearly spherical or irregularly oblate individuals. The peel can vary greatly in texture from smooth to extremely rough, warty and bumpy. The style abscises 3–5 weeks after fruit set, although in some varieties it persists and is strongly attached to the fruit. Sometimes the style seems to be an integral part of the fruit, with peel forming on the lower parts of the structure. Fruit size at maturity ranges from 10 cm long and 7 cm in diameter to 30 cm long × 20 cm wide. Weight can range from 100 to 1,500 g, although there are extreme cases of fruit weighing up to 3 or even 5 kg (Nicolosi et al. 2005). Citron has a great deal of phenotypic plasticity – many different shapes and sizes of fruits can be found on the same tree or even the same branch, such that the uninitiated can have a hard time distinguishing varieties by looking at fruits on a tree. One variety of citron, often called Sacred Citron or Finger Citron or Buddha's Fingers, is always deeply divided from the apex into 7–20 slender sections. The sections can be united forming a "fist" (often found in fruit that form in winter), or separated, forming an open "hand" (in summer-grown fruit).

The peel of young citrons appears dark green or almost black, although the green aspect is from inner layers of epidermal tissues that contain chlorophyll and the dark hue

Fig. 1 Citron fruits in a range of colors and sizes. All fruits shown are used in Jewish ritual

is actually very concentrated red pigments, mostly anthocyanins that are localized in outer epidermal cells (Klein, unpublished). As fruit mature and enlarge, the peel colour lightens to a medium-dark green, eventually becoming light green. At maturity, the peel turns yellow, initially at the calyx end of the fruit, then towards the stylar end. When over-ripe, the peel turns orange (Fig. 1). Some citrons have a pronounced "belt" or narrowing around the middle 1–3 cm of the fruit. The peel in this region is often smoother, and often turns yellow after the rest of the fruit. The developmental processes leading to belt formation are unknown, but may be related to non-lethal infections with exocortis viroid (Bitters et al. 1987; Gafny et al. 1995; Solel et al. 1993).

The albedo of citron is very pronounced in proportion to the flavedo, ranging from 0.5 to 2 cm in thickness. It can be acidic or sweet. Most varieties of citron have fleshy pulp with not much juice. The pulp is pale yellow, and is divided into 10–15 segments, with numerous monoembryonic seeds. The Yemenite citron variety has no pulp at all, just empty locules with seeds, while the Finger citron has neither pulp nor seeds, and is propagated only vegetatively. Seeds, where present, are ovoid, with a smooth cream-colored seedcoat and white endosperm.

Classification

Class: Magnoliopsida, subclass Rosidae
Order: Sapindales
Family: Rutaceae
Genus: Citrus

Species: *Citrus medica* L.
Subspecies: *Citrus medica* var. *sarcodactylis* (Hoola van Nooten) Swingle (Finger citron, also known as Buddha's Hand or Buddha's Fingers or Sacred Citron)

Citrus medica var. *dulcis* Risso & Poit. (Corsican)

Origin and Distribution

The species name medica indicates that Linnaeus thought that the origin of citron was in Medea/Persia (modern-day Iran), and does not, as is sometimes mistakenly thought, refer to the medicinal value of the fruit. More recently, it has been determined that citron is one of the three primordial citrus fruits, the others being mandarin and pommelo (Barkley et al. 2006; Barrett and Rhodes 1976). Interspecies crosses and intraspecies backcrosses between these three fruits have led to the development of the broad range of citrus fruits, from oranges to lemons to tangelos. The point of origin of all three founding citrus species is likely southwestern China/ northeastern India, although the subsequent distribution patterns of the three fruits and their descendants have varied. Alexander the Great is credited with bringing the citron from Persia to the Mediterranean Basin in about 300 BCE, probably for medicinal use (Morton 1987). However, there is also evidence that Jews living in Babylonia in the sixth century BCE used the citron in religious ritual, and that coreligionists (Nicolosi et al. 2005) or local Persian rulers in ancient Palestine/ Judea (Langgut et al. 2013) were responsible for introducing the fruit to the eastern Mediterranean.

Soil Requirements

Like most citrus, citron trees thrive in well-drained soil that is well-tilled. Sandy soils are preferable to loams or to rocky soils.

Climatic Requirements

The citron tree is by far the most sensitive to temperature of all the citrus trees. Temperatures below 12 °C will injure the leaves and fruit, while temperatures above 35 °C can cause abortion of flowers or young fruit (within 4 weeks of fruit set). Most citron orchards are located in coastal regions no more than 100 m above sea level. Exceptions are the Assads region in Morocco and the Saana region in Yemen, at 750 and 2,500 m, respectively. Interestingly, the Moroccan and Yemenite citron varieties grown in these two widely-separated locales are morphologically different from each other (being pulpy and pulpless, respectively) and from other citrons (both have fewer seeds than other varieties), but are much closer to each other genetically than they are to citrons from the Mediterranean Basin (Nicolosi et al. 2005). It should be noted that Chinese and Indian varieties can grow even in the lower reaches of the Himalayan foothills at 4,000 m.

Production Levels

Citron production for religious purposes is tallied on an individual fruit basis, rather than on a weight basis. Approximately 1.6 million fruit are marketed annually, of which close to 90 % is produced in Israel (A. Aftabi, E. Gorelik, and H. Kirschenbaum, personal communication). Because of the extremely high quality standards invoked for citrons used for ritual, it is estimated that only 10–15 % of the fruit produced in the orchard are actually harvested for that specialized market. A tiny percentage of the leftover fruit are used to make candied citron peel or slices but the vast majority are discarded.

Citron production worldwide for dried or candied peel is estimated at 7,000 t annually, from approximately 1,000 ha. Most of the production is divided between Puerto Rico and the Mediterranean region. Less than 10 ha of citron trees are grown in the Mediterranean region for alcoholic beverage manufacture.

Major Production Areas

The citron tree is probably the most cold-sensitive tree in the citrus family. Temperatures lower than 10 °C can cause significant injury to the tree and can destroy the current flower or fruit load. Therefore, citrons are only grown commercially in locations where warm temperatures can be guaranteed. Currently, citrons grown for religious use are produced in Israel along the Mediterranean coast (100 ha), Italy (10 ha), and Morocco (15 ha), with minor production in Greece. Plantings for processed fruit are found in Italy (Calabria and Sardinia), France (Corsica), and Greece (Rhodes and Naxos), as well as in Puerto Rico and Brazil (Morton 1987). Citrons are grown extensively in China, southeast Asia, and Korea, mostly for medicinal purposes and a little for ornamental use (Morton 1987).

Cultivars

Citron cultivars used for Jewish ritual include Diamante (also called Yanover, Genoa, or Calabri), Yemenite and Moroccan (both varieties originate in their eponymous countries, but are grown mostly in Israel, with some acreage in Morocco), and the Israeli-origin varieties Urdang (also called Barash), Halperin and Kivilevitch (both are also referred to as "Hazon Ish") and Braverman (Nicolosi et al. 2005).

Cultivars that are used in the food industry include Diamante, Poncire, and Corsican (also known as Citron of Commerce). The Finger citron is grown mostly in Asia.

Cultural Practices

Propagation

Although other citrus can be pollinated by citron pollen, the citron itself is considered to be receptive only to pollen of the same species (Barkley et al. 2006). According to Jewish religious law, citron fruit are acceptable for ritual only if they have grown solely from a citron tree, and not from interspecies grafting (Tractate Kilayim 1:7). Therefore, citrons grown for religious use are propagated by cuttings or occasionally by seed. Seedless varieties such as Finger citron, or those with few viable seeds such as Moroccan, are grown solely from cuttings. Citrons grown for other purposes can be grown on sour orange, Palestine lime, rough lemon, or other commercial rootstocks.

Soil Preparation

Citron trees require good drainage, so they cannot be planted in heavy soils. With sufficient irrigation, citron thrives in sandy soil, but can tolerate rocky soils, as well.

Planting

Citron seedlings and saplings in Israel are usually grown in plastic pots or transplant bags for 12–18 months before being transplanted with their root boll to the orchard. The Berber citron growers of Assads, Morocco, make a nursery by placing cuttings directly in the soil, and transplant the young trees once roots are established (Nicolosi et al. 2005; E. Goldschmidt, personal communication).

Fertilization

Citrus trees in general are fertilized more heavily during the active growing and flowering periods in spring and summer, and less in the more dormant fall and winter. In Israel, citrons are fertilized with approximately 1.5 t/ha NPK in the ratio

3-1-3, with additional K_2NO_3 in the last months of fruit development (E. Gorelik and H. Kirschenbaum, personal communication). Ammonium sulfate is often added in alkaline soils.

Irrigation

In order to produce fruits of adequate size (minimum 100 g) for ritual use during the Jewish holiday of Sukkot (Tabernacles), which can occur any time between mid-September to mid-October, citron growers supply up to 30 m^3/ha of water daily, beginning in August (E. Gorelik, personal communication). During spring and late summer, less water is needed, as fruits are still developing and there is often some groundwater left from winter rains. Fruit grown for confectionary use requires less water, especially since the growers prefer to have fruit with thick peels rather than fruit with heavy juice vesicles.

Weed Control

Since growing citrons can require large amounts of water, growers are very careful to prevent growth of weeds that compete with the trees for irrigation. Most growers spray herbicides at regular intervals, although some use plastic or other sheeting within rows to cover open spaces between trees. The herbicide picloram (a synthetic auxin) is used not only for maintaining weed-free rows, but is also applied in low concentrations early in the season to citrons grown for ritual use, since it has been found to aid in retention of the stigma and style (called a "pitam") on the fruit (Goldschmidt and Leshem 1971). Presence of the "pitam", which in most citron varieties abscises naturally, is considered by many consumers to contribute to the beauty and desirability of the fruit.

Pest Control

Citrons are susceptible to the range of citrus pests – thrips, mites, scale insects, and leafhoppers. Both systemic and specific (contact) pesticides are used in controlling insect pests. Pesticide application is particularly intensive for citrons used for Jewish ritual, since consumer standards for perfect fruit are extremely high. Some pesticide emulsifying agents cause staining of the citron peel (E. Gorelik, personal communication). Accordingly, the pesticides available to citron growers are limited. Applications are stopped approximately 2–3 weeks before harvest, according to manufacturers' recommendations.

Disease Control

One of the most problematic diseases of citron is mal secco, or Phoma tracheiphila, a vascular fungal disease that causes collapse of the tree. There is no known cure for this disease, although there is some evidence that the presence of viroids can protect against the fungus (Solel et al. 1993). Because the fungus can survive for a long time in the soil, growers usually do not replant in areas where trees have been removed because of mal secco. Phytophthera citrophthera foot rot is also a problem with citrons. There is a tendency for growers of citrons for ritual use to leave the roots exposed to a certain extent, so as to demonstrate that there is no graft union, as religiously required. Roots left exposed in this manner are often liable to fungal infections (A. Aftabi, personal communication). Standard antifungal compounds are applied regularly throughout the growing season. Fruit are often dipped in benzimidazole or similar compounds for postharvest fungal control.

Harvesting

Citron leaves, green twigs, flowers, and immature and mature fruit can be harvested at the appropriate season for production of flavorings. Flowers are harvested for extraction of volatile oils used in perfume, and young leaves and green twigs are harvested to make petitgrain oil, which is an aqueous extract used in cosmetic and medicinal creams, as well as for flavoring of alcoholic beverages. Fruits can be harvested while green for production of succade for confectionery and baking, and while yellow for production of essential oils used as flavoring for candy and for specialty alcoholic beverages.

Citrons grown for ritual purpose have to meet extremely high consumer standards of quality. Since fruit can be quite variable in shape, off-type citrons are removed during the growing season as soon as they are identified. This means that at harvest most citrons on the tree have style and stem aligned in a straight line, that the fruit is regular in shape (oval or pyriform), and that the "pitam", if not naturally abscised, is not broken in any way. Fruit should have no skin blemishes or scarring caused by rubbing against other fruit or against limbs, nor should they have any puncture wounds caused by thorns or insects. Similarly, the fruit skin should be free of off-color raised areas, often caused by leaf rubbing, insect activity, or spray residue.

In order to maintain fruit in acceptable condition during the growing season, fruit-bearing branches are often tied with twine to a frame surrounding the trees. This prevents excessive movement due to wind, and helps reduce incidence of blemishes caused by other fruit, limbs, thorns, and leaves. Many growers place protective sleeves of expanded polystyrene on fruit while they are still on the tree, especially if two high-quality fruits would otherwise touch and rub against each other. Fruit are clipped from the tree with a 1–2 cm stem, and are placed directly in citron-shaped foam rubber inserts (10 or 12 per box) for transport to the packing house.

Postharvest Handling

Part of Plant Harvested and Harvesting Techniques Used

Because most orchards of citrons for industrial use are relatively small (no more than 5 ha), fruit are harvested by hand. This is even more important in fruit harvested for ritual use, since only careful hand-harvesting will guarantee that fruit will remain blemish-free en route to the packing house. Citrons for ritual use can be marketed according to consumer demand in degrees of ripeness from medium-green to light green to yellow-green to yellow. All fruit are harvested green, and most are dipped in solutions of gibberellic acid (GA) to preserve the calyx and stem and keep them from yellowing in storage (Klein et al. 2013). Varying the GA concentration can also help control the degree of peel degreening. Some fruit are allowed to ripen naturally for marketing as medium or light green. Other harvested fruit are exposed briefly to ethylene in ripening rooms to ensure that they are already yellow when they reach the market. Care must be taken that over-exposure to ethylene does not trigger abscission of the calyx.

Grading

Flowers, leaves, and fruit used for food and flavoring are not graded, since the plant parts are processed. Citrons used for Jewish ritual are graded for size, shape, color, freedom of the skin from any sort of blemish, presence and straightness of a "pitam", central location and straightness of the stem at the bottom of the fruit and presence of a "belt". There are many subtle quality parameters within each type of citron used for ritual – degree of bumpiness of the skin, color gradation from yellow to green and location of the gradient on the fruit, and fruit scent are all used by buyers and by marketers in determining fruit quality. Overall, fruit are usually divided into four categories, from "extremely beautiful" to "beautiful", to "standard" to "acceptable". Some growers market sub-standard but ritually acceptable fruit as "student" fruit for demonstrating the ritual in classrooms, while other growers dispose of such fruit. Care must be taken when using ripening rooms that the fruit do not become unmarketable because of overly-yellow color or because of stem yellowing, senescence and abscission.

Packaging

Citrons grown for food and flavoring are not packaged until after processing, either as bottled essence or as barrels or boxes of candied or brined peel. In Asia, the "fingers" of finger citrons are sometimes packed in cans or jars as a "healthy condiment". Medicinal extracts are packaged in small vials.

Citrons used for ritual are often placed in individual low-density polyethylene bags to avoid water loss and to slow down peel yellowing to a certain extent. Fruit (with or without bags) are packed for retail sale in individual cartons. Padding the fruit is critical for maintenance of quality. Fruit are placed either in expanded polystyrene socks, such as described earlier to protect fruit on the tree, or they are placed in individual citron-sized foam rubber inserts that in turn fit into a cardboard carton. Fruit are also packed in larger foam inserts for wholesale in cardboard trays of single layers of 8–12 fruits each. The trays are constructed so that they can be stacked without the bottom of a tray being in contact with the fruit in the tray below.

Storage

Citrons are considered sensitive to cold-storage, and should therefore not be stored at temperatures lower than 10–12° (Klein, unpublished). In practice, however, only fruit used in Jewish ritual are stored at all, and those rarely for more than the 2–3 months from the earliest harvest in July to marketing in September. There are some varietal differences, with Diamante being able to withstand lower temperatures (even to 8°) than the Yemenite variety, which can get symptoms of chilling injury even at 12° (Klein, unpublished). Most ritual citrons that are shipped by sea from Israel to North America are held in containers at 16° for the 3–4 week voyage.

Marketing

Tradition plays a major role in marketing citrons and citron products, whether the fruit are used for confection or food flavoring, for medicine, or for ritual purposes. Most of the candy, succade, and processed citron peel is used as part of traditional foods, often with religious associations, such as Christmas fruit cakes. Alcoholic beverages such as cedratine made from Corsican citrons or Kitron liqueur made from the leaves of Greek citron trees are also associated with traditional local cuisines, and are marketed partly based on nostalgia for tradition. The use of citron extracts for medicinal purposes in the Mediterranean Basin is mostly based on folk practices and traditions, although the medicinal use of Finger citron in Asia is more in the medical mainstream. In the cases mentioned above, there are not many steps between the grower and the consumer, with the grower often processing or marketing the citron and its product himself. The exception to this perhaps is the manufacture of citron peel for the baking industry, where the processors in northern Europe buy fruit from growers in southern Europe, and in turn sell the product to bakeries for seasonal production of fruit cakes.

Growers of citrons for ritual purposes are often unable to sell their fruit unless they have a close connection with marketers who are trusted by the purchasing public. While production costs in Israel of a single citron for religious use can be as much as US$ 1.5 (Gorelik, personal communication), the retail cost of such fruit can range from US$ 5 to 100 or more, depending on the quality, with average prices in Israel of US$ 10–15, and twice that in export markets. In some religious communities, it is not unusual for someone to forego certain purchases throughout the year in order to be able to afford an expensive, but very high quality, citron. The retail customers trust the sellers to provide citrons of high quality and from sources with unquestioned reputation for observing the Jewish laws related to agricultural production. Since there is only a very narrow retail marketing window of 3–4 weeks at the time of the Jewish holidays, most citrons are sold at street markets, rather than as a standard item in stores. These markets often sell other items related to the seasonal holidays, as well. The personal connection and trust between the grower and the marketer, and between the marketer/seller and the consumer, is often a factor in completing the sale. Surprisingly, considering the economic pressure of supplying a product for a very short period of time, there is not much unity in citron marketing, with many relatively small-scale sellers, although growers will sometimes unite to supply the market.

Production Schedules

Citrons grown for confectionary, baking or medicinal use are harvested once or twice a year, in spring and summer. Leaves and flowers are harvested in spring for extraction of essential oils used in perfumes. Citrons for Jewish ritual use during the Sukkot holiday are harvested from mid-summer (mid-late July) until early fall (mid-September to mid-October), with the end of harvest depending on the timing of the holiday. Excess fruit are removed from trees in November and December. These fruit are often yellow and large (Fig. 2), and are used in confectionary, or else are thrown away. Fruit from the small flush of flowering in winter-early spring are usually removed by May or early June.

Since citron trees are susceptible to many biotic and abiotic stresses, in addition to having inherently weak wood and branch structure, they often are commercially productive for the relatively short time of 10–12 years. The trees can take 3–5 years to reach full productivity, so growers often have young plantings timed to succeed mature orchards, which are then pulled out and replanted to maintain the supply of fruit to the market.

Fig. 2 Citron fruits "Urdang" variety in an orchard in central Israel. Fruit photographed in winter, 3 months after the Sukkot holiday. White foam pads to protect fruit from limb rub are visible on branches

Utilization

Cosmetic

Essential oils from flowers or petitgrain oils from young twigs and leaves are used in the perfume industry (Lota et al. 1999; Menichini et al. 2011; Venturini et al. 2010).

Pharmaceutical and Therapeutic

Greek, Roman, Arab and Jewish sources from ancient to medieval times all refer to citron as a plant source of extensive medicinal value (Arias and Ramon-Laca 2005; Krispil 2000; Levey 1973; Nicolosi et al. 2005; Panara et al. 2013; Rosner 1968; Tolkowsky 1938). Extracts and preparations of the fruit and seeds were particularly valued for their antitoxic activity (Arias and Ramon-Laca 2005; Morton 1987), although whole fruit and leaves were also used as insect repellents (Morton 1987), which in turn could reduce insect-borne diseases. Maimonides (thirteenth century) in his Treatise on Healing mentions ground citron seeds, applied either fresh or dry in a poultice with colocynth root, as particularly efficacious against scorpion poison (Rosner 1968). His contemporary ibn Taysan (Saladin d'ascoli Salerno, c. twelfth century) recommended a reduction of citron juice to counteract poison in general, as well as to improve liver and heart function, and as an ingredient in eyewash (Levey 1973). Extracts of seeds, leaves and shoots are used as a vermifuge (Arias and Ramon-Laca 2005; Panara et al. 2013). The essential oils in the peel have antiseptic and antibiotic qualities. These aromatic oils also are the reason that citrons are placed in clothes closets and storage chests to protect textiles from moths and other insects (Morton 1987). In addition, the albedo is eaten to improve digestion (Krispil 2000), while fruit (and sometimes leaf) extracts help reduce symptoms of Alzheimer's disease (Conforti et al. 2007), cancer (Entezari et al. 2009) diabetes (Conforti et al. 2007; Sah et al. 2011), ulcers (Nagaraju et al. 2012), rheumatism and evil spirits (Morton 1987). Eating raw peel, sugared peel, or jam made from citrons used in the Sukkot holiday is considered to ease childbirth (Sofer 1969; Sperling 1982), although some consider only the pitam to have this beneficial effect. Eating the pitam is also considered to be a way of enhancing fertility (Sperling 1982).

In Asia, the Finger citron is preferred to the round citron for medicinal use. The stage of ripeness influences the concentration of bioactive compounds in the fruit, which in turn affects the efficacy of medicinal extracts (Liang et al. 2006). Extracts of citron have been found to enhance insulin secretion (Peng et al. 2009), to contain trigonelline, an alkaloid that may aid in alleviating diabetes symptoms (Mucci et al. 2013), to mimic estrogens (El-Alfy et al. 2012; Sharangouda 2008), to have anti-inflammatory activity (Negi et al. 2010), and to inhibit cholinesterase activity (Conforti et al. 2007). These specific metabolic and physiological activities provide a quantifiable base for further exploration of the medicinal effects and value of citron.

Food and Flavouring

Fresh citron fruit can be infused to flavor alcoholic drinks such as vodka, while the citron leaves are distilled to produce kitron, a specialty liquor made in Naxos, Greece. Essential oils from citron peel can be used in flavoring for many foods. The peel itself is processed by drying, or by brining, rinsing and candying for use in many kinds of baked goods, especially fruit cakes for Christmas. Candied peel (succade) is also marketed independently as a sweet, both plain and sometimes coated with chocolate. The albedo of the pulpless Yemenite citron is notably thick and non-acidic, and is sometimes eaten raw as a snack or condiment. The Finger citron also lacks pulp and can be eaten raw, but is more often used medicinally. In India and Asia, citron fruits are occasionally pickled and used as condiments.

Industrial

Citron products are not used for industrial non-food processes, with the exception of the perfume industry (see above).

Other

The major non-food use of citrons is for the Jewish ritual of waving the Four Species, which is enacted during the harvest festival of Sukkot (Tabernacles) in early autumn (Leviticus 23:40). The citron fruit is held along with a bound grouping of one palm branch, two willow branches, and three myrtle branches.

Safety Data

A heretical High Priest in the first century BCE did not perform a Sukkot ceremony properly in the Temple in Jerusalem. The congregation reacted by pelting him with citrons (Babylonian Talmud Sukka 48b), which, as mentioned earlier, can reach 1 kg in size (Nicolosi et al. 2005). There are no other known safety hazards associated with citron fruit or products.

Acknowledgements I thank the citron growers Hagai Kirschenbaum and Eliezer Gorelik for always making time to answer my extensive questions, extension agent (retired) Asaf Aftabi for his interest and help, Naif Abu-Muammar for always being gracious in helping me in the orchards, and Rabbi LY Sofer for his great interest and encouragement in citron research. Publication 111/2013 of the Agricultural Research Organization-Volcani Center.

References

Arias BA, Ramon-Laca L (2005) Pharmacological properties of citrus and their ancient and medieval uses in the Mediterranean region. J Ethnopharmacol 97:89–95

Barkley NA, Roose ML, Krueger RR, Federici CT (2006) Assessing genetic diversity and population structure in a citrus germplasm collection utilizing simple sequence repeat markers (SSRs). Theor Appl Genet 112:1519–1531

Barrett HC, Rhodes AM (1976) A numerical taxonomic study of affinity relationships in cultivated citrus and its close relatives. Syst Bot 1:105–136

Bitters WP, Duran-Vila N, Semancik JS (1987) Effect of exocortis viroid on flower and fruit structure and development on etrog citron. Plant Dis 71:397–399

Conforti F, Statti GA, Tundis R, Loizzo MR, Menichini F (2007) In vitro activities of *Citrus medica* L. cv. Diamante (Diamante citron) relevant to treatment of diabetes and Alzheimer's disease. Phytother Res 21:427–433

El-Alfy TH, Hetta MH, Yassin NZ, Rahman FA, Kadry EM (2012) Estrogenic activity of *Citrus medica* L. leaves growing in Egypt. J Appl Pharm Sci 2:180–185

Entezari M, Majd A, Falahian F, Mehrabian S, Hashemi M, Lajimi AA (2009) Antimutagenicity and anticancer effects of *Citrus medica* fruit juice. Acta Med Iran 47:373–377

Gafny R, Mogilner N, Nitzan Y, Ben-Shalom J, Bar-Joseph M (1995) The movement and distribution of citrus tristeza virus and citrus exocortis viroid in citrus seedlings. Ann Appl Biol 126:465–470

Goldschmidt EE, Leshem B (1971) Style abscission in the Citron (*Citrus medica* L.) and other Citrus species: morphology, physiology and chemical control with Picloram. Am J Bot 58:14–23

Klein JD, Cohen S, Hebbe Y, Shapovalov A, Shklyar G, Korol L (2013) Changes in peel color of citron fruits from different genetic origins in response to postharvest copper and gibberellic acid treatments. Acta Hortic 1012:385–390

Krispil N (2000) Medicinal plants in Israel and throughout the world: the complete guide. Hed Arzi Publishing House, Ohr Yehuda, p 59

Langgut D, Gadot Y, Porat N, Lipschits O (2013) Fossil pollen reveals the secrets of the Royal Persian Garden at Ramat Rahel, Jerusalem. Palynology 37:115–129

Levey M (1973) Early Arabic pharmacology. An introduction based on ancient and medieval sources. Brill Publishing, Leiden

Liang SJ, Wu H, Lun X, Lu DW (2006) Secretory cavity development and its relationship with the accumulation of essential oil in fruits of *Citrus medica* L. var. *sarcodactylis* (Noot.) Swingle. J Integr Plant Biol 48:573–583

Lota ML, de Rocca Serra D, Tomi F, Bessiere J-M, Casanova J (1999) Chemical composition of peel and leaf essential oils of *Citrus medica* L. and *C. limonimedica* Lush. Flavour Fragr J 14:161–166

Menichini F, Tundis R, Bonesi M, De Cindio B, Loizzo MR, Conforti F, Statti G, Menabeni R, Bettini R, Menichini F (2011) Chemical composition and bioactivity of *Citrus medica* L. cv. Diamante essential oil obtained by hydrodistillation, cold-pressing and supercritical carbon dioxide extraction. Nat Prod Res 25:789–799

Morton J (1987) Citron. In: Fruits of warm climates. Julia F. Morton, Miami, pp 179–182. http://www.hort.purdue.edu/newcrop/Morton,1987/citron.html#Food Uses

Mucci A, Parenti F, Righi V, Schenetti L (2013) Citron and lemon under the lens of HR-MAS NMR spectroscopy. Food Chem 141:3167–3176

Nagaraju B, Anand SC, Ahmed N, Chandra JNS, Ahmed F, Padmavathi GV (2012) Antiulcer activity of aqueous extract of *Citrus medica* L. fruit against ethanol-induced ulcer in rats. Adv Biol Res 6:24–29

Negi SA, Vijay VJ, Melkani AB (2010) Analgesic activity of fruit decoction of *Citrus medica* Linn. J Pharm Res 3:2119–2121

Nicolosi E, La Malfa S, El-Otmani M, Negbi M, Goldschmidt EE (2005) The search for the authentic citron (*Citrus medica* L.): historic and genetic analysis. HortSci 40:1963–1968

Ollitrault P, Terol J, Garcia-Lor A, Berard A, Chauveau A, Froelicher Y, Belzile C, Navarro RL, Brunel D, Talon M (2012) SNP mining in *C. clementina* BAC end sequences; transferability in the Citrus genus (Rutaceae), phylogenetic inferences and perspectives for genetic mapping. BMC Genomics 13:13

Panara K, Joshi K, Nishteswar K (2013) A review on phytochemical and pharmacological properties of *Citrus medica* Linn. Int J Pharm Biol Arch 3:1292–1297

Peng CH, Ker YB, Weng C-F, Peng C-C, Huang C-N, Lin L-Y, Peng RY (2009) Insulin secretagogue bioactivity of finger citron fruit (*Citrus medica* L. var. Sarcodactylis Hort, Rutaceae). J Agric Food Chem 57:8812–8819

Pessina D, Gentili R, Barcaccia G, Nicolè S, Rossi G, Barbesti S, Sgorbati S (2011) DNA content, morphometric and molecular marker analyses of *Citrus limonimedica*, *C. limon* and *C. medica* for the determination of their variability and genetic relationships within the genus Citrus. Sci Hortic 129:663–673

Rosner F (1968) Moses Maimonides' treatise on poisons. J Am Med Assoc 205:914–916

Sah AN, Joshi A, Juyal V, Kumar T (2011) Antidiabetic and hypolipidemic activity of *Citrus medica* Linn. seed extract in streptozotocin induced diabetic rats. Pharmacogn J 3:80–84

Sharangouda PSB (2008) Estrogenic activity of petroleum ether extract of seeds of *Citrus medica* on immature albino rats. Int J Green Pharm 2:91–94

Sofer YC (1969) Kaf HaChaim. M. Sofer pub, Jerusalem. Section 764:60 (Hebrew)

Solel Z, Mogilner N, Gafny R, Bar-Joseph M (1993) Induced tolerance to mal secco disease in Etrog citron and Rangpur lime by infection with citrus exocortis viroid. Plant Dis 79:60–62

Sperling AY (1982) Taamei HaMinhagim (Sources of [Jewish] customs). Eshkol Publishing, Jerusalem, p 67. (Hebrew)

Tolkowsky S (1938) Hesperides: a history of the culture and use of citrus fruits. J. Bale, Sons & Curnow, Limited, London

Venturini N, Curk F, Desjobert J-M, Karp D, Costa J, Paolin J (2010) Chemotaxonomic investigations of peel and petitgrain essential oils from 17 citron cultivars. Chem Biodivers 7:736–751

Therapeutic Use of Aleppo Pine (*Pinus halepensis* Mill.)

Gabriel Schiller

Abstract Aleppo pine has a circum-Mediterranean distribution. In coastal areas it grows from sea level up to 600 m a.s.l. while in the south (Morocco) it can be found at altitudes up to 2,600 m in the Atlantic mountains. The optimum climatic conditions for this species are 350–700 mm annual rainfall and between −2 and 10 °C absolute minimum temperatures. This species is a thermophilic, drought-tolerant species; it is probably one of the pines most tolerant to high temperatures and drought. It grows mainly on calcareous shallow soils which are the weathering product of limestone and chalk.

Early human settlement around the Mediterranean partly coincides with the areal distribution of Aleppo pine. Hence the use, from antiquity to modern times, of therapeutic chemicals (e.g. turpentine oil) produced by this tree species to treat a variety of illnesses as described by Hippocrates, Dioscorides, Maimonides and others. These conditions include: nasal discharge; stroke; lethargy; depression; pleurisy; hemorrhages (interior); external wounds (as an haemostatic agent); respiratory tract diseases (catarrh and common cold); tooth problems (toothaches) and wounds; diabetes prevention and treatment and also as an aphrodisiac agent.

Keywords Aphrodisiac agent • Depression • Diabetes prevention and treatment Dioscaorids • Hemorrhages • Haemostatic agent • Hippocrates • Lethargy • Maimonides • Nasal discharge • Pleurisy • Resin • Respiratory tract diseases • Terpentine oil • Terpenes • Toothaches • Stroke

Introduction

Studies on the genetic diversity, within and between circum Mediterranean natural populations of Aleppo pine (*Pinus halepensis* Mill.), revealed the similarity between an Aleppo pine forest growing in Umbria, Italy, and native Israeli populations in their genetic markers such as isosymes or resin monotrepene composition, and also

G. Schiller (✉)
Department of Natural Resources, Institute of Plant Sciences, Agricultural Research Organization, The Volcani Centre, Bet Dagan, Israel
e-mail: vcgabi@volcani.agri.gov.il

Z. Yaniv, N. Dudai (eds.), *Medicinal and Aromatic Plants of the Middle-East*,
Medicinal and Aromatic Plants of the World 2, DOI 10.1007/978-94-017-9276-9_11

some eco-physiological parameters (Schiller and Brunori 1992). The results evoked interest to understand the cause of these findings, which led to historical documented events from which it can be deduced that seed material or seedlings were transferred by monks from Palestine to Italy in the third or fourth century A.D. Monks usually obtained the common medical knowledge of their time, which means that possibly they knew about the therapeutic value of Aleppo pine and therefore they took with them seeds or seedlings when they were asked to come to Italy, a country whose medical plant flora was unknown to them.

According to Barceloux (2008), oil of turpentine extracted from pine resin has been used as an herbal treatment for a variety of illnesses since ancient times. Hippocrates (370–460 B.C.) mentioned the value of turpentine oil as an emmenagogue and inhibitor of nasal discharge. The Greek physician Dioscorides used turpentine oil as an aphrodisiac agent. The Romans used this oil for a wide variety of internal and external diseases including stroke, lethargy, depression and pleurisy. The use of Aleppo pine as a chemotherapy agent of infectious diseases in the Mediterranean area in the twelfth century was described in books by Maimonides (Casal and Casal 2004). Byzantine and Ottoman period medical treatments in the Middle East used Aleppo pine for treating hemorrhages (interior); as an Haemostatic agent (in external wounds); respiratory tract diseases (catarrh and common cold); tooth problems (toothaches) and wounds. The method of use was fumigation (inhaling, air-conditioning or exposing the concerned part of the body) (Lardos 2006). Diabetes prevention and treatment according to the Greco-Arab and Islamic-based natural products also used Aleppo pine (Ziad et al. 2011).

Description of the Plant (After Vidakovic 1991)

Pinus halepensis is a tree that according to site conditions can live up to 200 years, grow up to 25–30 m in height, and reach a circumference of more than 3 m.

General Description

1. **Bark**, on young trees is smooth and light gray; later on brown and fissured.
2. **Branches**, are spreading.
3. **Shoots**, are of 2–3 mm in diameter, ash-gray color.
4. **Buds**, are ovate, 5–10 cm in length, not resinous.
5. **Needles**, in 2's, seldom three on the short shoot, 6–10 cm in length, thin, slender light green (Yellowish green), at the top of the shoot in brush-like tufts; usually abscising between the 2nd and the 4th year; resin ducts marginal; leaf sheath 8 mm long.
6. **Flowers**, appear in April-May
7. **Cones**, are ovate, with thick and up to 2 cm long stalks, more or less turned back along the carrying branch or downwards. Cones are solitary or up to 3, seldom 4 in a whorl. Cones are 6–10 cm in length, up to 4 cm across, gray to reddish brown, ripening in September-October of the 2nd season. Cones open mostly in

the 3rd and 4th year under the influence of climate conditions (hot and dry spells) and shed an abundance of seeds (~70 seeds per cone). After seed shed, the empty cones persist on the tree for very many years, and these dry cones become a hazardous element in the tree canopy which enhances forest fires. Up to a third of the yearly cone production are serotinuous cones, i.e. cones that open and shed their seed under the heat created by forest fire. Apophysis of fertile scales is flat or raised, with a more or less conspicuous umbo, gray, without mucro

8. **Seed**, 6–7 mm long, darkly mottled, wing 18–28 mm long, dark brown.

As the result of its circum Mediterranean distribution, i.e. growing in many ecologically differing areas there is variation among the different geographic regions (ecotypes) not only in morphological traits but also in the Genetic structure, Terpene composition of the resin, Xylem and needles anatomy that influences the water economy of the plants, and more (Esteban et al. 2010; Korol et al. 2002; Mirov 1967; Schiller 1982, 2000a, b; Schiller et al. 1986).

Picture 1 Illustration of Aleppo pine morphological traits

Classification (Taxonomy)

According to Mirov, Vidaković and Quézel the taxonomy of *Pinus halepensis* Mill. is as follows:

1.	Division = Gymnosprmae	8.	Family = Pinaceae
2.	Subdivision = Conipherophytina	9.	Subfamily = Pinoideae (Pinus)
3.	Class = Ginkgoatae	10.	Genus = *Pinus*
4.	Order = Ginkgoales	11.	Subgenus = Pinus
5.	Family = Ginkgoaceae	12.	Section = Pinus
6.	Class = Pinatae	13.	Subsection = Sylvestres
7.	Subclass = Pinidae (Coniferae)	14.	Species = *Pinus halepensis*

According to Themis Nasopoulou (2004) the taxonomy of *Pinus halepensis* Mill. is as follows:

1.	Kingdom = Plantae	8.	Family = Pinaceae
2.	Subkingdom = Embryobiota	9.	Subfamily = Pinoideae
3.	Division = Tracheophyta	10.	Genus = *Pinus*
4.	Subdivision = Spermatophytina	11.	Subgenus = Pinus
5.	Class = Pinopside (Coniferopsida)	12.	Section = Pinea
6.	Subclass = Dicotyledonidae	13.	Subsection = Pinaster
7.	Order = Pinales (Conirerales)	14.	Species = *Pinus halepensis*

Origin and Distribution

Aleppo pine is a distinct Mediterranean species, the latest description of its natural occurrences shows that this species has a circum Mediterranean distribution. Natural forests cover about 2.5–3.0 million hectares located in geographic regions influenced by the Mediterranean climate on both sides of the western part of the Mediterranean basin, i.e., in Europe and North Africa. In the eastern part of the Mediterranean basin natural forests were decimated during the centuries and became relicts that still exist in: Turkey, Syria, Lebanon, Israel and Jordan where the most South-eastern border of distribution is reached. In the Balkans, Aleppo pine grows in scattered localities along the Adrian coast. It occurs abundantly in Greece mainly in the Peloponnese, Attica, Euboea and Khalkidhiki. Aleppo pine is not abundant anywhere in Italy, it occurs in a few localities in the southern part of the peninsula, on islands in the Adrian Sea, on Sicily, Sardinia and the Mediterranean Italian Riviera. The occurrence of Aleppo pine in France is mainly east of the Rhône valley in the Provence, where it reaches its most northerly limit in the Mediterranean region. Aleppo pine can be found along the Mediterranean region of the Spanish peninsula; and on the coastal ranges of Catalonia and Valencia region. There are also rare occurrences in the Andalucía province and the inland of the peninsula. Aleppo pine is abundant on the Balearic Islands. In North Africa Aleppo pine occurs in the Akhdar Mountains in Cyrenaica, Libya; this occurrence is a disjunct isolated one from the large and continuous forest area in Tunisia and Algeria that grows on all coastal ranges to the West of Algeria and on the high plateau of the Saharan Atlas Mountains. Aleppo pine is rather rare in Morocco; it grows in some isolated stands in a few steep valleys, on the margins of mountain ranges, protected from oceanic influences. It is quite abundant on the Mediterranean slopes of the central-eastern Rif, on the northern and southern flanks of the central eastern Middle Atlas. Aleppo pine becomes more frequent toward the Algerian boarder on the Beni Snassene, Oujda and Debdou Mountains.

In the last century, *Pinus halepensis* has been planted extensively with the aim to rehabilitate areas degraded as a result of erosion occurring after clear cut and/or large forest fires, large scale abandonment of Agriculture land etc. This species has been introduced into new regions in the world with Mediterranean type climate such as Australia, South Africa, Argentina and Chile, and California where it was planted on marginal and droughty land (Map 1).

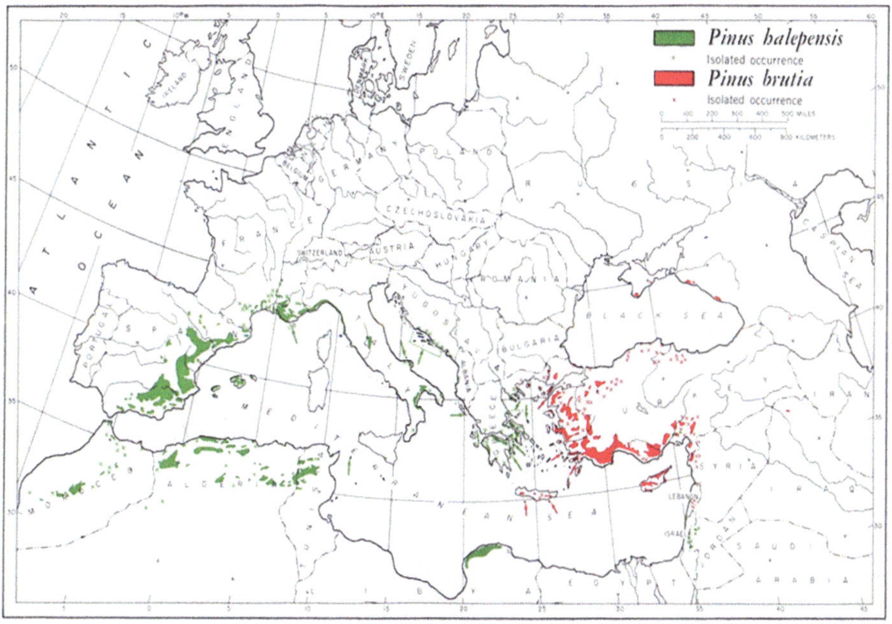

Map 1 Geographic distribution of Aleppo pine around the Mediterranean Sea (From Critchfield and Little 1966)

Ecological Requirements: Climate, Geology and Soils (After Quézel 2000)

Pinus halepensis natural forests occur within the circum Mediterranean low elevation broad leaf forest belt from sea level up to ± 1,000 m, i.e., the Thermo-Medit. vegetation belt which has a temperate to warm climate; mean minimal temp. 3 °C and more and annual average T of 17 °C and more. It occurs also within the Meso-Medit. vegetation belt which has a fresh climate with a mean minimal temperature of between 0 and 3 °C and an annual average T of between 13 and 17 °C. In North Africa the species grow also at higher elevations of between 1,000 and 2,500 m in the Atlas Mountains. Aleppo pine can endure temperatures as low as −12 °C without damage to the needles and the photosynthetic apparatus. This pine can be found over its entire distribution range in areas with a yearly average rainfall from 200 mm up to 1,500 mm, i.e., ranging from lower arid bio climate to the humid. This pine occurs most abundantly in the semi-arid and sub-humid zones, i.e., between 350 and 700 mm average yearly rainfall, and between 7 and 3 month duration of summer drought. Through its entire range of distribution Aleppo pine grows mainly on marly limestone and marls, it also grows on hard limestone and dolomite formations. The quality of growth, i.e., the site quality is strongly affected by both the lithological and moisture-retaining characters rather than by the geological age of the rock strata. The soils covering these bedrock formations are calcareous in varying degree and shallow up to 60 cm in depth.

Pinus halepensis is a pioneer species, occupying disturbed and abounded soils such as: abandoned agricultural areas, burned areas, eroded and flooded areas etc. This phenomenon is the result of a yearly large seed crop that is dispersed by wind to areas with bar soils with none or low competition by other vegetation. These characteristics enabled the natural occupation of tens of thousands of hectares during the twentieth century.

Major Chemical Constituents and Bioactive Compounds

A survey of scientific Ethno pharmacological literature published since 2000 revealed the use of Aleppo pine as a medical plant in many areas around the Mediterranean (Gonzalez-Tejero et al. 2008).

Crude Drug: The Crude Drug Used

The different chemicals which constitute Pines resin and the essential oil are the agents for curing different illness and discomfort because of their biological activity. According to "Plants for a future" "*turpentine is antiseptic, diuretic, rubefacient and vermifuge. It is a valuable remedy used internally in the treatment of kidney and bladder complaints and is used both internally and as a rub and steam bath in the treatment of rheumatic affections. It is also very beneficial to the respiratory system and so is useful in treating diseases of the mucous membranes and respiratory complaints such as coughs, colds, influenza and TB. Externally it is a very beneficial treatment for a variety of skin complaints, wounds, sores, burns, boils etc', and it is used in the form of liniment plasters, poultices, herbal steam bath and inhalers*".

Until today, Aleppo pine needles and seeds are used in the eastern region of the Mediterranean to cure diabetes, using a standard decoction of 50 g seeds or needles taken internally, 150 cc two times a day; and for sexual weakness, using 10–15 g seeds eaten daily (Said et al. 2002; Azaizah et al. 2006; Daoud 2008). Antioxidant activity analysis of some Jordanian medical plants used traditionally for the treatment of diabetes revealed that Aleppo pine pollen grain has only low antioxidant activity compared with other medical plants (Al-Mustafa and Al-Thunibat 2008). In the Calabria region in Italy use is made of Aleppo pine decoction of buds and needles to cure bronchitis, and cure joint inflammation by bath (Passalacqua et al. 2007). Ethno botanical data from Turkey revealed that *Pinus* species have been used against rheumatic pain and for wound healing; experimental studies revealed that essential oils from cones of *Pinus halepensis* display remarkable wound healing activity (Suntar et al. 2012). Antibacterial activity of Aleppo pine essential oil was tested against strains of *Staphylococcus aureus*, *Pseudomonas aeruginosa*, *Escherichia coli* and *Bacillus cereus*. The results revealed that this essential oil showed moderate activity against all the bacterial strains except *Pseudomonas aeruginosa* and *Escherichia coli* that were revealed as very resistant (Abi-Ayad et al. 2011).

Aleppo pine resin is used for the treatment of muscular pains, as a disinfectant of the respiratory and urinary tracts and as antifungal in North Africa (Boulaacheb 2009). Furthermore, according to Duke's et al., (2007) *Handbook of Medicinal Plants of the Bible* thin-shelled Aleppo pine seeds are added to festive cakes, ground seeds are sprinkled over Tunisian traditional pastries, seeds in honey are eaten first thing in the morning to augment sperm. Aleppo pine resin is used for wine fermentation, pills and suppositories. Crude sap is used in Lebanon internally for cold and coughs, externally for sores and venereal ailments. Astringent powdered bark is applied to wounds; the resin is used as antiseptic.

Collection Practice

The resin of Pines is obtained by tapping the trunk, or by destructive distillation of the wood. Different components of the resin can also be extracted using the decoction procedure with needles or seeds.

Chemical Composition of Aleppo Pine Essential Oil

Results of chemical analysis of the resin, as a genetic marker, revealed significant differences among geographic groups in the relative amount of chemicals present in the resin (Baradat et al. 1995; Schiller and Grunwald 1987; Dob et al. 2005). Out of the

100 different constituents present in Aleppo pine's resin: *α-pinene, Sabinene, Myrcene, d-3-Carene, p-cymene, Limonen, α-Terpinolene, γ-Terpinene and β-Caryophyllene* showed high variation in their relative amounts between geographic regions around the Mediterranean. Many of the other 90 constituents, regularly present in relative low quantities are difficult to analyze (Dob et al. 2005). There are significant differences in the relative quantity of constituents in the essential oil extracted from needles, branches or cones from the same tree (Macchionu et al. 2003). There are also significant differences between cones and seeds in their amounts of antioxidant compounds and antiradical properties (Dhibi et al. 2012).

Recently, Sterol in *Nigella sativa* L. (black cumin) and in *Pinus halepensis* Mill. seeds have gained importance due to the high level of *β-sitosterol* that could make it the most suitable and effective for lowering blood cholesterol and preventing coronary heart disease (Cheikh-Rouhou et al. 2008).

Conclusion

Since ancient times and up to date Aleppo pine is traditionally used as a therapeutic agent with the aim to cure human internal and external disorders such as: diabetes, sexual weakness, muscular pains, and disinfectant of the respiratory and urinal tracts, anti-fungal, bronchitis, joint-inflammation, tooth problems and wounds. Use was made of bark, resinous wood, needle and currant year sprouts, all containing resin, e.g. essential oils composed of many components with antioxidant and antibacterial activity. Decoction, inhalation, baths and salves are used according to the treatment needed.

References

Abi-Ayad M, Abi-Ayad FZ, Lazzouni HA, Rebiahi SA, Cherif C, Bessiere JM (2011) Chemical composition and antifungal activity of Aleppo pine essential oil. J Med Plant Res 5(22):5433–5436

Al-Mustafa AH, Al-Thunibat OY (2008) Antioxidant activity of some Jordanian medical plants used traditionally for treatment of diabetes. Pak J Biol Sci 11(3):351–358

Azaizah H, Saad B, Khalil K, Said O (2006) The state of art of traditional Arab herbal medicine in the eastern region of the Mediterranean: a review. eCAM 3(2):229–235

Baradat PH, Michelozzi M, Tonnetti R, Khouja ML, Khaldi A (1995) Geographic variation in the terpene composition of *Pinus halepensis* Mill. In: Baradat P, Adams WT, Muller-Starck G (eds) Population genetics and genetic conservation of forest trees. SPB Academic Publishing, Amsterdam, pp 5–28

Barceloux DG (2008) Chapter 104. In: Medical toxicology of natural substances: food, fungi, medical herbs, plants and venomous animals. Wiley Online Library. Published online 02.01.2008

Boulaacheb N (2009) La Resin de *Pinus halepensis* Mill. Usage Traditional par la Population de la Petite Kabylie (Algerie, Nord Afrique). Acta Hortic 853:435

Casal MT, Casal M (2004) Maimonides and the chemotherapy of infectious diseases. Rev Esp Quimioterap 17(3):289–294

Cheikh-Rouhou S, Besbes S, Longnay G, Blecker C, Deroanne C, Attia H (2008) Sterol composition of black cumin (*Nigella sativa* L.) and Aleppo pine (*Pinus halepensis* Mill.) seed oil. J Food Compos Anal 21:162–168

Critchfield WB, Little EL (1966) Geographic distribution of the Pines of the world, USDA forest service, miscellaneous publication 991. USDA, Washington, DC, 991

Daoud RTE (2008) Studies on folkloric medicinal plants used by Palestinians in the Qalqilia district. M.Sc. thesis, An-Najah national University, Nablus

Dhibi M, Mechri B, Brahimi F, Skhiri F, Alsaif M, Hammami M (2012) Fatty acid profiles' antioxidant compounds and antiradical properties of *Pinus halepensis* Mill. cones and seeds. J Food Agric 92(8):1702–1708. Wileyonlinelibrary.com. doi:10.1002/JSFA.5535

Dob T, Berramdan T, Chelgoum C (2005) Chemical composition of essential oil of *Pinus halepensis* Miller growing in Algeria. C R Chimie 8:1939–1945

Duke JA, Duke PAK, du Cellier JL (2007) Duke's hand book of medicinal plants of the bible; Allepo pine (*Pinus halepensis* Mill.). CRC Press, Boca Raton, pp 332–333

Esteban LG, Martin JA, de Palacios P, Fernández FG, López R (2010) Adaptive anatomy of *Pinus halepensis* trees from different Mediterranean environments in Spain. Trees 24:19–30

Gonzalez-Tejero MR, Casares-Porcel M, Sanchez-Rojas CP, Ramiro-Gutierrez JM, Molero-Mesa J, Pieroni A, Ciusti ME, Censorii E, de Psquale C, Dekk A, Paraskeva-Hadijchambi D, Hadjichambis A, Houmani Z, El-Demerdash M, El-Zayat M, Hmamouchi M, Eljohrig S (2008) Medical plants in the Mediterranean area: synthesis of the results of the project RUBIA. J Ethnopharmacol 116:341–357

Korol L, Shklar G, Schiller G (2002) Diversity among circum – Mediterranean populations of Aleppo pine and differentiation from Brutia pine in their isoenzymes: additional results. Silvae Genet 51:35–41

Lardos A (2006) The botanical *material medica* of the *Iatrosophikon* – a collection of prescriptions from a monastery in Cyprus. J Ethnopharmacol 104:387–406

Macchionu F, Cioni PL, Flamini G, Morelli G, Maccioni S, Ansaldi M (2003) Chemical composition of essential oils from needles, branches and cones of *Pinus pinea*, *P. halepensis*, *P. pinaster* and *P. nigra* from central Italy. Flavour Fragr J 18:139–143

Mirov NT (1967) The genus Pinus. The Ronald Press Co., New York

Nasopoulou Th (2004) Taxonomic portfolio. The Pinaceae family and *Pinus halepensis* Mill. M.Sc. thesis, University of Plymouth, School of Science

Passalacqua NG, Guarrera PM, De Fine G (2007) Contribution to the knowledge of the folk plant medicine in Calabria region (Southern Italy). Fitoterapia 78:52–68

Quézel P (2000) Taxonomy and biogeography of Mediterranean Pines (*P. halepensis* and *P. brutia*). In: Ne'eman G, Trabaud L (eds) Ecology, biogeography and management of *Pinus halepensis* and *P. brutia* forest ecosystems in the Mediterranean basin. Backhuys Publishers, Leiden, pp 1–12

Said O, Khalil K, Azaizah FH (2002) Ethnopharmacological survey of medicinal herbs in Israel, the Golan Heights and the West Bank region. J Ethnopharmacol 83:251–265

Schiller G (1982) Significance of bedrock as a site factor for Aleppo pine. For Ecol Manag 4:213–223

Schiller G (2000a) Eco-physiology of *Pinus halepensis* Mill. and *Pinus brutia* Ten. In: Ne'eman G, Trabaud L (eds) Ecology, biogeography and management of *Pinus halepensis* and *Pinus brutia* forest ecosystems in the Mediterranean basin. Backhuys Publishers, Leiden, pp 51–65

Schiller G (2000b) Inter-and-intra-specific genetic diversity of Pinus halepensis Mill and P. brutia Ten. In: Ne'eman G, Trabaud L (eds) Ecology, biogeography and management of *Pinus halepensis* and *Pinus brutia* forest ecosystems in the Mediterranean Basin. Backhuys Publishers, Leiden, pp 13–35

Schiller G, Brunori A (1992) Aleppo pine (*Pinus halepensis* Mill.) in Umbria (Italy) and its relation to native Israeli populations. Isr J Bot 41:123–127

Schiller G, Grunwald C (1987) Resin monoterpene in range-wide provenance trials of *Pius halepensis* in Israel. Silvae Genet 36:109–114

Schiller G, Conkle MT, Grunwald C (1986) Local differentiation among Mediterranean populations of Aleppo pine in their isozymes. Silvae Genet 35:11–19

Suntar I, Tuman I, Ustun O, Keles H, Kupeli-Akkol E (2012) Appraisal on the wound healing and anti-inflammatory activities of the essential oils obtained from the cones and needles of *Pinus* species by *in vivo* and *in vitro* experimental models. J Ethnopharamacol 139:533–540

Vidaković M (1991) CONIFERS, morphology and variation (The English version). Graficki Zavod Hrvatske, Zagreb

Ziad H, Said O, Hadich B, Kamil A, Saad B (2011) Diabetes prevention and treatment with Greco-Arab and Islamic-based natural products. JAMI'A 15:19–38

Pomegranate: Aspects Concerning Dynamics of Health Beneficial Phytochemicals and Therapeutic Properties with Respect to the Tree Cultivar and the Environment

Doron Holland and Irit Bar-Ya'akov

Abstract The pomegranate was recognized as an important fruit very early in human history dating back thousands of years ago. Since ancient times, pomegranate culture was developed along with human communities all over the world. Pomegranate provides tasty, healthy and aesthetic fruit and its importance in human diet is fully appreciated in recent years. The health promoting properties of pomegranate flowers, fruits and sprouting seedlings are now being extensively studied in many laboratories around the world revealing a wide spectrum of biological activities in human and animal cells. These, include among others, antioxidant activity, chemoprevention of cancer, protection against cardiovascular diseases and antimicrobial activities. In the following years it is expected that more and more important phytochemicals that are associated with the health therapeutic capabilities of pomegranate will be isolated and characterized from pomegranate. Unfortunately, most of the public and scientific interest in pomegranate is focused on its medicinal properties and much less effort was dedicated to study the biology of the pomegranate tree, its genetic variability and the relevance of the genetic and biochemical variability to the various medicinal activities displayed by pomegranates. In this chapter an attempt was made to focus on studies that demonstrated variability in the content of various phytochemicals among different pomegranate varieties and the relevance of these aspects to studies of the effect of pomegranate on human and animal health.

Keywords Pomegranate • Genetic variability • Health benefits • Phytochemicals • Pomegranate varieties • Anthocyanin • Punicalagin • Germplasm collection • Cultivars • Metabolites • Ellagitannins • Antioxidant activity

D. Holland (✉) • I. Bar-Ya'akov
Unit of Deciduous Fruit Tree Sciences, Institute of Plant Sciences, Newe Ya'ar Research Center, Agricultural Research Organization, P.O. Box 1021, Ramat Yishay 3009500, Israel
e-mail: vhhollan@volcani.agri.gov.il

Z. Yaniv, N. Dudai (eds.), *Medicinal and Aromatic Plants of the Middle-East*, Medicinal and Aromatic Plants of the World 2, DOI 10.1007/978-94-017-9276-9_12

Description of the Plant

The pomegranate tree (*Punica granatum* L.) is naturally grown as a shrub (Levin 2006). Under cultivating conditions the plant is often trained as a single trunk and grown as a tree which can reach up to five meters height (Holland et al. 2009). There are also dwarf pomegranate trees that do not exceed 1.5 m (Levin 2006). Most of the pomegranate varieties are deciduous trees. However, there are evergreen pomegranates that will not shed their leaves in warm winters, yet in cold winter even these varieties will shed their leaves and therefore should be regarded as conditionally deciduous (Holland et al. 2009). Flowering occurs in spring and lasts about a month. The tree bears basically two types of flowers, hermaphrodite flowers with fully developed reproductive organs (vase shape) and male flowers with poorly developed or undeveloped reproductive organs (bell shape). Intermediate stages between these two flowers types also exist. Only the vase shape flowers are fertile and set fruit (Goor and Liberman 1956; Holland et al. 2009). The flower has 5–8 separate pink-orange to orange-red petals alternate with 5–8 sepals fused in their base, forming a red fleshy vase shape. It has many long stamens and usually eight carpels superimposed in two whorls. The fruit is a nearly round fleshy berry crowned by a prominent calyx. It contains 2–3 lower chambers and 6–9 upper chambers separated by membranous walls and fleshy mesocarp. The chambers are filled with many arils that contain a juicy edible layer and a woody seed (Fahan 1976). The pomegranate is both self pollinated and cross pollinated by insects, mainly by bees. Following fruit set the peel color changes until it reaches its characteristic color as the fruit ripens. The fruit matures in late summer and autumn depending on the cultivar. The color of the fruit peel and arils strongly depends on the variety and on environmental conditions. The external and internal colors are not necessarily coordinated (Holland et al. 2009). External peel color of mature fruits ranges from yellow, green, or pink overlaid with pink to deep red or bourdeaux to fully red, pink or deep purple cover. The color of the edible juicy arils can vary from white to deep red. High variability is also observed in the size of the fruit which shows strong dependence on the variety (Holland et al. 2009). Thus, the range of mature fruit size can vary from less than 100 g to more than 1 kg.

The geographical origin of the pomegranate is considered to be somewhere in central Asia in an area which includes North of Iran, Turkmenistan and probably Caucasia (Goor and Liberman 1956; Morton 1987). It is estimated that pomegranate domestication initiated in the Neolithic Era (Levin 2006; Still 2006). Wild pomegranates are known today from North India, Iran and Central Asia (Levin 1981; Rana et al. 2007; Hajiahmadi et al. 2013), while culture escapees are very common in the wild through Asia and the Mediterranean basin (Holland et al. 2009). Only two species: *Punica granatum* L. and *Punica protopunica* Balf. F. belong to the Punicacea (Zukovskii 1950; Mars 2000; Levin 2006). *P. protopunica* is endemic to Sokotra Island and was suggested to be the ancestor of the genus based on xylem anatomy (Balfour 1888; Shilkina 1973). All of the cultivated pomegranate species around the world belong to *P. granatum*. Due to its high adaptability to various climatic and soil conditions the pomegranate tree is widely distributed in many countries around the globe. Pomegranate is cultivated today throughout the world in subtropical, tropical and dry areas in many different microclimatic zones. Commercial orchards of pomegranate trees are now grown in the Mediterranean basin, in central and south Asia, in North, Central and South America, in South Africa and in Australia (Holland et al. 2009; Verma et al. 2010).

One of the most interesting aspects of pomegranate culture is concerned with the origin and variability of domesticated pomegranate cultivars. More than 20 centers that maintain pomegranate germplasm collections were described in the literature (Still 2006; Holland et al. 2009; Verma et al. 2010). The largest collections were reported from Iran, Azerbaijan, China, India, Turkmenistan, Turkey, Ukraine, USA and Uzbekistan. However, care must be taken that not all accessions within a collection or among diverse collections are different accessions (Levin 2006; Zamani et al. 2007; Holland et al. 2009). Thus, accessions might represent duplications of the same cultivar with synonym names or variety landraces or accessions from crossbred varieties. Molecular markers were used to characterize the genetic variations observed in the various collections, but the numbers of the analyzed accessions in each of these studies did not usually exceed 100 (Yuan et al. 2007; Melgarejo et al. 2009; Ercisli et al. 2007, 2011a, b; Soriano et al. 2011; Hajiahmadi et al. 2013; Zamani et al. 2007, 2013). It is therefore estimated, based on those studies and the published literature, that the total number of pomegranate accessions that represent genetically different varieties is not more than several hundreds.

Most of the published data which is based on molecular markers such as SSR's, AFLP and RAPD indicate high variability among the cultivars (Ercisli et al. 2007, 2011a, b; Yuan et al. 2007; Zamani et al. 2007, 2013; Melgarejo et al. 2009; Soriano et al. 2011; Hajiahmadi et al. 2013). Since the data was derived from local accessions and only few publications compared between varieties originating from different known geographical sites it is difficult to precisely estimate the extent of variability among the world wide pool of pomegranate cultivars. Comparison of relatively small number of Indian, Israeli and Spanish originated accessions demonstrated that it was possible to distinguish between the cultivars according to their geographical origin by using SSR markers (Soriano et al. 2011). Very few pomegranate varieties of the species *P. granatum* have been designated as wild varieties. Wild pomegranate species were reported from India, Iran and Turkmenistan (Levin 1981; Rana et al. 2007; Hajiahmadi et al. 2013). However, only in the Indian report a detailed description of the wild phenotype is given. These wild accessions were collected in the Southern slopes of the Himalaya Mountains and are characterized by a relatively small fruit with greenish pink rind color, sour aril taste and hard seeds (Rana et al. 2007). Interestingly, these fruit characteristics are also very true for the various *P. granatum* var. Nana dwarf accessions which were not defined as wild accessions, although in one of the studies this variant was suggested as a third species of the genus *Punica* (Jbir et al. 2008). *P. protopunica* is a distinct *Punica* species and there is no question concerning its wild nature. The flower and fruit characteristics of *P. protopunica* appear to be similar to those of *P. granatum*. There are not yet available molecular comparison between *P. granatum* and *P. protopunica*.

Such comparison could be very valuable as *P. protopunica* could serve as a solid molecular reference for an unquestionable wild species of Punicacea and help solve the issue of the domesticated cultivars origin.

Reported Therapeutic Effects Attributed to Pomegranate

A vast range of therapeutic activities was attributed to pomegranate since ancient times (Levin 2006; Seeram et al. 2006; Jurenka 2008; Holland et al. 2009). The traditionally known therapeutic effects of pomegranate as well as the newly discovered pomegranate effects and the mechanisms of its action in human and animals were thoroughly described by relatively recent outstanding reviews (Seeram et al. 2006; Lansky and Newman 2007; Jurenka 2008; Viuda-Martos et al. 2010; Teixeira da Silva et al. 2013). Recent scientific experiments with human patients, animals, and human and animal cell cultures suggest that pomegranate derived substances have antioxidant, anti-carcinogenic, anti-inflammatory, antimicrobial and estrogenic properties (Seeram et al. 2006; Lansky and Newman 2007; Jurenka 2008; Viuda-Martos et al. 2010; Clinical Trials.gov 2013). Although therapeutic effects were found to be present in many tissues and organs of the pomegranate tree including the bark, roots, leaves, fruit (peel, juice, seed) and flowers (Lansky and Newman 2007; Viuda-Martos et al. 2010), most of the available data was generated by using pomegranate juice (Aviram et al. 2000; Seeram et al. 2006; Lansky and Newman 2007; Jurenka 2008; Stowe 2011; Basu and Penugonda 2009). The use of juice could be not only because of the outstanding effects of this particular pomegranate substance but rather because pomegranate juice is the most readily available and easily stored pomegranate substance.

The potential therapeutic properties of pomegranate include treatment and prevention of cancer diseases (skin, lung, breast, prostate and colon), diabetes, cardiovascular diseases (atherosclerosis, hyperlipidemia, hypoxia, ischemia), stroke, diarrhea and dysentery, cough and fever, erectile dysfunction, dental health, brain disorders, protection against UV radiation, bacterial diseases, fungal diseases (*Candida* spp.) and viral diseases (including AIDS) (Aviram et al. 2000, 2002; Esmaillzadeh et al. 2006; Jayaprakasha et al. 2006; Rosenblat et al. 2006; Seeram et al. 2006; Katz et al. 2007; Lansky and Newman 2007; Jurenka 2008; Adhami et al. 2009; Basu and Penugonda 2009; Endo et al. 2010; Viuda-Martos et al. 2010; Stowe 2011; Paller et al. 2012; Clinical Trials.gov 2013). The biological mechanisms found to be involved in therapeutic and prevention of the various diseases are mainly free radicals scavenging, macrophage oxidative stress decrease, lipid peroxidation decrease, cell proliferation inhibition, cell cycle disruption, suppression of tumor cell invasion, angiogenesis interference, induction of apoptosis and key enzyme inhibition (Seeram et al. 2006; Lansky and Newman 2007; Jurenka 2008).

Essentially, positive medicinal influence of pomegranate derived substrates was reported for a very wide range of illnesses and diseases and the question arises how can a single plant provide the substances for such a diverse range of mal functions of human beings and animals? Another related and important question is what are the mechanisms of action through which the medicinal effects of pomegranate are exerted? Several approaches to provide the answers to these difficult questions were undertaken by modern science. The first was based on thorough analysis of the chemical nature of the various metabolites which are produced in the various parts of the pomegranate plant (Navindra et al. 2006; Lansky and Newman 2007). The second approach involved discovering the target sites for pomegranate substances actions (Malik and Mukhtar 2006; Fuhrman and Aviram 2007; Lansky and Newman 2007; Jurenka 2008). Prominent advances in understanding the molecular and genetic impairments that cause human diseases such as atherosclerosis, neoplasia and diabetes have been crucial for unraveling the mechanism of action of pomegranate derived substances. It is possible that several seemingly different diseases share common factors which play an essential role in their etiology, for example chronic inflammation. Thus, the beneficial effect of pomegranate derived substances on diseases could be due to the anti-inflammatory effects of its phytochemicals (Lansky and Newman 2007). This could explain in part why pomegranate was found to have an effect on such a diverse range of ailments.

Pomegranate Phytochemicals and Their Activities

Several studies reported detailed analyses of the phytochemicals present in pomegranate (Navindra et al. 2006; Lansky and Newman 2007; Viuda-Martos et al. 2010). These include among others, ellagitannins and gallotannins, ellagic acid derivatives, catechins and procyanidins, anthocyanins and anthocyanidins, flavonols, organic acids, simple gallyol derivatives, fatty acids and triglycerides, sterols and terpenoids and alkaloids, sugars, minerals and vitamins (Seeram et al. 2006; Lansky and Newman 2007; Jurenka 2008; Viuda-Martos et al. 2010; Teixeira da Silva et al. 2013). Although many types of phytochemicals were shown to have bioactivity against human diseases this section will focus on polyphenols. Obviously, prominent bioactive phytochemicals that are present in the pomegranate are various polyphenolic compounds including, ellagitannins, and flovonoids. Some of the most potent polyphenolic substances predominate in the fruit, although not necessarily in the edible parts of the fruit. Some of the commercially available pomegranate juices are obtained by hydrostatic pressing process of whole fruits which also include non edible part such as the fruit peel. Such juices include among other two types of predominant polyphenolic compounds flavonoids and hydrolyzable tannins (Navindra et al. 2006).

One of the remarkable biochemical activities of pomegranate polyphenols is the high antioxidative activity reported by numerous studies. It was shown that the antioxidative effect of pomegranate is mainly due to the presence of very high amount of a wide range of polyphenols present in the fruit, leaves and bark (Seeram et al. 2006; Lansky and Newman 2007). Of particular interest are the ellagitanins present in high concentration in the fruit peel and carpellary membranes (Seeram et al. 2006; Lansky and Newman 2007). Comparative experiments demonstrated that the fruit peel has a much higher antioxidative activity as compared to the arils (Tzulker et al. 2007). While most of the antioxidative activity in the peel was attributed to the presence of punicalagin and other ellagitanins (hydrolyzable tannins), the antioxidative activity in arils juice was attributed to anthocyanins (Tzulker et al. 2007). Anthocyanins by themselves belong to the large family of flavonoids which are also polyphenols and were reported to have profound beneficial medical effects (Butelli et al. 2008; de Pascual-Teresa and Sanchez-Ballesta 2008). Antioxidative activity was suggested to have an important preventive role in cancer diseases development (Lansky and Newman 2007; Adhami et al. 2009) and atherosclerosis (Aviram and Rosenblat 2012).

Punicalagin, a complex ellagitanin that has a strong antioxidative activity, was suggested to be the main chemical component that conferred anti fungal activity against plant pathogenic fungi. When aqueous extracts of pomegranate peels were assayed *in vitro* for their antifungal activity against six rot fungi, that cause fruit and vegetable decay during storage, it was observed that the growth rates of *Alternaria alternata*, *Stemphylium botryosum*, and *Fusarium* spp. were significantly inhibited by the extracts. The growth rates were negatively correlated with the levels of total polyphenolic compounds in the extract and particularly to punicalagins, the major ellagitannins in pomegranate peels (Glazer et al. 2012). These results suggest that ellagitannins, and more specifically punicalagins, which are the dominant compounds in pomegranate peels, may be used as a control agent of storage diseases and to reduce the use of synthetic fungicides. Experiments with human pathogenic fungi showed that the active fraction of crude extract prepared from the fruit peel contains punicalagin. This fraction exhibited strong activity against *Candida albicans* and *Candida parapsilosis* (Endo et al. 2010).

Both, the pomegranate peel and arils contain substantial amounts of anthocyanins which are the main source of color contributing molecules to pomegranate fruit and flowers. Anthocyanins play an important biological role in the recruitment of pollinators and seed dispersers. They may also have significant roles in the signaling between plants and microorganisms, plant defense mechanisms, auxin transport and UV protection (Winkel-Shirley 2001; Koes et al. 2005). Anthocyanins also play a major role in protecting plant tissues against oxidation processes. Several recent experiments demonstrated that anthocyanins have health beneficial values (Wallace 2001; de Pascual-Teresa and Sanchez-Ballesta 2008; Wang and Stoner 2008). Tomatoes with exceptionally high anthocyanin content were shown to have anticancer properties (Butelli et al. 2008). Afaq et al. (2005) found that pomegranate fruit extract contains anthocyanins, ellagitannins and hydrolyzable tannins and possesses strong antioxidant and antiinflammatory properties. Their results provided clear evidence that the extract possesses antiskin-tumor-promoting effects in CD-1 mice. Since the same extract was capable of inhibiting conventional as well as novel biomarkers of induced tumor promotion, the authors suggested that it may possess chemopreventive activity in a wide range of tumor models. The pomegranate fruit contains six major types of anthocyanins composed from cyanidin, pelargonidin and delphinidin in their mono or di-glycozidic forms (Du et al. 1975). Both, the fruit skin and the arils contain all six forms of anthocyanins although the ratio of the various anthocyanin derivatives is different between the peel and arils. The content and the relative amount of anthocyanins in the pomegranate fruit may vary as it develops. In two Israeli commercial cultivars, 'Wonderful' and 'Rosh-Hapered', changes in the major chemical composition in arils and peels during fruit maturation were determined. In both cultivars, the levels of total phenolics, antioxidant activity and hydrolysable tannins were reduced in the peels during maturation, while the anthocyanin level increased. The results also show that the sugar content in the aril juice increased in both accessions while the levels of acidity and of citric acid decreased (Schwartz et al. 2009a). Fawole and Opara (2013) investigated fruits of 'Bhagwa' and 'Ruby' at five distinct stages of maturity. Total soluble solids, pH, titratable acidity, phenolic concentrations, antioxidant capacity and volatile constituents were studied and the results show that major compositional changes in the fruit are developmentally regulated. Significant increases in total soluble solids, coupled with significant decline in titratable acidity and total phenolics occurred with advancing maturity. Fruit at advanced maturity stages were characterized by intense pigmentation of peel and aril, which coincided with maximum accumulation of anthocyanins. This demonstrated that the timing of fruit harvest could have an influence on bioactivity of the pomegranate fruit.

Variations in the Content of Phytochemicals Among Pomegranate Varieties

Data accumulating from several recent studies that compared the amount and activity of phytochemicals between different pomegranate varieties indicate that there is high variability among pomegranate varieties with respect to the content of their phytochemicals. The variability in phytochemicals, includes among others: Ascorbic acid (Vitamin C), polyphenols (anthocyanins, ellagitannins etc.), fatty acids, sugars and organic acids. Some of the analyses showed changes between varieties in antioxidant capacity, antimicrobial properties or other therapeutic properties. However these differences were, not attributed to a specific phytochemical, but to the type of pomegranate extract.

The relationships between antioxidant activity, total polyphenol content, total anthocyanins content, and the levels of four major hydrolyzable tannins were studied in four different extracts prepared from 29 different accessions. The results showed difference of up to two fold in the level of antioxidant activity, total polyphenol and juice color index in between the accessions and one order of magnitude difference among the extreme accessions with respect to total anthocyanin (Tzulker et al. 2007). In another study that compared juices of old Italian accessions with a commercial cultivar it was found that the polyphenolic content varied significantly among accessions (Calani et al. 2013). From the data reported the 'ME8' old accession was found being particularly rich in ellagic acid derivatives, while the juice of the commercial cultivar 'Dente di Cavallo' was particularly rich in anthocyanins. Differences among accessions in antioxidant capacity, attributed to polyphenolic content were also found in ten Moroccan pomegranate varieties (Legua et al. 2012). The results indicate that the variety was the most influencing factor conditioning pomegranate sugar and organic acids profiles, antioxidant activity, and total phenolics. Such differences were also found in Iranian varieties (Hajimahmoodi et al. 2008; Tehranifar et al. 2010) and Turkish varieties (Ozgen et al. 2008). When juices of 15 different varieties were compared in Italy for their phenolic content and antioxidant capacity 'Wonderful', 'Hicaz' and 'G2' were ranked as the best ones. Juices of these cultivars showed variations in their ability to protect HepG2 cells from oxidative damages (Di Nunzio et al. 2013).

A study that compared seven cultivars grown in South Africa with respect to their antibacterial capabilities showed that methanolic peel extracts had strong broad-spectrum activity against Gram-positive and Gram-negative bacteria (Minimum inhibitory concentrations 0.2–0.78 mg/ml). This study also showed significant differences in radical scavenging activities between the varieties tested (Fawole et al. 2012). Extracts from arils of six pomegranate varieties grown in Turkey were tested for their antimicrobial properties by the agar diffusion and minimum inhibitory concentration methods against seven bacteria and three fungi. The extracts had antimicrobial effect on all microorganisms, giving ranging sizes of inhibition zones (Duman et al. 2009). Variations in antibacterial effects were also found among five varieties of pomegranate (Opara et al. 2009). In this study fruit fractions showed differences in the inhibition zone of *Staphylococcus aureus* and *Pseudomonas aeruginosa*. Variations were also found in vitamin C content (52.8–72.0 mg/100 g fresh weight for arils and 76.8–118.4 mg/100 g fresh weight for peels). Tehranifar et al. (2010) also reported varietal differences in vitamin C content.

Health beneficial fatty acids were reported to be present particularly within the oil extracted from seeds (Lansky and Newman 2007). A comparative study that determined the level and composition of fatty acids in seeds of six pomegranate varieties revealed that the varieties differ in their oil content ranging between 51 and 152 mg kg-dry matter. Sour cultivars had the highest oil content, while sour-sweet cultivars had the lowest oil content. Inter-variety differences in fatty acid composition were demonstrated (Melgarejo et al. 1995). Fatty acid composition of pomegranate leaves was determined in six varieties in Turkey by using gas chromatography. Differences in the composition were found among varieties (Ercisli et al. 2007).

The above mentioned experiments demonstrate that although most pomegranate varieties have similar composition of phytochemicals, the amount and sometimes the composition of their phytochemicals may depend on the variety. This variability could have profound effects on the bioactivity of the pomegranate extracts and should be considered when the therapeutic capabilities of pomegranate are discussed. When comparing the content of phytochemicals one should take great care that all cultivars compared were grown under the same climate and agrotechniques.

The data clearly points to the major role of genetic variation and the importance of cultivars when assessing any medical or therapeutic activity to pomegranate.

Effects of Environmental Conditions on the Accumulation and Activity of Phytochemicals

In addition to intrinsic differences among varieties, other factors may influence the content of phytochemicals in pomegranate as well. Phytochemical content was found to be dependent on the tissue source. Thus, their content and composition in the flower, leaf or fruit was found to be very different (Lansky and Newman 2007). Substantial effects of the environment on the content and composition of phytochemical were discovered in some studies (Borochov-Neori et al. 2009, 2011). Such effects were discovered in a study that investigated the influence of a wide range of temperature regimes (~7–40 °C) on pomegranates' aril anthocyanins content and composition. The study included two deciduous and two evergreen accessions as well as desert and Mediterranean climate orchards. RP-HPLC analysis of the arils' anthocyanins revealed mono- and di-glucosylated delphinidins and cyanidins as the major anthocyanins and pelargonidins as minor components. Anthocyanin accumulation changed inversely to the season's temperatures. Cyanidins were generally more abundant but delphinidin accumulation was enhanced in cooler season. Monoglucosylated anthocyanins prevailed at cooler temperatures and subsided during seasonal warming with a concomitant increase in di-glucoside proportion (Borochov-Neori et al. 2011). In another study different sensory properties and harvest seasons were analyzed in three cultivars at various dates throughout the harvest season, corresponding to different climatic conditions during fruit ripening. Values obtained varied with cultivar and ripening date. Arils of fruit ripening later in the season contained more soluble phenolics and exhibited a higher antioxidant activity (Borochov-Neori et al. 2009). Yet, another study described the differences in the chemical composition of major ingredients of the arils and peels of 11 accessions grown in Mediterranean and desert climates in Israel. In most of the accessions, the levels of antioxidant activity and content of total phenolics, total anthocyanins, total soluble solids, glucose, fructose, and acidity were higher in the aril juice of fruit grown in the Mediterranean climate compared to those grown in the desert climate. However, the peels of fruit grown in the desert climate exhibited higher antioxidant activity, and the levels of total phenolics, including the two hydrolyzable tannins, punicalagin and punicalin, were higher compared to those in the peels of fruit grown in the Mediterranean climate (Schwartz et al. 2009b).

To explore water quality effect, fruit from trees irrigated with a wide range of salinities, (0.5–9 dS m^{-1}), were studied. Anthocyanins were analyzed by RP-HPLC and phenolics content was determined by the Folin-Ciocalteau assay. Results show that water salinity considerably affected the composition of anthocyanins in the arils. Increased salinity enhanced phenolics accumulation but reduced that of anthocyanins in a cultivar dependent manner (Borochov-Neori et al. 2013).

The results of those studies indicate that environmental conditions significantly affect pomegranate fruit quality and health beneficial compounds. The studies imply that pomegranate fruit antioxidant and sensory quality traits can be enhanced by the choice of cultivar, climate and cultivation management and those environmental conditions have substantial effects on anthocyanin content and composition. Thus, when speaking about phytochemical activity and health beneficial effects of pomegranate ingredients and probably in other plants species as well, it is very important to take care to refer to environmental parameters.

Identification of the Specific Health Beneficial Phytochemicals

One of the most intriguing questions concerning the bioactivity of pomegranate is the precise chemical identity of the bioactive substance in the plant tissue. The answer to this question is important for many aspects, particularly, for the ability to formulate an effective synthetic drug to specific diseases and to choose the appropriate varieties, that produce most effectively the desired phytochemical. Only very few recent experiments report some progress on the chemical nature of pomegranate bioactive components (Gasmi and Sanderson 2010; Viuda-Martos et al. 2010; Rocha et al. 2012; Wang et al. 2012). One of the biggest problems in this respect is most probably the fact that pomegranate derived materials need to act synergistically in order to have their maximal effects (Lansky et al. 2005b; Mertens-Talcott and Percival 2005; Jurenka 2008).

A prominent feature of pomegranate is its antioxidative ability. It is known today that most of the antioxidant activity is derived from the presence of very potent polyphenols such as ellgitannins, catechins and other gallic acid derivatives. In some of the tissues such as fruit membranes and peel, ellgitannins may constitute the majority of the metabolites present within these tissues. Since ellagic acid and its derivatives are such potent antioxidants, many commercially available pomegranate extracts were standardized according to the content of ellagic acid (Jurenka 2008). However, ellagic acid standardization could cause exclusion of other therapeutically important pomegranate constituents (Lansky 2006; Jurenka 2008). Due to the above mentioned complexities, combinatorial approaches were experimented were several key components of pomegranate juice were co-administrated. Such experiments included the addition of ellagic acid with flavonoids such as quercetin (Mertens-Talcott and Percival 2005; Jurenka 2008), which demonstrated that ellagic acid potentiates the effect of quercetin on p53 and MAP-kinase and that ellagic acid and quercetin synergistically interact with resveratrol in the induction of apoptosis (Mertens-Talcott and Percival 2005). Other experiments involved the usage of a combination of ellagic acid, caffeic acid, luteolin and punicic acid. The four pure chemicals, which represent chemical classes from the seeds, peel and arils, and known to possess anticancer activities, were tested as potential inhibitors of *in vitro* invasion of human PC-3 prostate cancer. All compounds significantly inhibited invasion when used individually and when caffeic acid, punicic, and luteolin were equally combined at the same gross dosage as when the compounds were tested individually, a supradditive inhibition of invasion was observed (Lansky et al. 2005a). Wang et al. (2012) studied the *in vitro* effect of the same four components mentioned above on prostate cancer cell growth, adhesion, migration, and chemotaxis. They showed that luteolin, ellagic acid and punicic acid, but not caffeic acid, inhibit growth of hormone-dependent and hormone-refractory prostate cancer cells and inhibit their migration and their chemotaxis toward stromal cell-derived factor 1α, a chemokine that is important in prostate cancer metastasis to the bone. These components also increase the expression of cell adhesion genes and decrease expression of genes involved in cell cycle control and cell migration. Furthermore, they increase several well-known tumor-suppression microRNAs, decrease several oncogenic miRNAs, and inhibit a chemokines receptor. The authors suggest that chemical modification of these components could further enhance their bioavailability and efficacy of treatment. They also suggest that these pomegranate components may also be effective in the treatment of metastasis of other cancers, because the mechanisms of metastasis are similar for most cancers (Wang et al. 2012). Luteolin, ellagic acid and punicic acid were also tested using breast cancer cell lines, MDA-MB-231 cells and MCF7, and the non-neoplastic cell line MCF10A. It was shown that, in addition to inhibiting growth of the breast cancer cells, these compounds increase cancer cell adhesion and decrease cancer cell migration but do not affect normal cells. These treatments also inhibit chemotaxis of the cancer cells to SDF1α, a chemokine that attracts breast cancer cells to the bone and pro-inflammatory cytokines/chemokines are significantly reduced (Rocha et al. 2012). These findings, indicating inhibitory of metastatic processes in breast cancer cells as well as in prostate cancer cells, suggest that luteolin, ellagic acid and punicic acid are a potentially effective treatment to prevent cancer progression in general.

Another study investigated the growth inhibitory, antiandrogenic, and pro-apoptotic effects of increasing concentrations of 13 pure compounds found in the pomegranate in androgen-dependent LNCaP human prostate cancer cells. Four compounds, epigallocatechin gallate, delphinidin chloride, kaempferol, and punicic acid, were found to inhibit cell growth at concentrations of 10 μM and above. Apoptosis contributes to the observed decrease in cell growth, as epigallocatechin gallate, kaempferol, and, in particular, punicic acid, induced DNA fragmentation after a 24 h treatment. Punicic acid was further found to induce intrinsic apoptosis via a caspase-dependent pathway (Gasmi and Sanderson 2010).

Although, all of the data presented in this review is based on *in vitro* studies, it suggests that pomegranate defined phytochemicals or their synthetic analogs could potentially function as therapeutic agents.

Concluding Remarks

Pomegranate was suggested to have profound health beneficial effects on a wide range of serious human and animal diseases. These data encourage the onset of many experiments aimed to understand the molecular nature of the phytochemicals involved in the pomegranate health promoting activity and the molecular mechanisms of their action. In this review an emphasis was made on the profound effects of pomegranate genetics and environmental conditions on the content of pomegranate phytochemicals and their consequent potential effects on the therapeutic activity of pomegranate. So far, very little is known about the biology and biochemistry of pomegranate and an integrated effort will be required among the various countries that grow and culture pomegranate in order to exploit the maximum potential of the pomegranate tree.

References

Adhami VM, Khan N, Mukhtar H (2009) Cancer chemoprevention by pomegranate: laboratory and clinical evidence. Nutr Cancer 61:811–815

Afaq F, Saleem M, Krueger CG, Reed JD, Mukhtar H (2005) Anthocyanin- and hydrolyzable tannin-rich pomegranate fruit extract modulates MAPK and NF-kappaB pathways and inhibits skin tumorigenesis in CD-1 mice. Int J Cancer 113:423–433

Aviram M, Rosenblat M (2012) Pomegranate protection against cardiovascular diseases. Evid Based Complement Alternat Med 2012:382763

Aviram M, Dornfeld L, Rosenblat M, Volkova N, Kaplan M, Coleman R, Hayek T, Presser D, Fuhrman B (2000) Pomegranate juice consumption reduces oxidative stress, atherogenic modifications to LDL, and platelet aggregation: studies in humans and in atherosclerotic apolipoprotein E-deficient mice 1, 2. Am J Clin Nutr 71:1062–1076

Aviram M, Dornfeld L, Kaplan M, Coleman R, Gaitini D, Nitecki S, Hofman A, Rosenblat M, Volkova N, Presser D, Attias J, Hayek T, Fuhrman B (2002) Pomegranate juice flavonoids inhibit low-density lipoprotein oxidation and cardiovascular diseases: studies in atherosclerotic mice and in humans. Drugs Exp Clin Res 28(2–3):49–62

Balfour IB (1888) Punica. In: Balfour IB (ed) Botany of Socotra. Robert Grant & Sons, Edinburgh, pp 93–96

Basu A, Penugonda K (2009) Pomegranate juice: a heart-healthy fruit juice. Nutr Rev 67:49–56

Borochov-Neori H, Judeinstein S, Tripler E, Harari M, Greenberg A, Shomer I, Holland D (2009) Seasonal and cultivar variations in antioxidant and sensory quality of pomegranate (*Punica granatum* L.) fruit. J Food Compos Anal 22:189–195

Borochov-Neori H, Judeinstein S, Harari M, Bar-Ya'akov I, Patil BS, Lurie S, Holland D (2011) Climate effects on anthocyanin accumulation and composition in the pomegranate (*Punica granatum* L.) fruit arils. J Agric Food Chem 59:5325–5334

Borochov-Neori H, Lazarovitch N, Judeinstein S, Patil BS, Holland D (2013) Climate and salinity effects on color and health promoting properties in the pomegranate (*Punica granatum* L.) fruit arils. In: Tropical and subtropical fruits: flavors, color, and health benefits. ACS symposium series, 1129, American Chemical Society, Washington, DC, pp 43–61

Butelli E, Titta L, Giorgio M, Mock HP, Matros A, Peterek S, Schijlen EGWM, Hall RD, Bovy AG, Luo J, Martin C (2008) Enrichment of tomato fruit with health-promoting anthocyanins by expression of select transcription factors. Nat Biotechnol 26:1301–1308

Calani L, Beghè D, Mena P, Del Rio D, Bruni R, Fabbri A, Dall'asta C, Galaverna G (2013) Ultra-HPLC-MS(n) (Poly)phenolic profiling and chemometric analysis of juices from ancient *Punica granatum* L. cultivars: a nontargeted approach. J Agric Food Chem 61:5600–5609

Clinical Trials.gov (2013) Pomegranate studies. The U.S. National Institutes of Health. http://clinicaltrials.gov/ct2/results?term=pomegranate&Search=Search. Accessed 22 Dec 2013

de Pascual-Teresa S, Sanchez-Ballesta MT (2008) Anthocyanins: from plant to health. Phytochem Rev 7:281–299

Di Nunzio M, Toselli M, Verardo V, Caboni MF, Bordoni A (2013) Counteraction of oxidative damage by pomegranate juice: influence of the cultivar. J Sci Food Agric 93:3565–3573

Du CT, Wang PL, Francus FJ (1975) Anthocyanins of pomegranate, *Punica granatum*. J Food Sci 40:417–418

Duman AD, Ozgen M, Dayisoylu KS, Erbil N, Durgac C (2009) Antimicrobial activity of six pomegranate (*Punica granatum* L.) varieties and their relation to some of their pomological and phytonutrient characteristics. Molecules 14:1808–1817

Endo EH, Cortez DA, Ueda-Nakamura T, Nakamura CV, Dias Filho BP (2010) Potent antifungal activity of extracts and pure compound isolated from pomegranate peels and synergism with fluconazole against *Candida albicans*. Res Microbiol 161:534–540

Ercisli S, Agar G, Orhan E, Yildirim N, Hizarci Y (2007) Interspecific variability of RAPD and fatty acid composition of some pomegranate cultivars (*Punica granatum* L.) growing in Southern Anatolia region in Turkey. Biochem Syst Ecol 35:764–769

Ercisli S, Gadze J, Agar G, Yildirim N, Hizarci Y (2011a) Genetic relationships among wild pomegranate (*Punica granatum*) genotypes from Coruh Valley in Turkey. Genet Mol Res 10:459–464

Ercisli S, Kafkas E, Orhan E, Kafkas S, Dogan Y, Estiken A (2011b) Genetic characterization of pomegranate (*Punica granatum* L.) genotypes by AFLP markers. Biol Res 44:345–350

Esmaillzadeh A, Tahbaz F, Gaieni I, Alavi-Majd H, Azadbakht L (2006) Cholesterol-lowering effect of concentrated pomegranate juice consumption in type II diabetic patients with hyperlipidemia. Int J Vitam Nutr Res 76:147–151

Fahan A (1976) The leaf. The flower. The seed. In: Plant anatomy. Hakkibutz Hameuhad Publishing House Ltd, Jerusalem, pp 171–212, 321–394, 419–430

Fawole OA, Opara UL (2013) Seasonal variation in chemical composition, aroma volatiles and antioxidant capacity of pomegranate during fruit development. Afr J Biotechnol 12:4006–4019

Fawole OA, Makunga NP, Opara UL (2012) Antibacterial, antioxidant and tyrosinase-inhibition activities of pomegranate fruit peel methanolic extract. BMC Complement Altern Med 12:200

Fuhrman B, Aviram M (2007) Pomegranate and cardiovascular diseases: pomegranate juice polyphenolic antioxidants protect against oxidative stress and atherosclerosis development. Acta Hortic 744:205–216

Gasmi J, Sanderson JT (2010) Growth inhibitory, antiandrogenic, and pro-apoptotic effects of punicic acid in LNCaP human prostate cancer cells. J Agric Food Chem 58:12149–12156

Glazer I, Masaphy S, Marciano P, Bar-Ilan I, Holland D, Kerem Z, Amir R (2012) Partial identification of antifungal compounds from *Punica granatum* peel extracts. J Agric Food Chem 60:4841–4848

Goor A, Liberman J (1956) The pomegranate. State of Israel, Ministry of Agriculture, Agricultural Publishing Section, Tel-Aviv

Hajiahmadi Z, Talebi M, Sayed-Tabatabaei BE (2013) Studying genetic variability of pomegranate (*Punica granatum* L.) based on chloroplast DNA and barcode genes. Mol Biotechnol 55:249–259

Hajimahmoodi M, Oveisi MR, Sadeghi N, Jannat B, Hadjibabaie M, Farahani E, Akrami MR, Namdar R (2008) Antioxidant properties of peel and pulp hydro extract in ten Persian pomegranate cultivars. Pak J Biol Sci 11:1600–1604

Holland D, Hatib K, Bar-Ya'akov I (2009) Pomegranate: botany, horticulture, breeding. In: Janick J (ed) Horticultural reviews, vol 35(2). Wiley, Hoboken, pp 127–191

Jayaprakasha GK, Negi PS, Jena BS (2006) Antimicrobial activities of pomegranate from pomegranate ancient roots to modern science. In: Seeram NP, Schulman RN, Heber D (eds) Pomegranates: ancient roots to modern medicine. CRC Press, New York, pp 167–183

Jbir R, Hasnaoui N, Mars M, Marrakchi M, Trifi M (2008) Characterization of Tunisian pomegranate (*Punica granatum* L.) cultivars using amplified fragment length polymorphism analysis. Sci Hortic 115:231–237

Jurenka JS (2008) Therapeutic applications of pomegranate (*Punica granatum* L.): a review. Altern Med Rev 13(2):128–144

Katz SR, Newman RA, Lansky EP (2007) *Punica granatum*: heuristic treatment for diabetes mellitus. J Med Food 10:213–217

Koes R, Verweij W, Quattrocchio F (2005) Flavonoids: a colourful model for the regulation and evolution of biochemical pathways. Trends Plant Sci 10:236–242
Lansky EP (2006) Beware of pomegranates bearing 40 % ellagic acid. J Med Food 9:119–122
Lansky EP, Newman RA (2007) *Punica granatum* (pomegranate) and its potential for prevention and treatment of inflammation and cancer. J Ethnopharmacol 109:177–206
Lansky EP, Harrison G, Froom P, Jiang WG (2005a) Pomegranate (*Punica granatum*) pure chemicals show possible synergistic inhibition of human PC-3 prostate cancer cell invasion across Matrigel. Invest New Drugs 23:121–122
Lansky EP, Jiang W, Mo H, Bravo L, Froom P, Yu W, Harris NM, Neeman I, Campbell MJ (2005b) Possible synergistic prostate cancer suppression by anatomically discrete pomegranate fractions. Invest New Drugs 23:11–20
Legua P, Melgarejo P, Abdelmajid H, Martínez JJ, Martínez R, Ilham H, Hafida H, Hernández F (2012) Total phenols and antioxidant capacity in 10 Moroccan pomegranate varieties. J Food Sci 77:C115–C120
Levin GM (1981) Wild pomegranate (*Punica granatum* L.) in Turkmenistan. Izvestiia Akad Nauk Turkm SSR Ser Biol Nauk 2:60–64
Levin GM (2006) In: Baer BL (ed) Pomegranate roads: a Soviet botanist's exile from Eden. Floreat Press, Forestville
Malik A, Mukhtar H (2006) Prostate cancer prevention through pomegranate fruit. Cell Cycle 5:371–373
Mars M (2000) Pomegranate plant material: genetic resources and breeding, a review. Options Méditerr Sér Sémin Méditer 42:55–62
Melgarejo P, Salazar DM, Amoros A, Artes F (1995) Total lipids content and fatty acid composition of seed oils from six pomegranate cultivars. J Sci Food Agric 69:253–256
Melgarejo P, Martcnez JJ, HerncLndez F, Martcnez R, Legua P, Oncinab R, Martínez-Murcia A (2009) Cultivar identification using 18S-28S rDNA intergenic spacer-RFLP in pomegranate (*Punica granatum* L.). Sci Hortic 120:500–503
Mertens-Talcott SU, Percival SS (2005) Ellagic acid and quercetin interact synergistically with resveratrol in the induction of apoptosis and cause transient cell cycle arrest in human leukemia cells. Cancer Lett 218:141–151
Morton J (1987) Pomegranate. In: Fruits of warm climates. Julia F Morton, Miami, pp 352–355
Navindra P, Seeram NP, Zhang Y, Reed JD, Krueger CG, Vaya J (2006) Pomegranate phytochemicals. In: Seeram NP, Schulman RN, Heber D (eds) Pomegranates: ancient roots to modern medicine. CRC Press, New York, pp 3–29
Opara LU, Al-Ani MR, Al-Shuaibi YS (2009) Physico-chemical properties, vitamin C content, and antimicrobial properties of pomegranate fruit (*Punica granatum* L.). Food Bioprocess Technol 2:315–321
Ozgen M, Durgaç C, Serçe S, Kaya C (2008) Chemical and antioxidant properties of pomegranate cultivars grown in the Mediterranean region of Turkey. Food Chem 111:703–706
Paller CJ, Ye X, Wozniak PJ, Gillespie BK, Sieber PR, Greengold RH, Stockton BR, Hertzman BL, Efros MD, Roper RP, Liker HR, Carducci MA (2012) A randomized phase II study of pomegranate extract for men with rising PSA following initial therapy for localized prostate cancer. Prostate Cancer Prostatic Dis 16:50–55
Rana JC, Pradheep K, Verma V (2007) Naturally occurring wild relatives of temperate fruits in Western Himalayan region of India: an analysis. Biodivers Conserv 16:3963–3991
Rocha A, Wang L, Penichet M, Martins-Green M (2012) Pomegranate juice and specific components inhibit cell and molecular processes critical for metastasis of breast cancer. Breast Cancer Res Treat 136:647–658
Rosenblat M, Hayek T, Aviram M (2006) Anti-oxidative effects of pomegranate juice (PJ) consumption by diabetic patients on serum and on macrophages. Atherosclerosis 187:363–371
Schwartz E, Glazer I, Bar-Ya'akov I, Matityahu I, Bar-Ilan I, Holland D, Amir R (2009a) Changes in chemical constituents during the maturation and ripening of two commercially important pomegranate accessions. Food Chem 115:965–973

Schwartz E, Tzulker R, Glazer I, Bar-Ya'akov I, Wiesman Z, Tripler E, Bar-Ilan I, Fromm H, Borochov-Neori H, Holland D, Amir R (2009b) Environmental conditions affect the color, taste, and antioxidant capacity of 11 pomegranate accessions' fruits. J Agric Food Chem 57:9197–9209

Seeram NP, Schulman RN, Heber D (2006) Pomegranates: ancient roots to modern medicine. CRC Press, Taylor & Francis Group, Boca Raton

Shilkina IA (1973) On the xylem anatomy of the genus *Punica* L. Bot Zhurnal 58:1628–1630

Soriano JM, Zuriaga E, Rubio P, LlcLcer G, Infante R, Badenes M (2011) Development and characterization of microsatellite markers in pomegranate (*Punica granatum* L.). Mol Breed 27:119–128

Still DW (2006) Pomegranate: a botanical perspective. In: Seeram NP, Schulman RN, Heber D (eds) Pomegranates: ancient roots to modern medicine. CRC Press, New York, pp 199–210

Stowe CB (2011) The effects of pomegranate juice consumption on blood pressure and cardiovascular health. Complement Ther Clin Pract 17:113–115

Tehranifar A, Zarei M, Esfandiyari B, Nemati Z (2010) Physicochemical properties and antioxidant activities of pomegranate fruit (*Punica granatum*) of different cultivars grown in Iran. Hortic Environ Biotechnol 51:573–579

Teixeira da Silva JA, Ranac TS, Narzaryd D, Vermae N, Meshramf DT, Ranadeg SH (2013) Pomegranate biology and biotechnology: a review. Sci Hortic 160:85–107

Tzulker R, Glazer I, Bar-Ilan I, Holland D, Aviram M, Amir R (2007) Antioxidant activity, polyphenol content, and related compounds in different fruit juices and homogenates prepared from 29 different pomegranate accessions. J Agric Food Chem 55:9559–9570

Verma N, Mohanty A, Lal A (2010) Pomegranate genetic resources and germplasm conservation: a review. Fruit Veg Cereal Sci Biotechnol 4:120–125

Viuda-Martos M, Fernández-López J, Pérez-Álvarez JA (2010) Pomegranate and its many functional components as related to human health: a review. Compr Rev Food Sci Food Saf 9:635–654

Wallace TC (2001) Anthocyanins in cardiovascular disease. Adv Nutr 2:1–7

Wang LS, Stoner GD (2008) Anthocyanins and their role in cancer prevention. Cancer Lett 269:281–290

Wang L, Ho J, Glackin C, Martins-Green M (2012) Specific pomegranate juice components as potential inhibitors of prostate cancer metastasis. Transl Oncol 5:344–355

Winkel-Shirley B (2001) Flavonoid biosynthesis: a colorful model for genetics, biochemistry, cell biology, and biotechnology. Plant Physiol 126:485–493

Yuan Z, Yin Y, Qu J, Zhu L, Li Y (2007) Population genetic diversity in Chinese pomegranate (*Punica granatum* L.) cultivars revealed by fluorescent-AFLP markers. J Genet Genomics 34:1061–1071

Zamani Z, Sarkhosh A, Fatahi R, Ebadi A (2007) Genetic relationships among pomegranate genotypes studied by fruit characteristics and RAPD markers. J Hortic Sci Biotechnol 82:11–18

Zamani Z, Abadi M, Khadivi-Khub A (2013) Comparative analysis of genetic structure and variability in wild and cultivated pomegranates as revealed by morphological variables and molecular markers. Plant Syst Evol 299:1967–1980

Zukovskii PM (1950) Punica. In: Cultivated plants and their wild relatives. State Publishing House Soviet Science, Moscow, pp 60–61

Juniperus phoenicea from Jordan

Mohammad Sanad Abu-Darwish, Célia Cabral, and Lígia Salgueiro

Abstract *Juniperus phoenicea* L. is a widespread species throughout the Mediterranean Region. In Jordan, it is well known by its uses in traditional medicine, some of them already validated scientifically. This review aims to compile the information available in the literature about this species in Jordan. A state of knowledge of the taxonomy and ecology, as well as, the chemical composition of essential oils and other extracts, uses in traditional medicine, biological activities and cytotoxicity are referred under the perspective of its potential industrial use.

Keywords *Juniperus phoenicea* • Leaves • Berries • Jordan • Forest type • Essential oil • Bioactive compounds • Monoterpene • α-pinene • Traditional use • Pharmacology • Red list

The genus *Juniperus* L. belong to the family Cupressaceae and it is the second most diverse genus of the Conifers, with 67 species and 34 varieties (using the more widely accepted variety category instead of the subspecies category). All the taxa are restricted to the Northern Hemisphere, except *J. procera* Hochst. *ex* Endl., which also grows along the rift mountains in East Africa into the Southern Hemisphere (Adams 2008). The genus is divided into three sections: *Juniperus* (10 species), *Caryocedrus* Endlicher (1 species) and *Sabina* (Miller) Spach (56 species) (Adams 2008).

Juniperus phoenicea L. (section *Sabina*), commonly known as Phoenician juniper (English), phönizischer Wacholder (German), Genévrier rouge (French), araâr (Arabic), are evergreen monoecious or dioecious small trees or shrubs, with fleshy

M.S. Abu-Darwish (✉)
Department of Basic and Applied Sciences, Shouback University College, Al-Balqa Applied University, Al-Shouback 71911, Jordan
e-mail: maa973@yahoo.com

C. Cabral • L. Salgueiro
Faculty of Pharmacy, Center for Pharmaceutical Studies and Center for Neuroscience and Cell Biology, University of Coimbra, Azinhaga de Santa Comba, 3000-548 Coimbra, Portugal
e-mail: celiacabral@ff.uc.pt

Z. Yaniv, N. Dudai (eds.), *Medicinal and Aromatic Plants of the Middle-East*, Medicinal and Aromatic Plants of the World 2, DOI 10.1007/978-94-017-9276-9_13

Pictures 1 and 2 *Juniperus phoenicea* in Jordan (Taken by author)

seed cones in which the cone scales are fused resembling 'berries'. These seed cones are fully ripe only in the second year, globose to ovoid, blackish when very young, later green or yellowish and slightly pruinose, dark red when ripe (Adams 2008). *J. phoenicea* is divided in two subspecies: *J. phoenicea* subsp. *phoenicea* and *J. phoenicea* subsp. *turbinata* (Guss.) Nyman, the first characterized by globose female cones and obtuse or subacute scale-leaves, seed cones globose with a wider distribution; and the latter with seed cones turbinate and a more restrict distribution (Franco 1986; Meloni et al. 2006).

This work is focused on *J. phoenicea* subsp. *phoenicea*, the subspecies that occurs in Jordan, so throughout the text it will be named *J. phoenicea.*

In 2012, Al-Ramamneh et al. reported that *J. phoenicea* trees growing in Jordan produce small cones (mean cone weight equals 0.25 g), each containing 5 small seeds. The cones are green-colored and become reddish when mature. The shape of seeds in *J. phoenicea* is ovoid. However, seeds tend sometimes to be tapered at both ends. They are wrinkled and characterized by the presence of protrusions and furrow-like grooves along their surface (Pictures 1 and 2).

Geographic Distribution

Juniperus phoenicea subsp. *phoenicea* is distributed throughout the Mediterranean Region: Albania, Algeria, Andorra, Bosnia-Herzegovina, Bulgaria, Croatia, Cyprus, Egypt (Sinai), France, Greece, Italy, Jordan, Lebanon, Libya, Morocco, Portugal,

Romania, Spain, Tunisia, Turkey; Saudi Arabia (along Red Sea). It also occurs in Macaronesia: Canary Islands and Madeira Arquipelago (Adams et al. 2002). *Juniperus phoenicea* subsp. *phoenicea* is perhaps the most widespread and common juniper in the Mediterranean and therefore assessed as Least Concern by the IUCN Red List of threatened species (IUCN 2013a).

Juniperus phoenicea subsp. *turbinata* is widespread but scattered, and limited to the Mediterranean coast, where it often occurs in or close to urbanized coastal strips, where much of the habitat has been destroyed irreversibly. Rates of decline are not known, and it is uncertain if it has stopped, but it has probably slowed down. The area of occupancy is quite small, and certainly under 2,000 km^2 so it would qualify for Vulnerable if decline was continuing. It is therefore assessed as Near Threatened until more information is available as it almost qualifies for a threatened listing under criterion B2ab (iii,v) (IUCN 2013a).

In Jordan, *J. phoenicea*, is distributed in forest type throughout south mountain regions and characterizes the vegetation in Tafileh and Shoubak woodlands at high altitudes of 1,200–1,700 m (Al-Eisawi 1996; Al-Qura'n 2008, 2009). Al-Eisawi (1996) showed that *J. phoenicea* forest formation extends from Tafileh to Ras An-Naqab, where the best stand of this forest in Al-Mgarieh and Al-Mansoura in Shoubak and Bsiarah near Al-Rashadyiah in Tafileh. However, this stand now is known as Dana conservation area. Moreover, Bender (1974) and Karim and Al-Quran (1990), speculated that Ajlun and Tafileh mountains may have been the center of origin for *Juniperus* species from which they migrated to the lower peripheral regions south or west or even to the north through many ways especially via cap fruits and pollen grains.

Habitat and Ecology

J. phoenicea grows in areas characterized by persistent drought and arid climate with high temperature ranges (El-Bana et al. 2010). After the IUCN Red List of threatened species, *J. phoenicea* occurs in garrigue, maquis, or evergreen microphyllous woodland on dry, stony ground, limestone outcrops, or sand dunes at altitudes between 1 m and 2,400 m. This species is commonly associated with *Pinus halepensis, P. brutia, Quercus ilex, Pistacea lentiscus, Cistus* spp., *Olea europaea, Lavandula* spp., *Artemisia herba-alba*, and numerous other genera. There is a predominance of limestone, but granitic rock, sandstone, serpentine, volcanic rock, as well as sand dunes are also mentioned as substrates. Soils are usually rocky or skeletal and it can grow well from crevices in bedrock. The climate is Mediterranean, with dry and hot summers.

In Jordan, it grows at high altitudes, over 1,000 m, on sandy rocks (Al-Qura'n 2005). In 2012, Abu-Darwish and Ofir reported that *J. phoenicea* grows in soils with moderately basic pH (7.7–7.9). The samples of soil tested in which *J. phoenicea* grows, were collected from two locations in South of Jordan (Dana and Al-Hisha regions). The electrical conductivity values of the studied soils in the two locations

Table 1 Locality, climate conditions, altitudes and latitudes of two different geographical locations of Jordan where *J. phoenicea* grows

Growth location	Altitude (m)	Ambient temp. (Min-Max)	Mean of rainfall (mm)	Relative humidity (%)	Latitude (degrees)
Dana	1,260	11.8–23.4	237.6	23.30	30.4030° N
Al-Hisha	1,365	4.11–19.9	294.2	24.80	30.3100° N

were different, ranging from 1.06 mS/cm in the upper layer to 1.62 mS/cm in the lower soil layer in Dana and increased from 1.76 to 3.70 mS/cm in Al-Hisha region. Moreover, the sieve and hydrometer analysis of soil samples showed that the grain size particle was composed of 67.6–74.4 % sand, 3.3–3.4 % clay and 22.3–29.0 % silt in Dana region which represented a sandy loam texture. The soil samples from *Al-Hisha* region contained 31.1–31.4 % sand, 28.6–30.9 % clay and 38.0–39.9 % slit which represented a clay loam texture. The locality, climate conditions, altitudes and latitudes of two different geographical locations of Jordan where *J. phoenicea* grows are shown in Table 1.

Collection Practices

To collect the leaves and ripe berries of *J. phoenicea* to be used as crude drugs, the guidelines on good agricultural and collection practices (GACP) for medicinal plants should be followed (WHO 2003). A list of physicochemical and biological contaminants should be avoided and controlled through quality assurance measures of GACP for *J. phoenicea* and other medicinal plants. The content of heavy, organic and radioactive metals, mycotoxins and endotoxins, microbiological and agrochemical contaminants in the berries and leaves of *J. phoenicea* should be controlled and they must not exceed their limits estimated by WHO. However, as WHO recommended for all medicinal plants, also the used parts of *J. phoenicea* should not be collected in or near areas where high levels of pesticides or other possible contaminants are used or found, such as roadsides, drainage ditches, mine tailings, garbage dumps and industrial facilities which may produce toxic emissions.

After collection, the plant material should be subjected to appropriate preliminary processing, including elimination of undesirable materials and contaminants, washing (to remove excess soil), sorting and cutting. The plant material should be protected from insects, rodents, birds and other pests, and from livestock and domestic animals.

There are no conclusive data showing the best time in the year or the phenological stage to collect the used parts of *J. phoenicea*. The essential oils are the main secondary metabolites founded in the leaves and ripe berries of *J. phoenicea*. Ennajar et al. (2009a) recommended that to obtain higher essential oil yield, the leaves of *J. phoenicea* should dry in the sun and berries should be oven-dried. According to the authors, to obtain higher percentage of some special components such as α-pinene and δ-3-carene shade-drying was more suitable.

Chemical Composition of Essential Oils

The leaves and berries of *J. phoenicea* are a source of essential oils which are responsible to several biological activities. In some countries, in addition to leaves and berries, also the wood is used to obtain the essential oil from *J. phoenicea* (Chalchat et al. 1990).

Beside the essential oils obtained from leaves and berries of *J. phoenicea,* this species is a source of other biologically active primary and secondary metabolites such as lipids, minerals and phenolic compounds.

The chemical composition of the essential oils of the different parts of *J. phoenicea* from different provenances has been subject of various publications. The essential oil of leaves of *J. phoenicea* from Algeria was described by Mazari et al. (2010), from Corsica by Rezzi et al. (2001), from Egypt by Afifi et al. (1992) and El-Sawi et al. (2007), from Greece and Spain by Adams et al. (1996), from Italy by Angioni et al. (2003), from Morocco by Barrero et al. (2006), Achak et al. (2009a, b) and Derwich et al. (2010), from Tunisia by Bouzouita et al. (2008) and Ennajar et al. (2009a, b), and from Saudi Arabia by Boulos (1999). Also, there are some studies concerning the chemical composition of essential oil extracted from the berries of *J. phoenicea* originated from various locations of the world (Delitala 1980; Ramic and Murko 1983; Angioni et al. 2003; Cosentino et al. 2003; El-Sawi et al. 2007).

Essential oils from leaves of *J. phoenicea* have a wide chemical variation, however, most of the works published report predominance of monoterpene hydrocarbons, with α-pinene as major constituent, followed by oxygenated monoterpenes, such as, α-terpinyl acetate, δ-3-carene, myrcene, α-phellandrene and ß-phellandrene (Afifi et al. 1992; Rezzi et al. 2001; Angioni et al. 2003; El-Sawi et al. 2007; Ennajar et al. 2009a, b; Mazari et al. 2010) (Fig. 1).

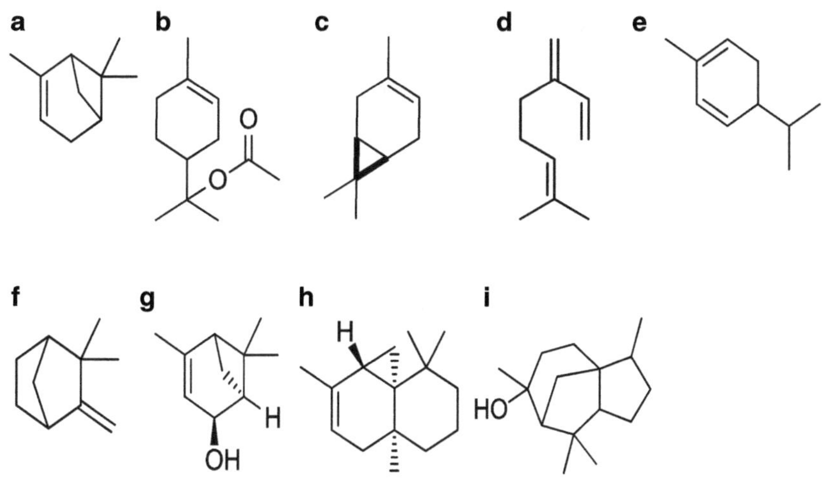

Fig. 1 Main monoterpenes and sesquiterpenes found in essential oils of *J. phoenicea.* (**a**) α-pinene; (**b**) α-terpinyl acetate; (**c**) δ-3-carene; (**d**) myrcene; (**e**) α-phellandrene; (**f**) camphene; (**g**) *trans*-verbenol; (**h**) thujopsene; (**i**) cedrol

Independently of the origin, the essential oils of *J. phoenicea* berries are composed mainly by monoterpenoids, followed by sesquiterpenoids, being characterized by a high content of α-pinene. Camphene, δ-3-carene and *trans*-verbenol were also detected in variable amounts in *J. phoenicea* berries oils (Delitala 1980; Lawrence 1989; Rezzi et al. 2001) (Fig. 1). The wood oil contains thujopsene and cedrol as major components (Adams 1991) (Fig. 1).

In 2013, Abu-Darwish et al. reported that berries essential oils of *J. phoenicea* originated from South of Jordan were compounded by 29 compounds in a total of 98.0 and 98.7 %. The oils were characterized by a high percentage of monoterpene hydrocarbons (82.6 and 90.5 %), being α-pinene the main compound (75.5 and 84.2 %). This group of compounds is followed by the oxygen containing monoterpenes (9.5 and 5.0 %). Among the sesquiterpenes, the hydrocarbons (2.8–1.2 %) were detected in lower concentration than the oxygenated ones (3.1–2.0 %).

Berries essential oils from Jordan *J. phoenicea* presents very high amount of α-pinene, what distinguish them from previous studies from other countries where α-pinene is also the main compound, however, in lower percentages (±40 %) (El-Sawi et al. 2007). Only the berries essential oil from Tunisia (Ennajar et al. 2009a, b), with 86.4 % of α-pinene maximum content is similar to our results. These results point the essential oil obtained from Jordan *J. phoenicea* berries as a good natural source of α-pinene (Abu-Darwish et al. 2013).

Chemical Composition of Other Extracts

To our knowledge, there are no reports about the chemical characterization of other secondary metabolites in *J. phoenicea* from Jordan, but there are some works from other countries. Aboul-Ela et al. (2005a, b) detected flavonoids, fatty acids, sterols and hydrocarbons from *J. phoenicea* grown in Egypt. Moreover, myricitrin, quercetin, cosmosin and quercitrin were among the detected flavonoids, which also were been isolated from the leaves of Egyptian *J. phoenicea* (Ali et al. 2010) .

The two lignans: desoxypodophyllotoxin and β-peltatin A methyl ether were detected by Cairnes et al. (1980) (Fig. 2).

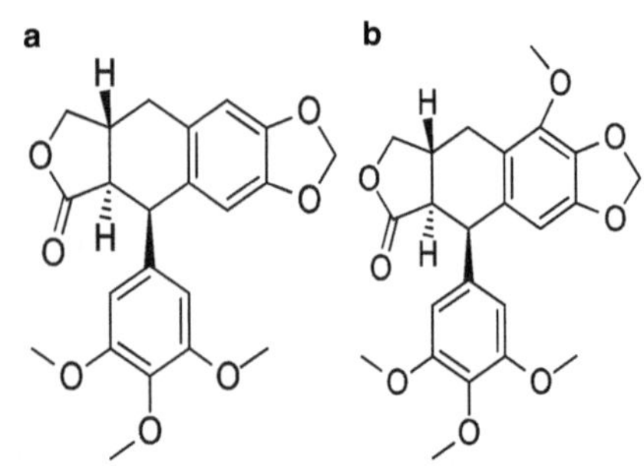

Fig. 2 Main lignans found in *J. phoenicea*. (**a**) desoxypodophyllotoxin; (**b**) β-peltatin A methyl ether

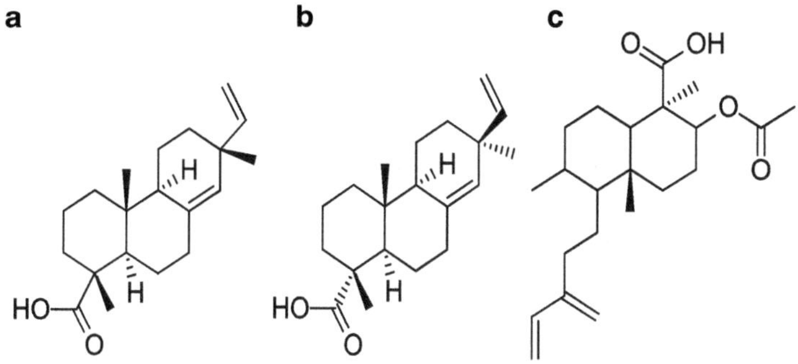

Fig. 3 Main diterpenes found in *J. phoenicea.* (**a**) sandaracopimaric acid; (**b**) pimaric acid; (**c**) juniperexcelsic acid

In *J. phoenicea* grown in France, furanon glycosides and phenylpropanoids were detected by Comte et al. (1996, 1997). Barrero et al. (2004) isolated isovalerate methyl ester derivatives of diterpenic acids, derivatives of p-methoxycinnamyl alcohol and linalool.

El-Sawi and Motawae (2008) isolated diterpenes from the berries of Egyptian *J. phoenicea*, which were identified as: sandaracopimaric acid, pimaric acid, juniperexcelsic acid, 3α-hydroxy-labda-8(17), 13(16)-14-triene-19-oic acid (isolated for the first time), 4-epi-abietic acid, 4-epi-abietol and 3β, 12-dihydroxyabieta-8,11,13-triene-1-one (Fig. 3).

J. phoenicea leaves and berries also contains other natural active compounds. Nasri et al. (2011) reported that the berries of Tunisian *J. phoenicea* contain 11 % of lipids, and minerals such as Na (63.8 mg per 100 gDW) or K (373.9 mg per 100 g DW) were also detected.

The trace elements content in berries of *J. phoenicea* grown in two different locations in South of Jordan (Dana area and Al-Hisha) were studied by Abu-Darwish and Ofir (2012). The authors found that heavy metal content was in variable concentrations in the order: Fe < Cu < Zn < Mn < Pb and Cd.

Cu concentration in *J. phoenicea* berries varied from 31.51 ppm in Dana to 71.86 ppm in AL-Hisha. It was higher than the maximum normal limits in plants (2.0–20 ppm). The concentrations of Fe, Zn, and Mn were very low and did not exceed their typical amount in non polluted plants. Toxic heavy metals Pb and Cd were not detected in the berries of *J. phoenicea.*

Uses in Traditional Medicine

In Jordanian and other traditional medicines, the leaves and berries of *J. phoenicea* are used individually or as mixed raw material to prepare infusions and other water extracts. Juniper has been used for centuries as a steam inhalant for bronchitis, and to control arthritis. In the German Pharmacopoeia, Juniper berry tea is listed as a digestive aid, both stimulating appetite as well as relieving flatulence. The oils also help increase the flow of digestive fluids, improve digestion and eliminate gas and stomach cramping (EL-Sawi et al. 2007). The Moroccan ethno pharmacological practice indicate powder leaves of *J. phoenicea* as diuretic and hypoglycemic agent (Bellakhder 1997). In Tunisian folk medicine, the leaves and berries are widely used in the treatment of diarrhea, rheumatism, acute gonococcal infection, eczema, dysmenorrhea and sunstroke. The water decoction of the leaves is used as bath for treatment of rheumatism, while the infusion of 30 g of *J. phoenicea* leaves in a liter of water for 15 min, is administrated orally in the doses of two glasses per day before meals (IUCN 2013b).

In Jordan, leaves and berries of *J. phoenicea* are used for therapy of diabetes mellitus (Hamdan and Afifi 2004), edema, and urinary tract diseases (Shkukani et al. 2008). In South Jordan, the decoction of *J. phoenicea* leaves is recommended in the treatment of rheumatism (Al-Qura'n 2009). Also, the Jordanian folk medicine recommended the use of aqueous extracts from leaves in the treatment of diarrhea and gout (Qnais et al. 2005).

The berries are probably better known as the unique flavoring agent of gin, an important component of the dry martini.

Biological Activities and Cytotoxicity

Due to the content in essential oils and other secondary metabolites by the leaves and berries of *J. phoenicea,* they exhibit various pharmacological activities. The essential oils isolated from leaves and berries of *J. phoenicea* from different countries could inhibit the growth of some pathogens such as *Candida albicans*, *Staphylococcus aureus*, *Escherichia coli*, *Pseudomonas aeruginosa*, *Klebsiella pneumonia, Bacillus subtilis, Streptococcus mutans*, *Saccharomyces cereviseae*, *Geotrichum candidum*, *Aspergillus flavus*, *Candida pseudotropicalis, C. albicans, Fusarium oxysporum*, *Maccrophomina phasioli, Mucor ramamnianus* and *Rhizopus stolonifer* (Stassi et al. 1996; Angioni et al. 2003; Cosentino et al. 2003; El-Sawi et al. 2007; Bouzouita et al. 2008; Ennajar et al. 2009b; Derwich et al. 2010; Mazari et al. 2010).

Recently, in 2013, Abu Darwish et al. evaluated the antifungal activity of the berries essential oils of *J. phoenicia* from South Jordan against dermatophytes (*Epidermophyton floccosum, Microsporum canis, M. gypseum, Trichophyton mentagrophytes*, *T. mentagrophytes* var. *interdigitale*, *T. rubrum*, *T. verrucosum*), yeasts (*Candida albicans*, *C. guillermondii, C. krusei*, *C. parapsilosis*, *C. tropicalis*, *Cryptococcus neoformans*) and *Aspergillus* species (*Aspergillus flavus, A. fumigatus* and *A. niger*). The essential oils exhibited higher antifungal activity against dermatophytes and *Cryptococcus neoformans* with MIC values ranging from 0.32 to 2.5 μL/mL.

The methanolic extract of *J. phoenicea* leaves from Egypt showed a hepatoprotective effect against carbon tetrachloride hepatotoxicity and remarkably enhanced liver and kidney functions (Ali et al. 2010). The administration of water extract of the leaves of Tunisian *J. phoenicea* in doses of 100 mg/kg reveled efficiency in reducing uric acid level in blood and the 200 mg/kg dose strongly reduced the activities of superoxide dismutase, catalase and glutathione peroxidase in liver and erythrocytes. Moreover, Medini et al. (2013) showed that the methanol and ethyl acetate extracts of *J. phoenicea* leaves from Tunisia exhibited potent scavenging activities towards ABTS and DPPH radical cations.

It has been stated that *J. phoenicea* exhibit cytotoxicity *in vitro*. In 2007, El-Sawi et al. demonstrated that berries and leaves essential oils of *J. phoenicea* from Egypt showed very high cytotoxicity against brain, cervix, lung, liver, and breast human cell lines. The cytotoxicity of berries essential oil is slightly higher than leaves essential oil against cells from lung (0.6 and 0.7 μg/mL, respectively), and from liver (0.7 and 0.9 μg/mL, respectively). The cytotoxicity of *J. phoenicea* essential oils could be explained due to the high content of total monoterpenes, α-pinene and β-pinene. Moreover, El-Sawi and Motawae (2008), found that the petroleum ether extract of *J. phoenicea* berries could exhibit a strong cytotoxicity against liver, lung and breast carcinoma cell lines.

Aqueous extracts of *J. phoenicea* leaves collected in Dana area, in Al-Tafilah, in South Jordan caused a dose dependent protection of rats against castor oil induced diarrhea and reduced castor oil induced enteropooling. This extract also caused a decrease in intestinal transit and showed a significant relaxant effect on rat ileal smooth muscle (Qnais et al. 2005). In 2008, Shkukani et al. showed that ethanol extract of Jordanian *J. phoenicea* possess antifertility effects in male albino rats. Also, pregnancy rate in females markedly reduced by 60 and 80 % after mating with males treated with administered single daily intra peritoneal injections of 400 or 800 mg/kg of ethanol extract of *J. phoenicea* berries, respectively. The arrest of spermatogenesis and the decrease in the number of mature sperms were markedly observed in treated rates.

Conclusion

Juniperus phoenicea is a very important medicinal plant from Jordanian flora, where its main parts used (leaves and berries) are widely used in folk medicine for treatment of various diseases. Its main secondary metabolites such as essential oils, flavonoids and lignans could exhibit various pharmacological activities, which indicate *J. phoenicea* as a promising medicinal plant for therapeutic and pharmaceutical industry.

References

Aboul-Ela M, El-Shaer N, El-Azim TA (2005a) Chemical constituents and hypoglycemic effect of *Juniperus phoenicea*: Part I. Alex J Pharm Sci 19(2):109–116

Aboul-Ela M, ElShaer N, ElAzim TA (2005b) Chemical constituents and antihepatotoxic effect of the berries of *Juniperus phoenicea* Part II. Nat Prod Sci 11(4):240–247

Abu-Darwish MS, Ofir R (2012) Heavy metals content and essential oil yield of *Juniperus phoenicea* L. in different origins in Jordan. Environ Eng Manag J (in press)

Abu-Darwish MS, Gonçalves MJ, Cabral C, Cavaleiro C, Salgueiro L (2013) Chemical composition and antifungal activity of essential oil from *Juniperus phoenicea* subsp. *phoenicea* berries from Jordan. Acta Aliment Hung 42(4):504–511

Achak N, Romane A, Alifriqui M, Adams RP (2009a) Chemical studies of leaf essential oils of three species of *Juniperus* from Tensift Al Haouz-Marrakech Region (Morocco). J Essent Oil Res 21:337–341

Achak N, Romane A, Dahbi A (2009b) Metals contents and organic composition in the leaf of three species of Cupressaceae from the Atlas Mountains, Marrakech (Morocco). Phys Chem News 47:98–102

Adams RP (1991) Cedar wood-oil-analysis and properties. In: Linskins HF, Jackson JF (eds) Modern methods of plant analysis: oils and waxes. Springer, Berlin, pp 159–173

Adams RP (2008) Junipers of the world: the genus *Juniperus*, 2nd edn. Trafford Publishing Company, Vancouver

Adams RP, Barrero AF, Lara A (1996) Comparisons of the leaf essential oils of *Juniperus phoenicea*, *Juniperus phoenicea* subsp. *eu-mediterranea* Lebr and Thiv and J. *phoenicea* var. *turbinata* (Guss) Parl. J Essent Oil Res 8:367–371

Adams RP, Pandey N, Rezzi S, Casanova J (2002) Geographic variation in the random amplified polymorphic DNAs (RAPDs) of *Juniperus phoenicea*, *J.p.* var. *canariensis*, *J.p.* subsp. *eu-mediterranea*, and *J.p.* var. *turbinata*. Biochem Syst Ecol 30:223–229

Afifi MS, El-Sharkawy SH, Maatoog GT, Sohly M, Rosazza JPN (1992) Essential oils of *Thuja occidentalis*, *Thuja orientalis*, *Cupressus sempervirens* and *Juniperus phoenicea*. Mandura J Pharm Sci 8:37–46

Al-Eisawi DM (1996) Vegetation in Jordan. UNESCO, Cairo

Al-Qura'n S (2005) Vegetation dynamics in the Tafilah woodlands of southern Jordan. Am J Environ Sci 1:46–49

Al-Qura'n S (2008) Comparison study of natural vegetation structure between Ajlun and Tafileh woodlands in Jordan. J Nat Prod 1:46–55

Al-Qura'n S (2009) Ethnopharmacological survey of wild medicinal plants in Showbak, Jordan. J Ethnopharmacol 123:45–50

Al-Ramamneh E, Dura S, Daradkeh N (2012) Propagation physiology of *Juniperus phoenicea* L. from Jordan using seeds and *in vitro* culture techniques: baseline information for a conservation perspective. Afr J Biotechnol 11(30):7684–7692

Ali SA, Rizk MZ, Ibrahim NA, Abdallah MS, Sharara HM, Moustafa MM (2010) Protective role of *Juniperus phoenicea* and *Cupressus sempervirens* against CCl4. World J Gastrointest Pharmacol Ther 1(6):123–131

Angioni A, Barra AT, Russo M, Coroneo V, Dessí S, Cabras P (2003) Chemical composition of the essential oils of *Juniperus* from ripe and unripe berries and leaves and their antimicrobial activity. J Agric Food Chem 51(10):3073–3078

Barrero AF, Quilez del Moral JF, Herrador MM, Akssira M, Bennamara A, Akkad S, Aitigri M (2004) Oxygenated diterpenes and other constituents from Moroccan *Juniperus phoenicea* and *Juniperus thurifera* var. *africana*. Phytochemistry 65(17):2507–2515

Barrero AF, Herrador MM, Arteaga P, Quílez del Moral JF, Sánchez-Fernández E, Akssira M, Aitigri M, Mellouki F, Akkad S (2006) Chemical composition of the essential oil from the leaves of *Juniperus phoenicea* L. from North Africa. J Essent Oil Res 18:168–169

Bellakhder J (1997) La pharmacopée marocaine traditionnelle. Édition Ibis Press, Paris

Bender F (1974) Geology of Jordan: contributions to the regional geology of the Earth. Gebruder Borntraeger, Berlin/Stuttgart

Boulos L (1999) Flora of Egypt. Alhadara Publishing, Cairo

Bouzouita N, Kachouri F, Ben Halima M, Chaaboun MM (2008) Composition chimique et activités antioxydante, antimicrobienne et insecticide de l'huile essentielle de *Juniperus phoenicea*. J Soc Chim Tunis 10:119–125

Cairnes DA, Ekundayo O, Kingston DGI (1980) Plant anticancer agents. X. Lignans from *Juniperus phoenicea*. J Nat Prod 43(4):495–497

Chalchat J, Garry R, Michet A, Peyron L (1990) Chemical composition of natural and empyreumatic oils and extracts from *Juniperus oxycedrus* and *Juniperus phoenicea* wood. J Essent Oil Res 2(23):1–236

Comte G, Allais DP, Chulia AJ, Vercauteren J, Delage C (1996) Two furanone glucoside derivatives from *Juniperus phoenicea*. Phytochemistry 41(5):1329–1332

Comte G, Vercauteren J, Chulia AJ, Allais DP, Delage C (1997) Phenylpropanoids from leaves of *Juniperus phoenicea*. Phytochemistry 45(8):1679–1682

Cosentino S, Barra A, Pisano B, Cabizza M, Pirisi FM, Palmas F (2003) Composition and antimicrobial properties of Sardinian *Juniperus* essential oils against food borne pathogens and spoilage microorganisms. J Food Prot 66(7):1288–1291

Delitala FL (1980) Ricerche chemiotassonomiche sul genere *Juniperus* Linnaeus. Riv Ital EPPOS 62:303–309

Derwich E, Benziane Z, Boukir A (2010) Chemical composition of leaf essential oil of *Juniperus phoenicea* and evaluation of its antibacterial activity. Int J Agric Biol 12:199–204

El-Bana M, Shaltout K, Khalafallah A, Mosallam H (2010) Ecological status of the Mediterranean *Juniperus phoenicea* L. relicts in the desert mountains of North Sinai, Egypt. Flora 205:171–178

El-Sawi SA, Motawae HM (2008) Labdane, pimarane and abietane diterpenes from the fruits of *Juniperus phoenicea* L. grown in Egypt and their activities against human liver carcinoma. Can J Pure Appl Sci 2(1):115–122

El-Sawi SA, Motawae HM, Ali AM (2007) Chemical composition, cytotoxic activity and antimicrobial activity of essential oils of leaves and berries of *Juniperus phoenicea* L. grown in Egypt. Afr J Tradit Complement Altern Med 4:417–426

Ennajar M, Bouajila J, Lebrihi A, Mathieu F, Savagnac A, Abderraba M, Raies A, Romdhane M (2009a) The influence of organ, season and drying method on chemical composition and antioxidant and antimicrobial activities of *Juniperus phoenicea* L. essential oils. J Sci Food Agric 90:462–470

Ennajar M, Bouajila J, Lebrihi A, Mathieu F, Savagnac A, Abderraba M, Raies A, Romdhane M (2009b) Chemical composition and antimicrobial and antioxidant activities of essential oils and various extracts of *Juniperus phoenicea* L. (Cupressacees). J Food Sci 74(7):M364–M371

Franco JA (1986) *Juniperus* L. In: Castroviejo S, Laínz M, López-González G, Montserrat P, Munoz-Garmendia F, Paiva J, Villar L (eds) Flora iberica 1. CSIC, Madrid, pp 181–188

Hamdan II, Afifi FU (2004) Studies on the *in vitro* and *in vivo* hypoglycemic activities of some medicinal plants used in treatment of diabetes in Jordanian traditional medicine. J Ethnopharmacol 93:117–121

IUCN (2013a) The IUCN red list of threatened species. Version 2013.2. http://www.iucnredlist.org. Downloaded on 28 March 2014

IUCN (2013b) Database for North African medicinal and aromatic plants. http://www.uicnmed.org/nabp/database/NA_Plants.htm. Downloaded on 28 March 2014

Karim F, Al-Quran S (1990) Toxic plants of Jordan (book assisted and published by Deanship of Research and Graduate Studies). Yarmouk University Press, Irbid

Lawrence BM (1989) Juniper berry oil. Essential oils (1987–1988). Allured Publishing Corporation, Wheaton

Mazari K, Bendimerad N, Bekhechi C, Xavier F (2010) Chemical composition and antimicrobial activity of essential oils isolated from Algerian *Juniperus phoenicea* L. and *Cupressus sempervirens* L. J Med Plant Res 4:959–964

Medini H, Elaissi A, Khouja ML, Chemli R (2013) Phytochemical screening and antioxidant activity of *Juniperus phoenicea* ssp. *phoenicea* L. extracts from two Tunisian locations. J Exp Biol Agric Sci 1(2):77–82

Meloni M, Perini D, Filigheddu R, Binelli G (2006) Genetic variation in five Mediterranean populations of *Juniperus phoenicea* as revealed by Inter-Simple Sequence Repeat (ISSR) markers. Ann Bot 97(2):299–304

Nasri N, Tlili N, Elfalleh W, Cherif E, Ferchichi A, Khaldi A, Triki S (2011) Chemical compounds from *Phoenician juniper* berries (*Juniperus phoenicea*). Nat Prod Res 18:1733–1742

Qnais EY, Abdulla FA, Abu Ghalyun YY (2005) Antidiarrheal effects of *Juniperus phoenicea* L. leaves extract in rats. Pak J Biol Sci 8:867–871

Ramic S, Murko D (1983) Chemical composition of *Juniperus species*. Arh Farm 33:15–20

Rezzi S, Cavaleiro C, Bighelli A, Salgueiro L, Proença da Cunha A, Casanova J (2001) Intraspecific chemical variability of the leaf essential oil of *Juniperus phoenicea* subsp. turbinata from Corsica. Biochem Syst Ecol 29:179–188, http://www.sciencedirect.com/science/journal/03051978

Shkukani HG, Salhab AS, Disi AM, Shomaf MS, Al Quadan F (2008) Antifertility effect of ethanolic extract of Juniperus phoenica (L.) in male albino rats. J Herb Pharmacother 7:179–189

Stassi V, Verykokidou E, Loukis A, Harvala A, Philianos S (1996) The antimicrobial activity of the essential oil of four *Juniperus* species growing wild in Greece. Flav Fragr J 11:71–74

WHO (2003) WHO guidelines on good agricultural and field collection practices (GACP) for medicinal plants. World Health Organization, Geneva

Myrtle (*Myrtus communis* L.) – A Native Mediterranean and Cultured Crop Species

Shmuel Zilkah and Eliezer E. Goldschdmidt

Abstract *Myrtus communis* L, 'Common myrtle' is a Mediterranean evergreen shrub, growing in the wild in hill regions from Morocco to Iran. The common myrtle has been used since ancient times as an ornamental and as a source of fragrance and medications. Its cultivation also dates back to the classic era. Major current uses include production of leaf essential oil for aromatherapy and perfume industry, production of liqueurs and the use of myrtle as one of the 'Four species' in the ritual of the Jewish feast of Tabernacles (Sukkot). Myrtle shoots for Tabernacles need to have a tricussate (three-leaved) phyllotactic arrangement. Wild myrtle as well as ornamental myrtle shrubs give rise to a small percentage of tricussate shoots. However, severe pruning either solely or followed by burning of myrtle shrubs during winter or early spring were found to dramatically increase the yield of tricussates; a technique currently used in cultivation of the tricussate myrtle in Israel. Detailed analyses of myrtle essential oils are currently available, revealing α-pinene as the principal constituent out of ca. 40 compounds. The increasing demand by the myrtle liqueur industry in recent decades created an ecological risk for the wild myrtle population of Sardinia. The survival potential and genetic variation among wild myrtle population was evaluated, prompting an extensive selection of the best cultivars, vegetative propagation and cultivation program. The liqueurs, Mirto rosso and Mirto bianco, are produced by hydro-alcoholic infusions of deep purple berries and yellow-white berries (with young shoots), respectively. *Myrtus communis'* medical potential focuses on its anti-oxidant as well as anti-hyperglycemic, antibacterial and analgesic properties. Of particular interest are the myrtucommulones, a unique group of myrtle compounds, showing anti-inflammatory and selective apoptosis-inducing activities.

Dedication: In loving memory of Dr. Tsvi Orlan, who dedicated a lifetime to the study and improvement of three-leaf myrtle shoots for Sukkot.

S. Zilkah (✉)
Institute of Plant Sciences, Agricultural Research Organization, The Volcani Center, P.O. Box 6, Bet Dagan 50250, Israel
e-mail: zilka@volcani.agri.gov.il

E.E. Goldschdmidt
The Robert H. Smith Faculty of Agriculture, Food and Environment, Institute of Plant Sciences and Genetics in Agriculture, The Hebrew University of Jerusalem, P.O. Box 12, Rehovot 76100, Israel

Z. Yaniv, N. Dudai (eds.), *Medicinal and Aromatic Plants of the Middle-East*, Medicinal and Aromatic Plants of the World 2, DOI 10.1007/978-94-017-9276-9_14

Keywords Anthocyanins • Antioxidants • Burning • Common myrtle • Ecological risk • Essential oils • 'Four Species' • Genetic diversity • Mirto liqueur • Myrtacea • Myrtucommulones • α-pinene • Pruning • The Feast Tabernacles • Tricussate phyllotaxis

Taxonomy and Plant Characteristics

The common myrtle (*Myrtus communis* L.) also called 'true myrtle' belongs to the family Myrtacea, order Myrtales, subclass Rosids, class Eudicots. The Myrtacea family is of about 130 genera and ca. 3,000 species of trees and shrubs (Stebbins 1974) growing mostly in tropical and subtropical floras. Myrtle has a diploid set of chromosomes (2n = 22) (Greco 1929). It is widespread and grows spontaneously in Mediterranean regions, including Southern Europe, North Africa and the Middle East, including Iran. It is commonly cultivated as an ornamental garden shrub, for its evergreen and sclerophyllous, leathery leaves, profuse flowering and pleasant odor.

The plant is an evergreen shrub or small tree. In the wild, myrtles form dense, several meters long bushes (Fig. 1). It is monoecious with hermaphrodite flowers in the leaf axil. The leaf is entire, 1–4 cm long, oval or oval-lanceolate shaped. Leaves are commonly arranged in an opposite position (=decussate phyllotaxis) and more rarely, in alternate (=spiral phyllotaxis) (Mabberley 1997), or three-leaved (=tricussate phyllotaxis) (Fig. 2). Leaves, flowers and fruit contain glands of essential oil

Fig. 1 Wild myrtle shrubs in Galilee Mountains, Northern Israel

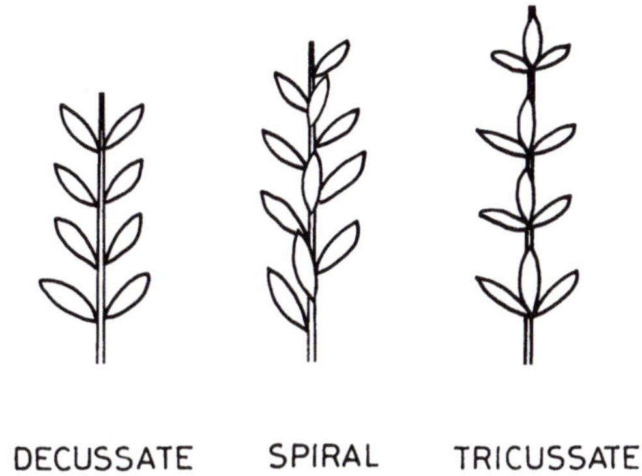

Fig. 2 Schematic illustration of the three phyllotactic myrtle shoot types; decussate, spiral and tricussate

Fig. 3 Mature, blue myrtle berries, of the kind used for the production of the 'Mirto rosso' liqueur

that renders the plant fragrant and spicy. Myrtle essential oil is used in the cosmetic and medical industries (Polunini 1984). The star-like flower has five petals and sepals, and numerous stamens. Petals are usually white. The fruit is edible, but astringent in taste, round, blue-black (Fig. 3) or white-amber in color, containing several seeds. Flowers are usually cross-pollinated by insects, but there are indications on self-pollination (Melito et al. 2013). Seeds are dispersed by birds that eat the berries. *M. communis* sub sp. *tarentina* is a smaller evergreen shrub than the common myrtle, with leaves similar in shape but smaller, with white flowers and berries about 12 mm in length (Mulas and Fadda 2004). Myrtle adapts to a broad range of Mediterranean climatic conditions but suffers under north European conditions. Myrtle grows best on neutral and sub-acidic soil, but is also adapted to other types of soils, showing relatively moderate tolerance to diseases and insects.

The Use of Myrtle in Worship and Ancient Cultures

Myrtle has been known and cultivated as an ornamental garden shrub since antiquity, praised for its beautiful, white flowers and delicate fragrance. Myrtle is mentioned by Theophrastus, *Enquiry into Plants* and Pliny the Elder, *The Natural History*. It plays a role in various scenes of Greek mythology and serves as a symbol of love and immortality throughout the Mediterranean and southern Europe. Myrtle in a wedding bouquet is a general European custom and its use as bridal thrones is also mentioned in ancient Jewish scripts. Myrtles buds were also used in Jewish funerals in ancient times. In the ceremony terminating the Sabbath (*Havdala*) a blessing is recited on a fragrant spice. For many generations it was customary to recite this blessing on a bud of myrtle, *Hadas* in Hebrew. This went to the extent that the article in which the fragrant spice was held was also called *Hadas*. According to Jewish Kabbalist customs a blessing on myrtle buds is recited also in other religious ceremonies. In the Jewish tradition, Myrtle is one of the ‘Four Species’ used in the worship ritual of ‘Sukkot’, the Feast of Tabernacles. Three myrtle shoots are held by the worshippers along with a palm branch, two willow shoots and a citron fruit. According to the Jewish law, myrtle shoots should be ‘three-leaved’, i.e. the leaves arranged along the shoot in a tricussate phyllotaxis, three leaves per node.

Propagation

For common gardening, the plant is propagated from seeds. There is great variability in the germplasm of this species. For commercial cultivation targeted towards crop industries, such as ornamental pot plants, ornamental cutting shoots, tricussate leaf shoots, essential oils for biochemical and medical purposes, and liqueur production, vegetative propagation of selected varieties is most advantageous. Selected genotypes can be propagated through rooting of cuttings but yields are relatively low (Canhoto et al. 1999). Rooting is performed by mean of softwood cuttings treated with IBA powder at 1 %, in a perlite growth medium and under mist irrigation (Mulas and Cani 1996; Mulas et al. 1998; Mulas 2012). Another study showed that rooting percentage of cuttings taken during December–February reached 70 %, while only 20 % of the cuttings taken during May–August rooted successfully (Klein et al. 2000). There was a high variability in rooting ability among myrtle ecotypes selected from germplasm spontaneously growing in the wild (Mulas and Cani 1996), as well as in morphological characters, production and vegetative growth (Mulas et al. 1998). The *in-vitro* multiplication through tissue culture (micropropagation) is advantageous in achieving rapid propagation, better homogeneity and higher rate of pathogen free plants. Consequently, protocols for *in vitro* propagation have been developed for ecotypes selected in various regions (Khoshkhui et al. 1984; Nobre 1994; Parra and Amo-Marco 1996; Scarpa et al. 2000). Variability among clones in rates of multiplication and rooting was found also in the *in vitro* systems (Ruffoni et al. 2003, 2010). Growth and rooting of myrtle cultures in aerated conditions was better than in closed vessels (Lucchesini et al. 2001, 2006). Rooting efficiency was influenced by level of sucrose and light intensity (Ruffoni et al. 2010). In order to improve *in vitro* plant regeneration in difficult-to-propagate species, a somatic embryogenesis has been induced from immature seeds, hypocotyls-cotyledons (Parra and Amo-Marco 1998) and mature zygotic embryos (Canhoto et al. 1999). These *in vitro* techniques could be potentially useful for genetic manipulation of myrtle.

Myrtle Cultivation

The removal of huge quantities on myrtle biomass from the wild for use in the various industries of drugs, perfumes and, in particular, liqueur caused an ecological threat to this species, to such an extent that it was listed as an endangered species. As a result, a long and progressive program of myrtle cultivation has been initiated in Sardinia. Unique, outstanding clones of myrtle were selected from the wild in order to suit the needs of liqueur industry (Mulas and Cani 1999; Mulas et al. 2002; Tuberoso et al. 2007). In the search of the better clones for liqueur production in Corsica, two distinguished groups of myrtle berries were characterized by their different concentrations of polyphenols in the alcoholic extracts (Barboni et al. 2010a, b). The extraction methods were optimized. It has been found that ethanol 80 % provides the extract with the best characteristics of anthocyanins and antioxidant activity (Snoussi et al. 2012). The anthocyanins in the hydro-alcoholic extracts of all the different selections were the same but varied in quantity (Tuberoso et al. 2006). The evolution of chemical composition was monitored during fruit development with the aim to identify integrated indices for the optimal harvest time (Fadda and Mulas 2010). The outcome of the intensive studies of about 30 years, the wild shrub of myrtle has been cultivated in Sardinia to a crop of about 200 ha which supports a liqueur industry of around 3.5 millions liters per annum (Mulas 2012). The cultivated fruits are better in yield, quality and supply security than the fruits harvested from the wild. This developed cultivated crop significantly contributed to mitigate the threat on the myrtle population growing spontaneously in the wild.

Genetic Variability

An alternative means for protecting the native Sardinian myrtle populations from the ecological risk, was to evaluate their survival potential through assessment of the genetic variation within and among populations in the wild (Agrimonti et al. 2007; Melito et al. 2013). The dispersion of myrtle in the Mediterranean Basin in regards to patterns of genetic diversity, mating system, and gene flow has been studied by using genetic markers, such as isozymes, AFLPs (Albaladejo et al. 2009; Bruna et al. 2007; González-Varo et al. 2009a, b) and microsatellites (Albaladejo et al. 2010). Genetic diversity of wild Tunisian *Myrtus* populations showed three ecological groups due to geographic distances, human activities and environmental factors (Messaoud et al. 2006). AFLP-based analysis suggested that Italy represents a botanical transition zone between the Western and Eastern Mediterranean region in myrtle (Bruna et al. 2007). This genetic diversity within populations and its variation between geographic groups should be taken into account in the conservation strategy of the myrtle in the wild. Nuclear ribosomal DNA and chloroplast DNA markers were used to explain the current distribution of the genus *Myrtus* in the Mediterranean Basin and in the contiguous isolated areas (Migliore et al. 2012). It was proposed that the myrtle populations have been able to survive the climatic fluctuations during the Pleistocene era probably in local refugia and by migration to adjacent regions (Melito et al. 2013). There are myrtle habitats where overall patch area is very low and spatial connectivity among patches is mostly lacking. These fragmented habitats might be threat to the plant reproduction because of the limited pollination. The effect of the pollinator assemblages in interaction with the patch sizes was studies with the aim to mitigate this problem within the fragmented habitats (González-Varo et al. 2009).

Use of Shoots with a Tricussate Phyllotactic Arrangement

As mentioned above, tricussate shoot myrtle is used in the celebration of the Jewish 'Sukkot' Holiday. There is a great demand to this special kind of shoots. However, in the wild, as well as in cultivated myrtle, the majority of shoots have an opposite (=decussate) leaf arrangement and only a small percentage of them are naturally in tricussate arrangement (Fig. 4). In the past, since there was no alternative, the

Fig. 4 Apical portion of a tricussate myrtle shoot

Fig. 5 A myrtle plantation destined for the Tricussate shoot industry. Emek Yizrael valley, Israel

tricussate shoots have been harvested in the wild of the Galilee, at the northern Israel. These shoots were highly variable in the rate of leaf tricussates, leaf size and shape, internode length and vegetative growth. The populations in the wild are often exposed to unpredictable environmental damages such as fires, excess grazing, insects and diseases. In this practice the harvested quantities could not supply the demand, the shoots were highly variable, the qualities were not satisfied and above all, the myrtle populations in the wild were endangered.

Consequently, cultivated crop of myrtle has been developed in Israel for producing high quality tricussate shoots (Fig. 5). High standard clones were selected from myrtle populations in the wild, gardens and seedling. The clone "Avot" was selected from seedlings and patented. This cultivar has three main leaves in a perfect tricussate arrangement. Additional smaller satellite leaves may evolve on both sides of each main leaf (Fig. 6). Nowadays the myrtle is cultivated in about 50 ha, producing around five million tricussate shoots per year.

Cultivation of Tricussate Shoot Crop

Although the physiological mechanism determining the phyllotactic arrangement is obscure, the traditional experience shows that pruning and/or burning myrtle shrubs in early spring increase the rate of the tricussate shoots (Orlan 1998). In practice, Israeli myrtle growers applied mechanical pruning to the ground level, which is sometimes followed by burning the trimmed shoots. This treatment reduces

Fig. 6 A tricussate shoot of the 'Avot' myrtle cultivar. Please note that each main leaf might be flanked by two smaller leaflets. Thus, a tricussate node could be, in a maximum of nine leaves

drastically the number of new shoots per shrub, thereby enabling the fewer shoots to grow more vigorously, to induce higher rate of tricussate shoots to eliminate flowering. The effects of several factors such as plant growth regulators, drought, day light and time of pruning on the development of high quality tricussate shoots have been studied (Grigorivker 2011; Orlan 1998).

Attempts to decipher the reasons for the scarcity of tricussate myrtle shoots had limited success (Halsall and Dourado 2001). However, extensive research by Orlan (1998) uncovered the gradual stages of the myrtle tricussate shoot development. The observations showed that almost every myrtle shoot starts as decussate. Then, as the shoot elongates, a phyllotactic transition commences; the decussate arrangement changes gradually to alternate-spiral and after a few more nodes the leaves are arranged to tricussate phyllotaxis, triple leaves in a whorl (Fig. 7).

The planting density is 10,000–20,000 plants/ha. The plants are propagated vegetatively from outstanding clones for tricussate shoot formation and leaf quality. The 'Sukkot' Holiday is celebrated usually during September. Therefore, the harvest time starts about one month ahead. For perennial plants, the drip irrigation starts on April (20 m^3/ha) and increases gradually to about 50 m^3/ha on August. The shoots at harvest time should be hardened, to avoid postharvest withering of the terminal leaves. Hence, the irrigation on heavy soils should be terminated 2–4 weeks before harvest; or gradually decreased on light soils. The shoot hardening could be enhanced by foliar application with a growth retardant such as paclobutrazol (Zilkah, unpublished data). This treatment could be applied in addition to the cessation of irrigation or as an alternative hardening treatment. The latter enables additional days for shoot elongation. It is recommended when the tricussate shoot sections are not expected to reach the required

Fig. 7 A common myrtle shoot, showing the phyllotactic transition from decussate (*bottom section*), to spiral (*middle section*), and to tricussate (*top section*). Arrows mark the transition zones

>30 cm in length for the superior quality. Fertilization with NPK is essential; 50–100 kg/ha for N and K; 25–50 kg/ha for P. Fe should be treated according to the degree of leaf greening (Steinmetz et al. 2013). All the lateral shoots have to be removed manually from the main stem. The premier tricussate shoots should be free of disease and insect symptoms, therefore the myrtle plants should be monitored carefully and accordingly controlled. The harvested shoots should be dipped in water (including antifungal and antibacterial agents), cooled for 12 h before sorting and kept at 4 °C, free of water, in horizontal position (Fig. 8), for several days until marketing (Steinmetz et al. 2013).

Essential Oils

Essential oils are natural, volatile, aromatic, hydrophobic components of plant organs produced by various plant classes, widely used in the perfume and food industries as well as in health and medicine. Essential oils play a major role in aromatherapy and as of recently, as antioxidants (Saleh et al. 2010). Best known,

Fig. 8 Harvested tricussate myrtle shoots ready for the market

perhaps, are the essential oils of citrus, but essential oils of mentha, lavender, clover and eucalyptus are also commercially available. Myrtle essential oil has been known and used since antiquity; the typical myrtle fragrance derives from its volatile, essential oil. Detailed chemical analyses of *Myrtus communis* essential oil from different geographical regions surrounding the Mediterranean are currently available, revealing slight regional, genotype differences (Mulas and Melis 2011). However, differences in extraction methods must also be taken into account. A study from Tunisia (Wannes et al. 2010) identified 44 compounds belonging to several chemical classes; mainly monoterpene hydrocarbons and oxygenated monoterpenes (66.45 and 26.01 % in leaf essential oil, respectively). Constituents of myrtle essential oil from Iran (Weyerstahl et al. 1994) and Spain (Boelens and Jimenez 1992) have also been analyzed. Leaves, stems and flowers differ in their essential oil composition; stems essential oil contains 52.43 % oxygenated monoterpenes, while flowers have 14.13 % phenyl propanoids (Wannes et al. 2010). The twenty principal myrtle leaf essential oil constituents and their concentrations, as determined in Sardinia, are listed in Table 1. Quantitatively, α-Pinene, Limonene, 1,8-Cineole and Geranyl acetate are the most dominant; however, other, minor constituents might still be major determinants of fragrance and aroma. The medical aspects of specific myrtle constituents will be discussed in a subsequent section. The myrtle essential oil fraction should not be confused with the hydro-alcoholic infusions used as a basis for the myrtle liqueurs 'Mirto rosso' and 'Mirto bianco', which have a somewhat different chemical composition (Barboni et al. 2010a, b; Mulas and Melis 2011).

Table 1 Concentrations (g.kg^{-1} Dry Weight) of principal *Myrtus communis* leaf essential oil constituents

Constituent	Concentration (g.kg^{-1} dry weight)
α -Pinene	518.4
Camphene	0.0
β -Pinene	5.7
Isobutylbutyrate	4.6
Limonene	191.4
1,8-Cineole	102.3
p-Cymene	3.1
Terpinolene I	16.2
Terpinolene II	9.3
Linalool	14.5
Lynalyl acetate	13.0
Bornyl acetate	1.2
α -Humulene	5.5
Myrtenyl acetate	0.9
α-Terpinol	10.1
Neryl acetate	1.4
Geranyl acetate	49.9
Myrtenol	0.9
Geraniol	2.5
Methyl eugenol	2.8

Compiled from Mulas and Melis (2011)
Averages from nine Sardinian myrtle cultivars; Spring 2008

The highest essential oil content obtained at bloom in a yield of 0.3–0.5 % from the fresh wt. of young shoots (Mulas and Cani 1999). Essential oil composition of myrtle leaves in the winter was quite similar to that from spring, but in a lower yield (Mulas and Melis 2011). The essential oils obtained from plants grown in culture *in vitro* were different from plants harvested at natural locations (Savikin-Fodulovic et al. 2000).

Production of Liqueur

Myrtle is an aromatic plant with various important essential oils, tannins and anthocyanins. It is a plant source for production of special liqueurs in the islands of Sardinia (Italy) and Corsica (France). The myrtle red liqueur ('Mirto rosso') is produced by cold hydro-alcoholic infusion of deep purple berries, while the white liqueur ('Mirto bianco') is produced from yellow-white berries and young shoots in spring time (Mulas and Cani 1999). The berries and the plant biomass used to be collected manually from the wild to an annual yield of 4,200 t (Nuvoli and Spano 1996). The liqueur industry has increased in recent decades to a conspicuous production of about 3.5 million liters. In spite of the intensive harvest from wild habitats it was impossible to supply the whole quantity demanded by the industry and the standards of the liqueur quality could not be stabilized, due to the large variability within the germplasm in the wild (Mulas 2012). Furthermore, excessive harvesting from the wild habitats endangered the survival of the natural myrtle ecosystem, leading to the genetic diversion studies and selection of cultivars for agricultural cultivation (Mulas 2012).

Medicinal Uses

Myrtle has been widely used in Herbal folklore medicine since ancient times (Sumbul et al. 2011). Egyptian medical texts recommend myrtle extracts as a remedy for urinary disorders and other ailments (Ozkan and Guray 2009). Myrtle occupies a prominent place in the medical writings of Greek, Roman and Arabic scholars. Myrtle extracts have been reported to possess antihyperglycemic, antibacterial and analgesic properties (Aleksic and Knezevic 2014; Sumbul et al. 2011). The latter review articles summarize the vast, rapidly increasing biomedical research literature dedicated to *Myrtus communis*. Commercial myrtle essential oil preparations are widely used in Aromatherapy. Recent reports focus on the antioxidant activity of myrtle extracts (Romani et al. 2004; Wannes et al. 2010) and attempt to identify the active ingredients. In addition to the major essential oil components (Table 1), myrtle extracts contain a series of polyphenol tannins with considerable radical scavenging activity (Yoshimura et al. 2008). However, most attention has recently been paid to the unique oligomeric, nonprenylated acylphloroglucinols of myrtle, the myrtucommulones (Fig. 9). Myrtucommulones and their antibacterial activity were first described by an Israeli team in 1974 (Rotstein et al. 1974) and then further investigated by Appendino et al. (2002) and Feisst et al. (2005). Myrtucommulones were shown to have significant anti-inflammatory activity as well as highly selective apoptosis-inducing activity (see Mueller et al. 2010). Their regular implementation in modern therapy awaits further research.

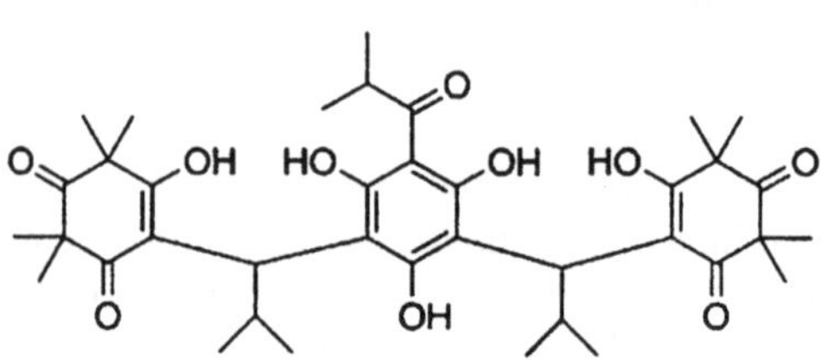

Fig. 9 A general formula of the myrtucommulones (After Rotstein et al. (1974) and Mueller et al (2010))

References

Agrimonti A, Bianchi R, Bianchi A, Ballero M, Poli F, Marmiroli N (2007) Understanding biological conservation strategies: a molecular-genetic approach to the case of myrtle (*Myrtus communis* L.) in two Italian regions: Sardinia and Calabria. Conserv Genet 8:385–396

Albaladejo RG, Carrillo LF, Aparicio A, Fernandez-Manjarres JF, Gonzalez-Varo JP (2009) Population genetic structure in *Myrtus communis* L. in a chronically fragmented landscape in the Mediterranean: can gene flow counteract habitat perturbation? Plant Biol 11:442–453

Albaladejo RG, Sebastiani F, González-Martinez SC, González-Varo JP, Vendramin GG, Aparicio A (2010) Isolation of microsatellite markers for the common Mediterranean shrub *Myrtus communis* (Myrtaceae). Am J Bot 97(5):23–25

Aleksic V, Knezevic P (2014) Antimicrobial and antioxidative activity of extracts and essential oils of *Myrtus communis* L. Microb Res 169(4):240–254

Appendino G, Bianchi F, Minassi A, Sterner O, Ballero M, Gibbons S (2002) Oligomeric acylphloroglucinols from myrtle (*Myrtus communis*). J Nat Prod 65:334–338

Barboni T, Cannac M, Massi L, Perez-Ramirez Y, Chiaramonti N (2010a) Variability of polyphenol compounds in *Myrtus Communis* L. (Myrtaceae) berries from Corsica. Molecules 15(11):7849–7860

Barboni T, Venturini N, Paolini J, Desjobert JM, Chiaramonti N, Costa J (2010b) Characterization of volatiles and polyphenols for quality assessment of alcoholic beverages prepared from Corsican *Myrtus communis* berries. Food Chem 122:1304–1312

Boelens MH, Jimenez R (1992) The chemical composition of Spanish myrtle oil. Part II. J Essent Oil Res 4(4):349–353

Bruna S, Portis E, Cervelli C, De Benedetti L, Schiva T, Mercuri A (2007) AFLP-based genetic relationships in the Mediterranean myrtle (*Myrtus communis* L.). Sci Hortic 113:370–375

Canhoto JM, Lopes ML, Cruz GS (1999) Somatic embryogenesis and plant regeneration in myrtle (Myrtaceae). Plant Cell Tissue Organ Cult 57:13–21

Fadda A, Mulas SM (2010) Chemical changes during myrtle (*Myrtus communis* L.) fruit development and ripening. Sci Hortic 125(3):171–544

Feisst C, Franke L, Appendino G, Werz O (2005) Identification of molecular targets of the oligomeric nonprenylated acylphloroglucinols from *Myrtus communis* and their implication as anti-inflammatory compounds. J Pharm Exp Ther 315:389–395

González-Varo JP, Albaladejo RG, Aparicio A (2009a) Mating patterns and spatial distribution of conspecific neighbours in the Mediterranean shrub *Myrtus communis* (Myrtaceae). Plant Ecol 203:207–215

González-Varo JP, Arroyo J, Aparicio A (2009b) Effects of fragmentation on pollinator assemblage, pollen limitation and seed production of Mediterranean myrtle (*Myrtus communis*). Biol Conserv 142:1058–1065

Greco R (1929) 'Notizie preliminarisull' embriologia e la cariologia del '*Myrtus communis* L.'. Nuovo Giorn Bot Italy 36(1):57–59

Grigorivker G (2011) Improvement of myrtle (*Myrtus communis* L.) branch quality: increasing tricussate phyllotactic arrangement and reducing lateral shoot sprouting. M.Sc. dissertation, The Hebrew University of Jerusalem

Halsall LJ, Dourado AM (2001) Germplasm evaluation of spontaneous myrtle (*Myrtus communis* L.) for cultivar selection and crop development. New Plantsman 8:224–226

Khosh-khui M, Shekafandeh A, Azarakhsh H (1984) Micropropagation of myrtle. Sci Hortic 22:139–146

Klein JD, Cohen S, Hebbe Y (2000) Seasonal variation in rooting ability of myrtle (*Myrtus communis*L.) cuttings. Sci Hortic 83(1):71–76

Lucchesini M, Mensuali-Sodi A, Massai R, Gucci R (2001) Development of autotrophy and tolerance to acclimatization *of Myrtus communis* transplant cultured *in vitro* under different aeration. Biol Plantar 44(2):167–174
Lucchesini M, Monteforti G, Mensuali-Sodi A, Serra G (2006) Leaf ultrastructure, photosynthetic rate and growth of myrtle plantlets under different *in vitro* culture conditions. Biol Plantar 50(2):161–168
Mabberley DJ (1997) The plant book, 2nd edn. Cambridge University Press, New York
Melito S, Chessa I, Erre P, Podani J, Mulas M (2013) The genetic diversity of Sardinian myrtle (*Myrtus communis* L.) populations. Electron J Biotechnol 16(6). ISSN:0717–3458. doi:10.2225/vol16-issue6-fulitext-13
Messaoud C, Khoudja MK, Boussaid M (2006) Genetic diversity and structure of wild Tunisian *Myrtus communis* L. (Myrtaceae) populations. Genet Res Crop Evol 53:407–417
Migliore J, Baumel A, Juin M, Medail F (2012) From Mediterranean shores to central Saharan mountains: key phylogeographical insights from the genus *Myrtus*. J Biogeogr 39:942–956
Mueller H, Paul M, Hartmann D, Huch V, Blaesius D, Koeberle A, Werz O, Jauch J (2010) Total synthesis of myrtucommulone A. Angew Chem Int Ed 49:2045–2049
Mulas M (2012) The myrtle (*Myrtus communis* L.) case, from a wild shrub to a new fruit crop. Acta Hortic 948:235–242
Mulas M, Cani MR (1996) Variability of rooting ability of softwood cuttings in myrtle germplasm. In: Pank F (ed) Proceedings of the international symposium on breeding resistance on medical and aromatic plants, Quedlinburg (Germany) vol 2(1), pp 191–194
Mulas M, Cani MR (1999) Germplasm evaluation of spontaneous myrtle (*Myrtus communis* L.) for cultivar selection and crop development. J Herbs Spices Med Plants 6(3):31–49
Mulas M, Fadda A (2004) First observations on biology and organ morphology of myrtle (*Myrtus communis* L.) flower. Agric Mediterr 134(3–4):223–235
Mulas M, Melis RAM (2011) Essential oil composition of myrtle (*Myrtus communis*) leaves. J Herbs Spices Med Plants 17:21–34
Mulas M, Cani MR, Brigaglia N (1998) Characters useful to cultivation in spontaneous populations of *Myrtus communis* L. Acta Hortic 457:271–277
Mulas M, Dias Francesconi AH, Perinu B (2002) Myrtle (*Myrtus communis* L.) as a new aromatic crop: cultivar selection. J Herbs Spices Med Plants 9(2–3):127–131
Nobre J (1994) In vitro shoot proliferation of *Myrtus communis* L., from field grown plants. Sci Hortic 58:253–258
Nuvoli F, Spano D (1996) Amalisi e prospettive economiche dell'utilizzazione industrial del mirto. Rivista Italiana EPPOS 7(19):231–236
Orlan T (1998) Factors affecting differentiation and growth of shots in myrtle (*Myrtus communis* L.) with an emphasis on shoots with a tricussate phyllotactic arrangement. PhD dissertation, The Hebrew University of Jerusalem
Ozkan AMG, Guray CG (2009) A Mediterranean: *Myrtus communis* L. (Myrtle). In: Morel JP, Mercuri AM (eds) Plants and culture: seeds of the cultural heritage of Europe. Edipuglia, Bari, pp 159–168
Parra R, Amo-Marco JB (1996) Effect of plant growth regulators and basal media on *in vitro* shoot proliferation and rooting of *Myrtus communis* L. Biol Plantar 38:161–168
Parra R, Amo-Marco JB (1998) Secondary somatic embryogenesis and plant regeneration in myrtle (*Myrtus communis* L.). Plant Cell Rep 18:325–330
Polunini O (1984) Guia Práctica Ilustrada de la Flora. Editorial Blume, Barcelona, 142
Romani A, Coinu R, Carta S, Pinelli P, Galardi C, Vinciery FF, Franconi F (2004) Evaluation of antioxidant effect of different extracts of *Myrtus communis* L. Free Radic Res 38:97–103
Rotstein A, Lifshitz A, Kashman Y (1974) Isolation and antibacterial activity of acylphloroglucinols from *Myrtus communis*. Antimicrob Agents Chemother 6:539–542
Ruffoni B, Airò M, Fascella G, Mascarello C, Zizzo G, Cervelli C (2003) Rooting and acclimatization of ornamental Myrtle genotypes. Acta Hortic 616:255–258

Ruffoni B, Mascarello C, Savona M (2010) *In vitro* propagation of ornamental Myrtus (*Myrtus communis*). In: Mohan Jain S, Ochatt SJ (eds) Methods in molecular biology, protocols for *in vitro* propagation of ornamental plants, vol 589. Humana Press, New York, pp 257–269

Saleh MA, Clark S, Woodward B, Deolu-Sobogun SA (2010) Antioxidant and free radical scavenging activities of essential oils. Ethn Dis 20:S1-78–S1-82

Savikin-Fodulovic KP, Bulatovic VM, Menkovic NR, Grubisic DV (2000) Comparison between the essential oil of *Myrtus communis* L. obtained from naturally grown and *in vitro* plants. J Essent Oil Res 12:75–78

Scarpa GM, Milia M, Satta M (2000) The influence of growth regulators on proliferation and rooting of *in vitro* propagated myrtle. Plant Cell Tiss Org Cult 62(3):175–179

Snoussi A, Hayet BHK, Essaidi I, Zgoulli S, Moncef CM, Thonart P, Bouzouita N (2012) Improvement of the composition of Tunisian Myrtle berries (*Myrtus communis* L.) alcohol extracts. J Agric Food Chem 60(2):608–614

Stebbins GL (1974) Flowering plants. Evolution above the species level. Belknap Press, Cambridge, MA, 399 pp

Steinmetz Y, Tamari Y, Gotlib Y, Zilkah S (2013) Directives for myrtle growth in Israel. http://shaham.moag.gov.il/ProfessionalInformation/Pages/hanchayot_gidul_adasim_2013.aspx

Sumbul S, Ahmad MA, Asifi M, Akhtar M (2011) *Myrtus communis* Linn. – a review. Indian J Nat Prod Res 2:395–402

Tuberoso CIG, Barra A, Angioni A, Sarritzu E, Pirisi FM (2006) Chemical composition of volatiles in Sardinian myrtle (*Myrtus communis* L.) alcoholic extracts and essential oils. J Agric Food Chem 54:1420–1426

Tuberoso CIG, Melis M, Angioni A, Pala M, Cabaras P (2007) Myrtle hydroalcoholic extracts obtained from different selections of *Myrtus communis* L. Food Chem 101:806–811

Wannes WA, Mhamdi B, Sriti J, Ben Jemia M, Ouchikh O, Hamdaoui G, Kchouk ME, Marzouk B (2010) Antioxidant activities of the essential oils and methanol extracts from myrtle (*Myrtus communis* var. *italica* L.) leaf, stem and flower. Food Chem Toxicol 48:1362–1370

Weyerstahl P, Marschall H, Rustaiyan A (1994) Constituents of the essential oil of *Myrtus communis* L. from Iran. Flavour Fragr J 9(6):333–337

Yoshimura M, Amakura Y, Tokuhara M, Yoshida T (2008) Polyphenolic compounds isolated from the leaves of *Myrtus communis*. J Nat Med 62:366–368

Eruca sativa, a Tasty Salad Herb with Health-Promoting Properties

Oz Barazani and Jotham Ziffer-Berger

Abstract *Eruca sativa* (Brassicaceae), commonly known as salad rocket or arugula, is an annual plant, naturally occurring around the Mediterranean Basin. The species has been gaining worldwide popularity as a salad vegetable with health-promoting properties, and, as in other members of the cabbage family, the main secondary metabolites, i.e., glucosinolates, give *E. sativa* a unique, rich aroma and taste. Early documentation of *E. sativa* as a culinary and medicinal herb dates back to Ancient Greece, and presumably these uses had been known in much earlier times. Among its therapeutic uses in traditional medicine, it has been valued by both ancient and modern scholars as a digestive and diuretic agent, to cure kidney and skin disorders, as an aphrodisiac, among many others. This chapter highlights the available information on the chemical composition of *E. sativa* and its biological activity in association with the reported health-promoting properties of the species.

Keywords Antigenotoxic • Antioxidant • Aphrodisiac • Arugula • Brassicaceae • Digestive agent • *Eruca sativa* • *Eruca vesicaria* • Glucosinolate • Glucosativin • Isothiocyanates • Kaempferol • Rocket • Sulforaphane • Traditional medicine

Introduction

Eruca sativa Mill. (Brassicaceae), commonly known as salad rocket, or arugula, is an annual herb, cultivated and appreciated worldwide. *E. sativa* is an erect herb with glabrous or sparsely pilose pinnatifid leaves and relatively large (1.5–2.0 cm in diameter) flowers with dark-veined, cream to yellow clawed petals (Fig. 1) (Zohary 1966). The species is self-incompatible, i.e., it relies on cross-pollination, performed by insects. Its elongate siliques dehisce when dry and release about 20 seeds.

O. Barazani (✉)
Institute of Plant Sciences, Agricultural Research Organization,
P.O. Box 6, Bet Dagan 50250, Israel
e-mail: barazani@agri.gov.il

J. Ziffer-Berger
Herbarium, the National Natural History Collections, The Hebrew University of Jerusalem,
Edmond J. Safra Campus, Giv'at Ram, Jerusalem 91904, Israel

Z. Yaniv, N. Dudai (eds.), *Medicinal and Aromatic Plants of the Middle-East*,
Medicinal and Aromatic Plants of the World 2, DOI 10.1007/978-94-017-9276-9_15

Fig. 1 *Eruca sativa* – *dark* veined petals and young fruit of an individual from a wild population in Israel (*left*) and fresh rocket leaves of the commercial cultivar at the market (*right*) (Photos by Oz Barazani)

The species occurs naturally around the Mediterranean Basin, from North Africa and Spain to the Levant. Unlike other crucifer crops, the common cultivated variety of *E. sativa* did not undergo a breeding process (Bennett et al. 2007) and, therefore resembles the wild types. Through its agricultural use as leaf vegetable and as a source of seed oil it has been introduced to all continents and became naturalized in many countries (Zohary 1966).

Beside *E. sativa* there are two additional annual rocket species, namely *E. vesicaria* (L.) Cav. and *E. pinnatifida*, both occurring in the Iberian Peninsula and North Africa; they can be distinguished from *E. sativa* in having obtuse anthers and sepals that persist into the young fruit stage, whereas in *E. sativa* the sepals are deciduous after fertilization and the anthers are rather acute (Schulz 1919). Some taxonomists recognized these three taxa only as subspecies in *E. vesicaria* (Thellung, Year not given). However, interbreeding experiments that resulted in low interfertility between *E. sativa* and *E. vesicaria* (Sobrino Vesperinas 1995) reaffirmed the existence of several different annual biological species of *Eruca*. In addition, a genetic AFLP (Amplified Fragment Length Polymorphism) analysis of 184 accessions (Warwick et al. 2006), supported the three-species concept. Besides the annuals, two perennial species – *E. loncholoma* and *E. setulosa* – occur exclusively in North Africa, suggesting that this area is the genus diversification center and the origin of annual rocket. Such diversity within the genus indicates the importance of *Eruca* genetic resources with respect to its agronomical potential (Padulosi 1994).

Traditional Uses of *E. sativa*

The medicinal properties of rocket were recognized by the ancient Greeks and Romans; Pliny (23–77 CE) in *Historia Naturalis* refers to rocket as an antihelminthic, as a treatment for eye diseases and as an aphrodisiac; Dioscorides (40–90 CE) in *Materia Medica* mentioned its use in Roman and Egyptian cultures, referred to the use of *E. sativa* as a diuretic and a digestive agent, and also mentioned the consumption of seeds either as a corn or as a relish, similarly to mustards. It is interesting to note that Dioscorides mentioned a wild race of rocket originating from the Iberian Peninsula, which differed from the cultivated one in having a more peppery taste and better diuretic effect; this wild race may be associated with the aforementioned *E. vesicaria*. Rabbinical literature (third and sixth centuries BCE) indicates the use of *E. sativa* for stimulation of salivation and for improving vision (Löw 1928).

Medieval physicians mentioned a range of therapeutic effects of *E. sativa*: against dog bites, as a repellent of parasites, and a cure for kidney and liver problems; cosmetic applications included use as a deodorant and as a cure for acne (Yaniv 1997; Yaniv et al. 1998). Later, the Renaissance physician, Laguna (1499–1559) referred to the efficiency of rocket in treating scorpion stings, as an analgesic, in removing marks from the skin, and as an expectorant (Font Quer 2000). More information on traditional uses of *E. sativa* can be found in a review of literature of the Levant from the Muslim conquest to the Ottoman period; the plant was used in treating skin diseases, psychological problems and epilepsy, stomach and intestinal disorders, as well as hemorrhoids and sexual diseases (Lev 2002).

In herbal medicine, *E. sativa* is used against edema and as a rubefacient and an antiscorbutic (Burkill 1985; Thellung Year not given). Saad et al. (2008), in discussing Arab herbal medicine, refer to the use of *E. sativa* to treat psoriasis and as aphrodisiac; and Afifi and Abu-Irmaileh (2000) and Ertuğ (2000) describe its current use in Jordan and Turkey, respectively, to treat hypercholesteremia and to promote blood coagulation. It is also used against hair loss and mange in Iran (Rajaei and Mohamadi 2012), and against diabetes in North Africa (Jouad et al. 2001). In Central Asia, rocket seed oil, called taramira oil or jamba oil, is used for massage (Sastry 2003), hair treatment, and as a medicine against influenza (Ahmad et al. 2008).

Secondary Metabolites in *E. sativa*

In the Brassicaceae, the most prominent secondary metabolites are the sulfur-containing compounds glucosinolates (GLs), which have a structure of a β-D-glucopyranose residue linked to a (Z)-N-hydroximinosulfate ester via sulfur (Wittstock and Halkier 2002; Halkier and Gershenzon 2006). The GLs are derived from amino acids and are grouped into three categories: (1) aliphatic GLs – derived from methionine, alanine, valine, leucine, isoleucine; (2) aromatic GLs – derived from tyrosine, phenylalanine; and, (3) indolic GLs derived from tryptophane. These amino acids determine the final structure of the residue groups of the various GLs that are formed after several elongation and transformation steps.

About 120 different GLs have been identified in plants, mostly in the Brassicaceae, as well as in closely related plant families of the order Brassicales (Fahey et al. 2001;

Mithen et al. 2010). In *E. sativa*, the main GL is the aliphatic methionine-derived 4-mercaptobutyl GL (glucosativin), followed by lower concentrations of methylthiobutyl GL (glucoerucin) and methylsulfinylbutyl GL (glucoraphanin) (Bennett et al. 2006, 2007; Kim and Ishii 2006; Pasini et al. 2012). Other compounds found in *E. sativa* include the indolic 3-indolylmethyl GL (glucobrassicin) (Kim and Ishii 2006; Bennett et al. 2007; Cataldi et al. 2007; D'Antuono et al. 2008; Pasini et al. 2012) and 4-methoxy-3-indolylmethyl GL (4-methoxyglucobrassicin) (Bennett et al. 2002, 2006; Kim and Ishii 2006; Cataldi et al. 2007; D'Antuono et al. 2008). Another compound found in *E. sativa* is the 4-(β-D-glucopyranosyldisulfanyl)butyl GL (Kim et al. 2004; Kim and Ishii 2006; Cataldi et al. 2007), which contains two glucose moieties and an S–S bond (Cataldi et al. 2007). A dimeric form of glucosativin was also found in most studies on *E. sativa*, this disulfide GL is formed during the extraction steps (Bennett et al. 2002).

Differences were found among GL profiles from different tissues of *E. sativa* (Bennett et al. 2006; Kim and Ishii 2006): whereas glucosativin was the main GL in leaves, it was scarce in roots, which contained mostly glucoerucin and glucoraphanin (Barillari et al. 2005; Bennett et al. 2006; Kim and Ishii 2006). The GLs in seeds of *E. sativa* comprised mostly glucoerucin (94 %) and glucoraphanin (6 %), and the major GL in flowers was glucoraphanin (Bennett et al. 2006).

Several studies have revealed variations in GL profiles and concentrations among various accessions of *E. sativa* (Bennett et al. 2007; D'Antuono et al. 2008; Pasini et al. 2012; Villatoro-Pulido et al. 2012a). Total GLs in dry weight (DW) of leaves of various Italian commercial accessions ranged from 300 to more than 3,000 mg kg^{-1} (D'Antuono et al. 2008), and Bennett et al. (2007) reported variations in total GLs among accessions originating from Europe, Asia and North Africa, demonstrating the importance and the potential of genetic resources; GL concentrations in DW of sprouts ranged from 1.1 to 53.5 μ mol g^{-1}, depending on the sprouts' accession and age (Bennett et al. 2007). In comparison, DW of the edible parts of other crucifer crops, such as broccoli, cabbage, cauliflower and kale, were reported to contain total GLs of 12.8, 10.9, 15.1 and 15.0 μ mol g^{-1}, respectively (Kushad et al. 1999). Ontogenic profiling of GLs, and comparison among plant tissues demonstrated that seeds (Kim and Ishii 2006), sprouts (Bennett et al. 2006, 2007), and roots (Bennett et al. 2006; Kim and Ishii 2006) contained higher concentrations of GLs than leaves.

In addition to GLs, leaves of *E. sativa* contain considerable amounts of phenolic compounds, mainly the flavonoid kaempferol (Bennett et al. 2006; Martinez-Sanchez et al. 2007; Pasini et al. 2012). Moreover, Arabbi et al. (2004) showed that concentrations of kaempferol in leaves of *E. sativa* were higher than those in other vegetable crops such as lettuce, bell pepper, onion and chicory (Arabbi et al. 2004). In *E. sativa* leaves, flavonoids are present as complex polyglycosylated and/or polyglycosylated flavanols, and young seedlings contain acylated flavanols as well (Bennett et al. 2006). Three quercetin triglucosides and their acylated sinapoyl derivates were identified in leaves of *E. sativa* (Weckerle et al. 2001), and Bennett et al. (2006) have shown that *E. sativa* contained glycosides of kaempferol, quercetin, and isorhamnetin, which were present in considerable amounts in young seedlings, leaves and flowers, but in very small amounts (<1 μg g^{-1} in fresh weight) in roots (Bennett et al. 2006). Recently, two new malonylated kaempferol – glycoside [kaempferol 3-O-(2″-O-malonyl-β-D-glucopyranoside)-4′-O-β-D glucopyranoside and rhamnocitrin 3-O-(2″-O-methylmalonyl-β-Dglucopyranoside)- 4′-O-β-D-glucopyra-noside] were isolated and indentified in ethanolic extract of *E. sativa* (Michael et al. 2011). Flowers of *E. sativa* contained, in addition to the core flavonol aglycones, a complex of at least 16 different anthocyanins (Bennett et al. 2006).

Biological Activity of *E. sativa*

Hydrolysis of GLs by the enzyme myrosinase releases biologically active compounds (Rask et al. 2000). GLs are separated from myrosinase at the cellular or subcellular level; therefore the hydrolysis process can start as they come into contact following tissue damage. Consequently, the hydrolysis of the glucose moiety releases unstable aglycones, which can be rearranged to form isothiocyanates (ITCs), nitriles, etc. (Wittstock and Halkier 2002; Halkier and Gershenzon 2006). As discussed below, these hydrolysis products account for the main biological activity of *E. sativa* as well as that of other members of the Brassicaceae (Matusheski and Jeffery 2001; Kim and Ishii 2006; Moreno et al. 2006; Khoobchandani et al. 2010; Villatoro-Pulido et al. 2012a). The nature of the hydrolyzed GLs depends mainly on the structure of the residue, but also on pH, concentration of ferrous ions, and presence of specifier proteins (Hopkins et al. 2009; Winde and Wittstock 2011). The specifier proteins account for the diversity of hydrolyzed products of GLs, and lead to formation of nitriles instead of isothiocyanates (ITCs) that are spontaneously formed from any GL (Lambrix et al. 2001).

The ITCs form the major proportion of the volatiles of *E. sativa*, as revealed by solid-phase microextraction (SPME) (Jirovetz et al. 2002) and analysis of the essential oil (Miyazawa et al. 2002). Moreover, Miyazawa et al. (2002) detected 67 different volatile aroma compounds, which they categorized as ITCs, nitriles, n-alkanes and alcohols, which comprised 61.4, 11.5, 7.0 and 4.9 %, respectively, of the essential oil. The ITCs in *E. sativa* included: the 4-(methythio)butyl-ITC, a principal hydrolysis product of glucoerucin (Fenwick et al. 1983); bis-(4-isothiocyanatobutyl) disulfide, which is formed by oxidation of 4-(mercapto) butyl ITC (Cerny et al. 1996); sulforaphane (1-isothiocyanato-4-methylsulfinylbutane), typical of broccoli, was also detected in *E. sativa* (Liang et al. 2006), and so was erucin (Barillari et al. 2005). The last two ITCs were formed by hydrolysis of glucoraphanin (4-methylsulfinylbutyl GL) and glucoerucin, respectively, by myrosinase. Erysoline (4-methylsulfonylbutyl) ITC, released in hydrolysis of glucoraphanin, was detected by Lamy et al. (2008), who also found erucin, sulforaphane and phenylethyl ITC. A similar profile was determined by SPME analysis of oil extracted from *E. sativa* seeds, which contained 3-butenyl ITC and allyl ITC in addition to the above – i.e., bis(4-isothiocyanatobutyl) disulfide, sulforaphane, 2-phenylethyl ITC (Khoobchandani et al. 2010).

Indolic GLs contribute the bitter taste to crucifer vegetables (Engel et al. 2002), whereas ITCs are known to contribute to their aroma and flavor (Wittstock and Halkier 2002; Halkier and Gershenzon 2006). The unique flavor and aroma of *Eruca* are associated with the presence of a thiol group in the glucostavin GL and its corresponding ITC hydrolyzed compound (Bennett et al. 2007). Indeed, sensory analysis of *E. sativa* has shown that the perception of aroma and pungency of *E. sativa* was strongly associated with GL content (D'Antuono et al. 2009; Pasini et al. 2011), especially of the dimeric glucostavin (Pasini et al. 2011). However, its palatability was attributed to kaempferol glucoside and not only to GLs (Pasini et al. 2011).

GLs and their corresponding hydrolyzed forms act as toxic and deterrent compounds against herbivores (Hopkins et al. 2009; Muller et al. 2010; Wittstock and Burow 2010; Kos et al. 2012), and various studies have shown that nitriles are less toxic to insects than ITCs (Winde and Wittstock 2011 and references therein). But, not only do ITCs protect plants against herbivore damage; they offer potential in promoting human health (Mithen 2001; Traka and Mithen 2009). Among the biological activities of ITCs, cancer prevention and protection are attracting the most attention; these effects of ITCs are associated with induction of phase II enzymes, glutathione transferases and NAD(P)H:quinone reductase, which are important protectants in detoxification of electrophiles (Zhang et al. 1992; Keum et al. 2004). More recent findings indicate that cancer chemopreventive properties of ITCs, and especially sulforaphane, are associated with several mechanisms, including suppression of cytochrome P450 enzymes, induction of apoptotic pathways, suppression of cell cycle progression, and inhibition of angiogenesis (Juge et al. 2007; Traka and Mithen 2009).

Among the various ITCs, sulforaphane attracts considerable attention, in light of its biological activity: it is considered to exhibit high capacity for inducing phase II enzymes (Barillari et al. 2005) and shows strong antioxidant activity and thereby to protect cells against damage by reactive oxygen species (Fahey and Talalay 1999). Indeed, *E. sativa* and sulforaphane showed ability to counteract the genotoxic effects of hydrogen peroxide and to increase the life span of *Drosophila melanogaster* (Villatoro-Pulido et al. 2012a), and in human cell culture the antigenotoxic effect of *E. sativa* extract was attributed to the synergistic effects of erucin and erysolin (Lamy et al. 2008). Oxidative stress is strongly associated with cancer development and reactive oxygen species; hydrogen peroxide in particular is produced by cancer cells to enhance their survival and proliferation. Thus, the above studies provide further evidence of the beneficial health-promoting effects of *E. sativa*.

The potential of *E. sativa* as a functional food was also attributed to the antioxidant activity of glucoerucin and its corresponding hydrolyzed product, erucin (Barillari et al. 2005). In addition, isolation of desulfo-GLs and chemiluminescence measurement of antioxidant activity provided evidence that 4-methoxyglucobrassicin, 4-(β-D-glucopyranosyldisulfanyl)butyl GL, and the dimeric 4-mercaptobutyl GL, exhibited strong antioxidant activity (Kim et al. 2004; Kim and Ishii 2006). It was further suggested that not only glucoerucin, but also flavonoids in ethanolic extract of *E. sativa* were responsible for its *in vitro* antioxidant properties and its ability to reduce renal toxicity of mercury ($HgCl_2$) (Alam et al. 2007). Thus, apart from ITCs and GLs, flavonoids in *E. sativa*, i.e., quercetin and kaemferol, also have been shown to exhibit antioxidant activity (Madsen et al. 2000). The potential of quercetin as an antitumor agent was attributed to its effect on tyrosine kinase and its interaction with type II estrogen-binding sites, as well as to its low toxicity and easy absorption by humans (Lamson and Brignall 2000). Among their various biological activities, the anti-inflammatory effects of ITCs and their antibacterial properties against *Helicobacter pylori* – a bacterium associated with peptic ulcer – were indirectly associated with cancer prevention (Traka and Mithen 2009 and references therein). Further experimental systems have also confirmed the anti-ulcer activity of ethanolic extract of *E. sativa*, which significantly reduced gastric ulceration and ulcer activity in rats exposed to noxious ulcer-inducing chemicals (Alqasoumi et al. 2009).

The antibacterial effects of *E. sativa* extracts and of its seed oil were proved to be effective against several gram-negative and -positive bacteria, with the oil exhibiting the strongest activity (Khoobchandani et al. 2010; Gulfraz et al. 2011). In pharmacological experimental systems, extracts of leaves and of seeds, respectively, were shown to stimulate diuretic effects and to increase excretion of Na^+, K^+ and Cl^- in dogs (Mahran et al. 1991). Antihepatotoxic properties of *E. sativa* were also confirmed in experimental system with rats: liver functions improved under a diet supplemented with its ethanolic extract (Hussein et al. 2010). In addition to these effects, extracts of *E. sativa* seeds were proved to be efficient in reducing the toxic effects of aflatoxin in rabbits: they restored normal blood values, improved semen motility, and reduced the toxic pathological effects of the toxin on liver and kidneys (Hanafi et al. 2010). In addition, *E. sativa* was found to contain considerably higher content of ascorbic acid (vitamin C) about 1.5 mg g^{-1} in FW (Kim and Ishii 2007) – than other fruits and vegetables such as cabbage, cauliflower, spinach and also orange fruit, which contain about 0.5–0.8 mg g^{-1} (Davey et al. 2000). Furthermore, *E. sativa* is a good source of minerals such as iron, potassium, magnesium, copper and calcium (Villatoro-Pulido et al. 2012b). All of these characteristics indicate the wholesomeness of *E. sativa* for humans.

References

Afifi FU, Abu-Irmaileh B (2000) Herbal medicine in Jordan with special emphasis on less commonly used medicinal herbs. J Ethnopharmacol 72:101–110

Ahmad M, Khan MA, Zafar M (2008) Traditional herbal cosmetics used by local women communities in district Attock of Northern Pakistan. Indian J Tradit Knowl 7:421–424

Alam MS, Kaur G, Jabbar Z, Javed K, Athar M (2007) *Eruca sativa* seeds possess antioxidant activity and exert a protective effect on mercuric chloride induced renal toxicity. Food Chem Toxicol 45:910–920

Alqasoumi S, Al-Sohaibani M, Al-Howiriny T, Al-Yahya M, Rafatullah S (2009) Rocket "*Eruca sativa*": a salad herb with potential gastric anti-ulcer activity. World J Gastroenterol 15:1958–1965

Arabbi PR, Genovese MI, Lajolo FM (2004) Flavonoids in vegetable foods commonly consumed in Brazil and estimated ingestion by the Brazilian population. J Agric Food Chem 52:1124–1131

Barillari J, Canistro D, Paolini M, Ferroni F, Pedulli GF, Iori R, Valgimigli L (2005) Direct antioxidant activity of purified glucoerucin, the dietary secondary metabolite contained in rocket (*Eruca sativa* Mill.) seeds and sprouts. J Agric Food Chem 53:2475–2482

Bennett RN, Mellon FA, Botting NP, Eagles J, Rosa EAS, Williamson G (2002) Identification of the major glucosinolate (4-mercaptobutyl glucosinolate) in leaves of *Eruca sativa* L. (salad rocket). Phytochemistry 61:25–30

Bennett RN, Rosa EAS, Mellon FA, Kroon PA (2006) Ontogenic profiling of glucosinolates, flavonoids, and other secondary metabolites in *Eruca sativa* (salad rocket), *Diplotaxis erucoides* (wall rocket), *Diplotaxis tenuifolia* (wild rocket), and *Bunias orientalis* (Turkish rocket). J Agric Food Chem 54:4005–4015

Bennett RN, Carvalho R, Mellon FA, Eagles J, Rosa EAS (2007) Identification and quantification of glucosinolates in sprouts derived from seeds of wild *Eruca sativa* L. (salad rocket) and *Diplotaxis tenuifolia* L. (wild rocket) from diverse geographical locations. J Agric Food Chem 55:67–74

Burkill HM (1985) The useful plants of west tropical Africa, vol 1. Royal Botanic Gardens, Kew

Cataldi TRI, Rubino A, Lelario F, Bufo SA (2007) Naturally occurring glucosinolates in plant extracts of rocket salad (*Eruca sativa* L.) identified by liquid chromatography coupled with negative ion electrospray ionization and quadrupole ion-trap mass spectrometry. Rapid Commun Mass Spectrom 21:2374–2388

Cerny MS, Taube E, Battaglia R (1996) Identification of bis(4-isothiocyanatobutyl) disulfide and its precursor from Rocket salad (*Eruca sativa*). J Agric Food Chem 44:3835–3839

D'Antuono LF, Elementi S, Neri R (2008) Glucosinolates in *Diplotaxis* and *Eruca* leaves: diversity, taxonomic relations and applied aspects. Phytochemistry 69:187–199

D'Antuono LF, Elementi S, Neri R (2009) Exploring new potential health-promoting vegetables: glucosinolates and sensory attributes of rocket salads and related *Diplotaxis* and *Eruca* species. J Sci Food Agric 89:713–722

Davey MW, Van Montagu M, Inze D, Sanmartin M, Kanellis A, Smirnoff N, Benzie IJJ, Strain JJ, Favell D, Fletcher J (2000) Plant L-ascorbic acid: chemistry, function, metabolism, bioavailability and effects of processing. J Sci Food Agric 80:825–860

Engel E, Baty C, le Corre D, Souchon I, Martin N (2002) Flavor-active compounds potentially implicated in cooked cauliflower acceptance. J Agric Food Chem 50:6459–6467

Ertuğ F (2000) An ethnobotanical study in central Anatolia (Turkey). Econ Bot 54:155–182

Fahey JW, Talalay P (1999) Antioxidant functions of sulforaphane: a potent inducer of phase II detoxication enzymes. Food Chem Toxicol 37:973–979

Fahey JW, Zalcmann AT, Talalay P (2001) The chemical diversity and distribution of glucosinolates and isothiocyanates among plants. Phytochemistry 56:5–51

Fenwick GR, Heaney RK, Mullin WJ (1983) Glucosinolates and their breakdown products in food and food plants. Crit Rev Food Sci Nutr 18:123–201

Font Quer P (2000) Plantas Medicinales, el Dioscórides Renovado, 2nd edn. Península, Barcelona

Gulfraz M, Sadiq A, Tariq H, Imran M, Qureshi R, Zeenat A (2011) Phytochemical analysis and antibacterial activity of *Eruca sativa* seed. Pak J Bot 43:1351–1359

Halkier BA, Gershenzon J (2006) Biology and biochemistry of glucosinolates. Annu Rev Plant Biol 57:303–333

Hanafi EM, Hegazy EM, Riad RM, Amer HA (2010) Bio-protective effect of *Eruca sativa* seed oil against the hazardus effect of aflatoxin B1 in male rabbits. Int J Acad Res 2:67–74

Hopkins RJ, van Dam NM, van Loon JJ (2009) Role of glucosinolates in insect-plant relationships and multitrophic interactions. Annu Rev Entomol 54:57–83

Hussein J, Salah A, Oraby F, El-Deen AN, El-Khayat Z (2010) Antihepatotoxic effect of *Eruca sativa* extracts on alcohol induced liver injury in rats. J Am Sci 6:381–389

Jirovetz L, Smith D, Buchbauer G (2002) Aroma compound analysis of *Eruca sativa* (Brassicaceae) SPME headspace leaf samples using GC, GC-MS, and olfactometry. J Agric Food Chem 50:4643–4646

Jouad H, Haloui M, Rhiouani H, El Hilaly J, Eddouks M (2001) Ethnobotanical survey of medicinal plants used for the treatment of diabetes, cardiac and renal diseases in the North centre region of Morocco (Fez–Boulemane). J Ethnopharmacol 77:175–182

Juge N, Mithen RF, Traka M (2007) Molecular basis for chemoprevention by sulforaphane: a comprehensive review. Cell Mol Life Sci 64:1105–1127

Keum YS, Jeong WS, Kong ANT (2004) Chemoprevention by isothiocyanates and their underlying molecular signaling mechanisms. Mutat Res-Fundam Mol Mech Mutagen 555:191–202

Khoobchandani M, Ojeswi BK, Ganesh N, Srivastava MM, Gabbanini S, Matera R, Iori R, Valgimigli L (2010) Antimicrobial properties and analytical profile of traditional *Eruca sativa* seed oil: comparison with various aerial and root plant extracts. Food Chem 120:217–224

Kim SJ, Ishii G (2006) Glucosinolate profiles in the seeds, leaves and roots of rocket salad (*Eruca sativa* Mill.) and anti-oxidative activities of intact plant powder and purified 4-methoxyglucobrassicin. Soil Sci Plant Nutr 52:394–400

Kim SJ, Ishii G (2007) Effect of storage temperature and duration on glucosinolate, total vitamin C and nitrate contents in rocket salad (*Eruca sativa* Mill.). J Sci Food Agric 87:966–973

Kim SJ, Jin S, Ishii G (2004) Isolation and structural elucidation of 4-(beta-D-glucopyranosyldisulfanyl)butyl glucosinolate from leaves of rocket salad (*Eruca sativa* L.) and its antioxidative activity. Biosci Biotech Biochem 68:2444–2450

Kos M, Houshyani B, Wietsma R, Kabouw P, Vet LE, van Loon JJ, Dicke M (2012) Effects of glucosinolates on a generalist and specialist leaf-chewing herbivore and an associated parasitoid. Phytochemistry 77:162–170

Kushad MM, Brown AF, Kurilich AC, Juvik JA, Klein BP, Wallig MA, Jeffery EH (1999) Variation of glucosinolates in vegetable crops of *Brassica oleracea*. J Agric Food Chem 47:1541–1548

Lambrix V, Reichelt M, Mitchell-Olds T, Kliebenstein DJ, Gershenzon J (2001) The *Arabidopsis* epithiospecifier protein promotes the hydrolysis of glucosinolates to nitriles and influences *Trichoplusia ni* herbivory. Plant Cell 13:2793–2807

Lamson DW, Brignall MS (2000) Antioxidants and cancer, Part 3: quercetin. Altern Med Rev 5:196–208

Lamy E, Schroder J, Paulus S, Brenk P, Stahl T, Mersch-Sundermann V (2008) Antigenotoxic properties of *Eruca sativa* (rocket plant), erucin and erysolin in human hepatoma (HepG2) cells towards benzo(a)pyrene and their mode of action. Food Chem Toxicol 46:2415–2421

Lev E (2002) Reconstructed materia medica of the Medieval and Ottoman al-Sham. J Ethnopharmacol 80:167–179

Liang H, Yuan QP, Dong HR, Liu YM (2006) Determination of sulforaphane in broccoli and cabbage by high-performance liquid chromatography. J Food Compos Anal 19:473–476

Löw I (1928) Die Flora der Juden, vol 1. R. Löwit Verlag, Wien/Leipzig

Madsen HL, Andersen CM, Jorgensen LV, Skibsted LH (2000) Radical scavenging by dietary flavonoids. A kinetic study of antioxidant efficiencies. Eur Food Res Technol 211:240–246

Mahran GH, Kadry HA, Isaac ZG, Thabet CK, Alazizi MM, Elolemy MM (1991) Investigation of diuretic drug plants. 1. Phytochemical screening and pharmacological evaluation of *Anethum graveolens* L, *Apium graveolens* L, *Daucus carota* L and *Eruca sativa* Mill. Phytother Res 5:169–172

Martinez-Sanchez A, Llorach R, Gil MI, Ferreres F (2007) Identification of new flavonoid glycosides and flavonoid profiles to characterize rocket leafy salads (*Eruca vesicaria* and *Diplotaxis tenuifolia*). J Agric Food Chem 55:1356–1363

Matusheski NV, Jeffery EH (2001) Comparison of the bioactivity of two glucoraphanin hydrolysis products found in broccoli, sulforaphane and sulforaphane nitrile. J Agric Food Chem 49:5743–5749

Michael HN, Shafik RE, Rasmy GE (2011) Studies on the chemical constituents of fresh leaf of *Eruca sativa* extract and its biological activity as anticancer agent in vitro. J Med Plant Res 5:1184–1191

Mithen R (2001) Glucosinolates – biochemistry, genetics and biological activity. Plant Growth Regul 34:91–103

Mithen R, Bennett R, Marquez J (2010) Glucosinolate biochemical diversity and innovation in the Brassicales. Phytochemistry 71:2074–2086

Miyazawa M, Maehara T, Kurose K (2002) Composition of the essential oil from the leaves of *Eruca sativa*. Flavour Fragr J 17:187–190

Moreno DA, Carvajal M, Lopez-Berenguer C, Garcia-Viguera C (2006) Chemical and biological characterisation of nutraceutical compounds of broccoli. J Pharm Biomed 41:1508–1522

Muller R, de Vos M, Sun JY, Sonderby IE, Halkier BA, Wittstock U, Jander G (2010) Differential effects of indole and aliphatic glucosinolates on lepidopteran herbivores. J Chem Ecol 36:905–913

Padulosi S (1994) Rocket genetic resources network. Report of the first meeting 13–15 November 1994, IPGRI, Lisbon, Portugal

Pasini F, Verardo V, Cerretani L, Caboni MF, D'Antuono LF (2011) Rocket salad (*Diplotaxis* and *Eruca* spp.) sensory analysis and relation with glucosinolate and phenolic content. J Sci Food Agric 91:2858–2864

Pasini F, Verardo V, Caboni MF, D'Antuono LF (2012) Determination of glucosinolates and phenolic compounds in rocket salad by HPLC-DAD-MS: evaluation of *Eruca sativa* Mill. and *Diplotaxis tenuifolia* L. genetic resources. Food Chem 133:1025–1033

Rajaei P, Mohamadi N (2012) Ethnobotanical study of medicinal plants of Hezar Mountain allocated in south east of Iran. Iran J Pharm Res 11:1153–1167

Rask L, Andreasson E, Ekbom B, Eriksson S, Pontoppidan B, Meijer J (2000) Myrosinase: gene family evolution and herbivore defense in Brassicaceae. Plant Mol Biol 42:93–113

Saad B, Azaizeh H, Said O (2008) Arab herbal medicines. In: Watson RR, Preedy VR (eds) Botanical medicine in clinical practice. CAB International, Cambridge, pp 31–39

Sastry ED (2003) Taramira (*Eruca sativa*) and its improvement – a review. Agric Rev 24:235–249

Schulz OE (1919) Cruciferae – Brassiceae Pars Prima. In: Das Pflanzenreich, vol IV. Wilhelm Engelmann, Leipzig, p 105

Sobrino Vesperinas E (1995) Diferencias morfologicas e interfertilidad entre las especies arvenses *Eruca vesicaria* (L.) Cav. Y *Eruca sativa* Miller. In: Altoaragoneses IE (ed) Congreso 1995 de la Sociedad Española de Malherbología. Actas, Huesca, pp 153–156

Thellung A (Year not given) Dicotyledones II Teil. In: Hegi G (ed) Illustrierte Flora von Mitteleuropa, vol 4(1). J. F. Lehmanns Verlag, Muenchen

Traka M, Mithen R (2009) Glucosinolates, isothiocyanates and human health. Phytochem Rev 8:269–282

Villatoro-Pulido M, Font R, Saha S, Obregon-Cano S, Anter J, Munoz-Serrano A, De Haro-Bailon A, Alonso-Moraga A, Del Rio-Celestino M (2012a) In vivo biological activity of rocket extracts (*Eruca vesicaria* subsp *sativa* (Miller) Thell) and sulforaphane. Food Chem Toxicol 50:1384–1392

Villatoro-Pulido M, Moreno Rojas R, Munoz-Serrano A, Cardenosa V, Amaro Lopez MA, Font R, Del Rio-Celestino M (2012b) Characterization and prediction by near-infrared reflectance of mineral composition of rocket (*Eruca vesicaria* subsp. *sativa* and *Eruca vesicaria* subsp. *vesicaria*). J Sci Food Agric 92:1331–1340

Warwick SI, Guge RK, Gómez-Campo C, James T (2006) Genetic variation in *Eruca vesicaria* (L.) Cav. Plant Genet Resour 5:142–153

Weckerle B, Michel K, Balazs B, Schreier P, Toth G (2001) Quercetin 3,3′,4′-tri-O-beta-D-glucopyranosides from leaves of *Eruca sativa* (Mill.). Phytochemistry 57:547–551

Winde I, Wittstock U (2011) Insect herbivore counter adaptations to the plant glucosinolate-myrosinase system. Phytochemistry 72:1566–1575

Wittstock U, Burow M (2010) Glucosinolate breakdown in *Arabidopsis*: mechanism, regulation and biological significance. In: The *Arabidopsis* book, vol 8. American Society of Plant Biologists, Rockville, p e0134. doi:10.1199/tab.0134

Wittstock U, Halkier BA (2002) Glucosinolate research in the *Arabidopsis* era. Trends Plant Sci 7:263–270

Yaniv Z (1997) Tradition, uses and research on rocket in Israel. In: Padulosi S, Pignone D (eds) Rocket: a Mediterranean crop for the world. IPGRI, Rome, pp 76–80

Yaniv Z, Schafferman D, Amar Z (1998) Tradition, uses and biodiversity of rocket (*Eruca sativa*, Brassicaceae) in Israel. Econ Bot 52:394–400

Zhang YS, Talalay P, Cho CG, Posner GH (1992) A major inducer of anticarcinogenic protective enzymes from broccoli – isolation and elucidation of structure. Proc Natl Acad Sci U S A 89:2399–2403

Zohary M (1966) Flora Palaestina, vol 1. The Israel Academy of Sciences and Humanities, Jerusalem

Oil Rose (*Rosa damascena*)

Kemal Hüsnü Can Baser and Neset Arslan

Abstract Rose oil of commerce is obtained by water distillation of fresh rose flowers of *Rosa damascena* which is a cultivated species. 3.5–4 t of flowers are needed to get 1 kg of this highly priced oil. Top ranking cultivators of oil rose and producers of rose oil are Turkey and Bulgaria producing evenly the annual world demand of 3 t of oil.

Due to very specific soil and climatic requirements, oil rose cultivation for a profitable crop is confined to Kazanlik region of Bulgaria and Isparta region of Turkey. Apart from rose oil, rose concrete (rose flower extract), rose water (rose hydrosol), rose jam, rose syrup, dried rose buds, rose cosmetics are also produced commercially from oil rose.

The chapter covers the cultivation practices, pest and weed control, production and uses of oil rose in a comprehensive manner.

Keywords Rose oil • Oil rose • *Rosa damascena* • Rose cultivation • Rose oil distillation • Rose water • Rose concrete • Rose products • Turkey • Bulgaria

General Aspects

Description of the Plant

Roses are perennial bushy plants with taproot containing hairy roots. Oil rose (*Rosa damascena* Miller) can attain a height of 1–3 m. Stem is cylindric, dark coloured and multibranched. Branches bear numerous bristles and prickles. Stem contains

K.H.C. Baser (✉)
College of Science, Department of Botany and Microbiology, King Saud University, Riyadh, Saudi Arabia
e-mail: Khcbaser@gmail.com

N. Arslan
Faculty of Agriculture, Ankara University, Ankara, Turkey
e-mail: Neset.Arslan@agri.ankara.edu.tr

Z. Yaniv, N. Dudai (eds.), *Medicinal and Aromatic Plants of the Middle-East*, Medicinal and Aromatic Plants of the World 2, DOI 10.1007/978-94-017-9276-9_16

buds which will give rise to leaves or shoots. Nodes bear dark green oval dentate 3–4 cm long 3–7 leaves with waxy and slippy surface and hairy underneath. Flowers are solitary, rarely in corymbs, hermaphrodite. Sepals 5, reflexed in full bloom. Petals 30, semi-double, oval, pink or pale pink, white spotted at base and highly fragrant. Numerous stamens and styles are placed in a bowl shaped receptacle. Rose flowers have a high rate of foreign pollination. Pollination is by bees and wind. Receptacle fatten by time and assumed red colour, and contains hairy nuts (Kurkcuoglu 1988; Ozcelik et al. 2009; Baydar and Kazaz 2010).

Classification

Ca. 100–250 species of roses are estimated to exist. This is due to self hybridization of rose taxa. The genus Rosa is separated to four subgenera *Hulthemia* (Dumort.) Focke, *Platyrhodon* (Hurst) Rehder, *Hesperhodos* and *Rosa*, however, these subgenera are monotypic and correspond to one or two species. All other species belong to subgenus *Rosa*. Subgenus Rosa is divided into 10 sections, *Banksianae* Lindl. (2 spp.), *Bracteatae* Thory (1–2 sp.), *Caninae* (DC.) Ser. (*ca.* 50 spp.), *Carolinae* Crep. (5 spp.), *Cinnamomeae* (DC.) Ser. (*ca.* 80 spp.), *Indicae* Thory (3 spp.), *Laevigatae* Thory (1 sp.), *Pimpinellifoliae* (DC.) Ser. (*ca.* 15 spp.), *Rosa* (1 sp.) and *Synstylae* DC. (*ca.* 25 spp.). (Anon. 2013g).

24 species are recorded in the Flora of Turkey (Nilsson 1972). Rose cultivars are classified in a different way as rose societies divide them into several distinct groups. In Turkey, there are 24 garden roses, eight being natural. There are an estimated more than 18.000 (45.000 to some sources) rose cultivars. New studies increase the number. Contrary to garden cultivars, improvement of oil rose cultivars is not common (Baytop 2001; Cheers 1999; Özçelik et al. 2009; Ritz 2005).

Origin and Distribution

Recent fossil records suggest that rose has been in existence in the World for the last 35 million years. Roses grow naturally in the northern hemisphere. Gene centre is believed to be Asia. Ancient Chinese, Assyrian and Sanskrit literature mention about the fine fragrance of roses. Petrified rose wreaths were discovered in ancient Egyptian tombs. Egyptian queen Cleopatra is said to love roses and rose fragrance. It also has importance in the ancient Greek and Roman cultures. Rose oil and rose water has been used in religious ceremonies for at least two millenia. Rose water production by immersing rose flowers in warm water and olive oil macerated rose oil have been known for centuries but there is no reliable early record of rose oil by distillation. Earliest records of rose oil distillation can be found in the works of Ibn Haldun (1332–1406). He mentioned that the most precious rose oil was obtained by distillation, and it was traded as a commodity in eighth and ninth centuries in India and China. This means that distilled rose oil was already known in Asia earlier. Distillation was known by the Indus Valley civilization 5,000 years ago since artefacts of an early earthenware distillation apparatus was discovered in Harappa. Traditional attar production has been in existence in Kannauj, India and rose attar is one of their products.

Rose oil and rose water distillation was performed in Edirne during the Ottoman Empire. Katip Chelebi wrote in his book "The Geography of Thrace and Bosnia" that roses were grown in large gardens and rose water was produced in early seventeenth century in Edirne, Turkey.

There was a gülhane (rose works) in the Ottoman palace and a rose distillation facility at a hospital in Edirne in the fifteenth century. The culture of rose oil and rose water have existed in the Ottoman society. A gülhane was also built in Istanbul, the new Capital of the Ottoman Empire after 1453, at Topkapi palace between 1460 and 1478. Evliya Chelebi, an Ottoman travelogue, mentions that there were 300 shops in Istanbul selling rose water in the seventeenth century. In the eighteenth century, Kazanlik, Zagra, Karlova and Plovdiv in Bulgaria (then Ottoman land) became a centre for rose cultivation and rose oil and rose water production. Ottomans, then, started getting their rose supplies from those areas. Oil rose was brought back to Turkey after the 1877–1878 Balkan war along with Turkish immigrants fleeing into Turkey from the Balkans. Those coming from Kazanlik area were accommodated in the Cavusbasi Farm of the Ottoman Sultan Abdulhamit II (1842–1918) in Istanbul and started growing oil roses and rose oil distillation was carried out in 1880s. With royal decree, oil rose cultivation was encouraged in Bursa, Adana, Ankara, Aydın, Diyarbakır, Edirne, Kastamonu, Konya, Kayseri, Erzurum, Trabzon provinces of Turkey by distributing rose plants to farmers free of charge. However, due to availability of insufficient number of stills and poor yields of rose oil, finally rose cultivation was confined to Isparta, Burdur provinces where oil rose cultivation was initiated unilaterally by Ismail Efendi (1840–1915) (Arslan 2002; Baser et al. 2012; Baydar and Kazaz 2010; Baytop 2001; Karabulut 1993; Kürkçüoğlu 1988; Özçelik et al. 2009).

Soil Requirements

Roses prefer fresh, loamy soils rich in organic matter. Such soils are sticky, fast-drying and easy to treat soils. Generally roses can be grown in all kinds of garden soils. However, they do not grow only in extreme conditions such as clayey, sandy or too calcareous soils. As rose is a perennial plant soil preference is important. According to the soil characteristics and location, a plantation can be beneficial for 20–25 years with rejuvenescence pruning between 7 and 12 years. *Rosa damascena* is not so selective for soil. However, it does not like too clayey and calcareous soils. Root system of the rose is sensitive. Rose shows good development in slightly sandy, deep, penetrable soils rich in organic matter to enable absorption of nutrients. Trace amounts of iron salts and slight moisture are beneficial. Rose plantations in

Isparta, Turkey are generally in deep and sandy soils. However, plantations may also occur in very variable soil types. Rose plantations should not be established in too clayey calcareous, not so deep, hydrophilic soils where root rot and chlorosis may occur. Loamy and salt-free soils with pH 6–8 are ideal for rose planting (Açatay 1969; Anon. 1987, 2009a; Arslan 1994; Baydar and Kazaz 2010; Bilir 2010; Işık 1953; Kürkçüoğlu 1988; Özcan 1998; Temurcin 2004; Yalcin et al. 1992; Yılmaz 1992).

Climatic Requirements

Rose is a natural and ornamental plant grows in all parts of the northern hemisphere except for areas with very harsh unfavourable climate. Rose planting is not so common in the southern hemisphere. Important countries planting roses for rose oil production are mainly Turkey and Bulgaria, and Iran, Russia, Morocco, India and China. 80 % of oil rose cultivation in Turkey is in Isparta and the rest in Burdur and Afyon provinces. A small amount of cultivation occurs in Karacasu of Aydın province.

Oil rose cultivation occurred in some central and eastern Anatolian provinces before the 1st world war. Rose likes a temperate climate. It does not like too hot and too cold occasions. It grows well in sunny environments. Plantations should be in open and sunny areas. Rose prefers areas where there is adequate rain, morning dew is frequent during flowering and devoid of drought, frost and white frost in spring months. Temperature and moisture influence flowering time and period. High temperature and drought during flowering in the months of May and June result in oil losses from the petals. Air temperature is preferred to be between 10 and 15 °C during the flowering period for better oil yields. Night temperatures below 5 °C and day temperatures above 20 °C decrease oil yield. Very hot days accelerate the rate of flowering. Deciduous oil rose plants are resistant to frost in winter months. However, they are highly sensitive to low temperatures and frost in budding period which starts in March. Early whipping type pruning result in damage due to frost in pruning sites. Grave frost damage may occur in young shoots due to spring frost in some years. Shoots with flower buds occur in the nodes of annual branches in spring. Therefore, early spring climatic conditions are crucial for the number and development of flower buds. High temperature, low moisture, dry and hot wind force buds to flower before full maturity resulting in light pink flowers with low oil yield. Regular rains in spring and early summer prevents plants to suffer water stress. Ideal relative humidity over 60 % (preferably ca. 70 %) contributes positively to oil accumulation in flowers in these months. However, heavy rains hamper flower harvest and inflict physical damage to flowers incurring oil loses. Cool light winds ensure slow opening of flowers and increase oil yield. On the contrary, heavy storms damage buds and negatively affect yield.

Isparta has appropriate climatic conditions to grow oil rose (Table 1). Isparta is located in the transition area between the Mediterranean and continental (central

Table 1 Some average climatic data for Isparta province (1960–2012)

Months	Jan	Feb	Mar	Apr	May	Jun	Jul	Aug	Sep	Oct	Nov	Dec
Average temp (°C)	1.9	2.8	6.1	10.7	15.6	20.2	23.6	23.2	18.6	13.0	7.4	3.5
Average highest temp (°C)	6.4	7.6	11.7	16.5	21.7	26.6	30.4	30.5	26.5	20.5	13.9	8.1
Average lowest temp (°C)	−2.0	−1.4	0.9	4.7	8.4	12.3	15.3	14.9	10.6	6.5	2.2	−0.4
Average sunny period (hour)	3.5	4.5	5.6	6.5	8.4	10.4	11.4	11.1	9.5	7.1	5.3	3.3
Average rainy days	12.0	11.3	10.7	10.9	10.2	6.4	3.0	2.2	3.7	6.4	7.6	12.4
Monthly average total rain (kg/m²)	72.7	64.8	54.3	56.1	50.1	29.5	13.5	10.5	15.6	36.4	46.1	85.6
Average monthly relative humidity (%)	77	74	66	60	60	52	45	44	50	62	71	77
Lowest and highest values over the years (1960–2012)												
Highest temperature (°C)	17.6	19.0	26.8	28.4	32.0	36.2	42.3	41.2	35.6	31.0	25.4	20.0
Lowest temperature (°C)	−19.2	−21.0	−18.5	−7.7	−1.2	4.5	4.9	7.0	−0.8	−4.2	−11.5	−15.0

Ref. : General Directorate of Meteorology

Anatolia) climates (Anon. 1987, 2009a; Arslan 1994; Baydar and Kazaz 2010; Bilir 2010; Özcan 1998; Temurçin 2004; Yalçın et al. 1992; Yılmaz 1992).

Therefore, the province enjoys both climatic conditions. Summers are not as warm as Aegean and South Anatolia and winters are not as harsh as Central Anatolia. Average temperature rarely exceeds 30 °C in summer and hardly drops below −15 °C in winter months. Precipitation has a typical transitionary character and annual average rain fall is 500–600 mm. Distribution of precipitation according to seasons is 41 % in winter, 29 % in spring, 9 % in summer and 21 % in autumn. Another characteristics of the region is that it is located in Lakes District possessing 6 lakes and humidity is 65–70 %. Average temperatures in the flowering season in May is 15.3 °C in Isparta, 16.3 °C in Burdur and 15.2 °C in Afyon (Anon. 1987, 2009a, 2013b; Baydar and Kazaz 2010; İkiz 2011; Özcan 1998; Temurçin 2004; Yılmaz 1992).

Flowering season is May and June in Bulgaria and Turkey. Mean temperature in Kazanlık is 15.6 °C in May and 19.2 °C in June (Yılmaz 1992).

Table 2 World production of rose flowers (ton)

Year	Turkey	Bulgarıa
2003	6,000	5,000
2004	6,000	5,000
2005	6,500	5,500
2006	7,500	6,000
2007	6,500	5,000
2008	8,400	6,000
2009	6,500	5,500
2010	6,000	5,000
2011	6,000	3,500
2012	6,500	4,500

Source: Gülbirlik

Production Levels

Production of Rose Flower in the World

Main countries producing rose flower in the world are Turkey and Bulgaria. Both countries supply rose oil and rose concrete to the world perfumery and cosmetic industries. Production figures of rose flowers in the last 9 years are shown in Table 2.

Apart from these countries, Iran, Morocco, India, Afghanistan, Pakistan, France, Saudi Arabia, China also cultivate oil rose. Iran has been cultivating it for long. Cultivation areas in Iran are 1,610 ha, production 5,483 t and the yield is 3,401 kg/ha. Production is mainly in Isfahan and western Azerbaijan, and also in eastern Azerbaijan, Horasan, Kashan, Semnan, Gilan, Merkezi and Yazd provinces. 90 % of the rose fields and 91.5 % of the production is in Isfahan and western Azerbaijan. Although rose production is as good as Turkey and Bulgaria, its position in the world rose oil trade is very limited and instead rose water production and utilization is more frequent. In Morocco, companies of French origin cultivate roses in fields rented or purchased, and process them in their factories at low cost. Indian, Chinese and Saudi Arabian rose oils are mainly consumed in domestic markets. Those countries with low production volumes do the same. In Afghanistan, rose cultivation is considered as an alternative to illicit opium poppy cultivation (Anon. 1987, 1991, 2009a, 2011, 2012, 2013a, c, d; Arslan 1994, 2002; Baydar 2006; Baydar and Kazaz 2010; Baytop 1963, 2001; Javani et al. 2013; Kovacheva et al. 2010; Kovacheva 2011).

Production in Turkey

Rosa damascena (Damask rose), an old cultivar whose origin is unknown, has been cultivated in and around Isparta. 80 % of oil rose cultivation is in Isparta and the rest in Burdur and Afyon provinces. A small amount is cultivated in Denizli and Karasu town of Aydın. In some other provinces oil roses are also grown for ornamental purposes and for rose jam production. Rose cultivation areas, production and yield figures are given in Table 3.

Table 3 Rose fields, rose flower production and yield in Turkey

Year	Cultivated fields (Ha)	Production (Ton)	Yield (Kg/Ha)
2004	1,200	6,000	5,000
2005	1,300	6,500	5,000
2006	1,500	7,500	5,000
2007	1,500	6,500	4,300
2008	1,700	8,400	5,500
2009	1,700	6,500	5,000
2010	1,600	6,000	4,000
2011	1,800	6,000	3,000
2012	1,900	6,500	3,430

Source: Gülbirlik (Anon. 2011, 2012)

Cultivation of oil rose (*Rosa damascena*) is common in Lakes District of South-West Anatolia. Important cultivation areas in Isparta are located in Keçiborlu, Gönen and Central townships. Furthermore, Eğirdir, Uluborlu, Atabey, Aksu and Sütçüler are other production centres. Central township, Ağlasun and Çeltikçi in Burdur and Başmakçı in Afyon are also centres of production (Anon. 1987, 2009a, 2011, 2012, 2013a, c, d; Arslan 1994, 2002; Baydar 2006; Baydar and Kazaz 2010; Baytop 1963, 2001; Bilir 2010; İkiz 2011).

Major Production Areas in Africa

Morocco is the major rose growing area in Africa. Studies are ongoing for rose cultivation in South Africa and Ethiopia (Anon. 2009b, 2013e; Frederick et al. 2002).

Cultivars

Main Rose species used in rose oil and rose water production in the world are *Rosa damascena* Mill., *Rosa gallica* L., *Rosa centifolia* L. *Rosa alba* L. *Rosa moschata* Mill. *R. bourboniana* L. (Edward rose). The former four are also the oldest garden roses. Most important of them is *Rosa damascena* Mill. which is known as Pembe yağ gülü, Şam gülü, Isparta gülü or Damask rose. It is believed to be a hybrid of *Rosa gallica* L. and *Rosa phoenicia* Boiss. Although ornamental roses have numerous registered cultivars, sorts of oil roses are quite limited. This is probably due to growing oil rose vegetatively and mainly by traditional methods. No oil rose cultivar is registered in Turkey. Research has indicated that roses cultivated in Turkey have genetical resemblance (Baydar et al. 2004). However, there are researchers who claim that there are 15 different oil rose cultivars is Turkey. Isparta rose was registered as geographical indication in 2006. Similar is also indicated in Bulgaria. However, there are also registered oil rose cultivars in Bulgaria.

Research in Iran and Pakistan has indicated that there are genotypic variations among oil rose cultivars in these countries. ***Rosa gallica*** L. (French rose) is naturally distributed in Turkey, South and Central Europe, Caucasia and Iraq (Dehghan et al. 2012). Gallica group constitutes taxonomical basis of oil roses. In the past, it was used for rose water production. Its dried flowers are used to flavour teas. ***Rosa centifolia*** L. (May rose) is a rose species used for rose water production and for ornamental purposes. It is also used as rootstock. It has highly fragrant pink flowers. *ca.* 10 cultivars are believed to exist in Turkey. It used to be cultivated in Burdur as oil rose. Flowers are heavier and easier to harvest compared to Isparta rose. It is grown in Turkey for landscaping and also for household food and fragrance materials. It is preferred by florists. It is grown in France and Morocco as oil rose. ***Rosa alba*** L. (White rose) which is a hybrid of *R. corymbifera x gallica* is one of the oldest garden roses. This fragrant rose is found among oil roses cultivated in Kılıç (Keçiborlu) and Yakaören (Isparta). It is easily distinguished by its leaves and habitus even outside the flowering season.

It is grown in areas where *Rosa damascena* cultivation is not economical, for rose water and rose jam production at smaller scale. ***Rosa moschata*** Mill. (Musk rose) is a recent record for the flora of Turkey. Its distribution is quite narrow. Baytop (2001) mentions that it has not been found but believed to exist in Turkey according to some records. It is known to be grown in Iran. ***R. bourboniana*** Desf. (Edward rose, Bourbon rose) is a hybrid of *R. chinensis* and *R. damascena*. It is a recent record for Turkey and is cultivated in Isparta, Denizli, Istanbul and Edirne. Since it is showy and fragrant, it is grown and distributed by florists. It is preferred in graveyards, tombs, mosque gardens, parks and home gardens. According to Roberts et al. (2003), 20 cultivars exist in Turkey. It is an old native garden rose of Turkey. Its flowers are used for rose water and rose jam production in some parts of Turkey. It is locally named as jam rose or fragrant rose. Most of the roses in seven flowering or 11 month group belong to this taxon. It is grown in India as oil rose.

Rosa damascena appears to be the most appropriate among the above roses in terms of oil yield and quality. However, its flowering period is short and flowers only once. There is a seven flowering variety of this species, *R. damascena* var. *semperflorens* (Yediveren Şam gülü or Kral gülü) which is grown in Anatolia as a garden rose. It is also rich in oil. Other species have varieties which flower 2–3 times or ever flower (Anon. 1987, 2009a, 2013a; Gora et al. 1995; Arslan 2002; Baydar and Baydar 2005; Baydar 2006; Baydar and Kazaz 2010; Baytop 1963, 2001; Bilir 2010; İkiz 2011; Frederick et al. 2002; Dobreva and Kovacheva 2010; Hussain and Khan 2004; Kaul et al. 2009; Kovacheva 2011; Kovacheva et al. 2010; Özçelik et al. 2009; Farooq 2011; Nedkov et al. 2009; Rusanov et al. 2005, 2009; Zeinali et al 2009; Riaz et al 2012).

Cultivation Practices

Propagation

Rose is propagated vegetatively via cuttings. During the establishment of a rose plantation, rose cuttings are obtained from 7 to 12 year old roses via rejuvenational pruning. This is named as “cutting” technique. 400 kg (1,200–1,400) rose cuttings are required to establish 1 decare rose plantation. Lignified thick cuttings are cut from 4 to 5 cm of soil level. They must be pruned and 100–150 cm in length. This is a traditional technique with propagation rate of ca. 1:8 – 1:10. Although tested and proven, in practice, due to planting of cuttings without further processing there is a risk to transfer diseases to the new plantation and having gaps between the plants. A selection is out of question. Furthermore, layering is also employed. In that, one branch is submerged in soil without cutting to initiate rooting. Propagation rate is 1:10 – 1:15. It is possible to select good rootstocks and plants propagate earlier. Another method is grafting oil rose to rose rootstocks (*e.g., Rosa canina*). Propagation rate with this method is very high and it allows grafting from good quality roses. Many studies are ongoing for the propagation of roses. In addition, rooted wood and green cuttings are also used in propagation. This is also applied in Turkey to a lesser extent (Anon. 1987, 2009a, Arslan 1994, 2002; Baydar and Kazaz 2010; Baytop 1963, 2001; Bilir 2010; İkiz 2011; Frederick et al 2002; Dobreva and Kovacheva 2010; Ginova et al. 2012; Hussain and Khan 2004; Özçelik et al. 2009; Pati et al. 2012).

Soil Preparation

Establishment of a plantation: Fields must be spacious, airy, sunny and well lit. They must be closed to cold and strong winds which may affect rose yield. In order not to be affected negatively from frost, fields not shallow, slightly leaning and south facing must be preferred.

Preparation of the soil: Before planting, soil must be plowed deeply and levelled. September and October are proper for soil processing in Isparta (Anon. 1987, 2009a; Arslan 1994, 2002; Baydar and Kazaz 2010; Baytop 1963; Bilir 2010; İkiz 2011; Özçelik et al. 2009; Sarıbaş and Aslancan 2011).

Planting

Oil rose is generally planted in Autumn months. 15–20 days before, parallel ditches 40 cm wide and 50 cm deep are opened. Ditches must be perpendicular to the slope in sloppy fields, and in North-South direction on flat terrain. These ditches are filled with topsoil 10–15 cm. Sometimes this is not done. Rose cuttings are laid down in these ditches end to end and filled with 10 cm soil mixed with farm manure. Cuttings are then pressed down with a roller or hind wheel of a tractor. When saplings emerge in spring months ditches are freed of weeds and extra soil is added to the bottom of the saplings. This is continued until ditches are levelled with the soil. Soil between ditches is ploughed to get rid of weeds and to aerate the soil (Anon. 1987, 2009a; Arslan 1994, 2002; Baydar and Kazaz 2010; Sarıbaş and Aslancan 2011).

Pruning

One of the most important maintenance work in oil rose is pruning. Pruning in oil rose is very different from that of ornamental roses. Separation, whisking (top pruning) and rejuvenation pruning are the three different pruning methods applied in oil roses. Dry extraction is removing dried or broken branches from the plants each year regularly between November and March. Whisking type pruning is performed in end-February and beginning of March before bud opening by leaving 5–7 eyes on young branches. Timing of whisk pruning is very crucial. Early pruning may bear the risk of frost. Late pruning however bears the risk of low yield. According to an Iranian study, flower yield was 29 % higher in pruning 75 days before flowering compared to 54 days before flowering (711.1 kg/day and 551.7 kg/day, resp.), and 11 % (642.1 kg/day) compared to unpruned plants (Anon. 1987, 2009a; Arslan 1994, 2002; Baydar and Kazaz 2010; Hussain and Khan 2004; Saffari et al. 2004).

If pruning is delayed one way or another, pruning is better not to be done in dry and hot years, however, this must not be repeated the following year. Whisking brings about an ergonomical adjustment of 60–80 cm table length for easy harvesting of flowers. Furthermore, old branches are stimulated to generate young saplings. Within the first few years of the establishment of rose plantation, no pruning is required since the plant has not reached enough height and flower density, or a light pruning to shape the plant may be resorted to. 3 year old rose plants produce a good number of shoots and flowers. Based on soil character and productivity, manuring, irrigation and climatic conditions, each year old branches give rise to young shoots. If not pruned, each year shoots may reach 30 cm and roses in ditches may attain the height of a man. Therefore, regular whisking pruning is necessary to keep the height of roses evenly and to facilitate harvesting of flowers. If whisking is not done, plant becomes bushy and productivity drops (Anon. 1987, 2009a; Arslan 1994, 2002; Baydar and Kazaz 2010).

Pruning is performed with pruning shears or hedge trimmers. Pruning gears must be sterilized regularly to avoid transmitting diseases to healthy plants. Ten percent sodium hypochloride and hydrogen peroxide can be used for sterilization. After pruning, plants must be applied with Bordeaux mixture (prepared by mixing 2 kg copper sulphate and 1 kg quicklime in 100 L water. Quicksilver must be added first) (Anon. 2009a; Baydar and Kazaz 2010).

Rejuvenation pruning: Although, oil rose plants are perennial and with a long life span, they are economically productive 6–10 years. From 10th year on their yield drops considerably. Therefore, rejuvenation is necessary once in every 10 years by cutting old oil roses just below the soil level to rejuvenate in autumn months before winter after the leaves are shed. Healthy cuttings may be used for replanting. Plants subjected to rejuvenation pruning begin producing healthy and strong shoots in the following spring. They reach economical productivity in 2 years in irrigated fields, and at most 3 years in non-irrigated fields. Some oil rose farmers avoid rejuvenation pruning taking into account losses in the following few years. However, rejuvenation pruning makes it possible to harvest roses economically for 20–25 years from a rose plantation (Anon. 1987, 2009a; Arslan 1994, 2002; Baydar and Kazaz 2010; Sarıbaş and Aslancan 2011).

Fertilisation

As oil rose plants are perennial and long lived, they absorb nutrients from the soil regularly. According to a Bulgarian study, oil rose absorbs 6.4 kg N, 0.87 kg P and 3.65 kg K from 1 day; 100 kg flower takes up 3 kg N, 0.34 kg P and 1 kg K. Fertilization is necessary for high yield and quality production in oil rose. Micronutrients required according to soil and plant analysis are given to rose via the soil or the leaf like in other cultivated plants. The required micronutrients can be determined especially after leaf analysis during the budding period. Lack of nitrogen turns leaves yellow, reduces the number of fresh shoots and buds. Dark red and purple spots appear near the centre of dark green leaves in phosphorus deficiency. Leaf edges turn dark and begin drying in the deficiency of potassium. According to soil analysis taken from 43 different areas of rose cultivation in Isparta, soils are poor in terms of organic matter, nitrogen and iron, generally rich in available phosphorus, rich in exchangeable calcium, magnesium, potassium, zinc and mangane, sufficient in copper. Even after nitrogen fertilization, a significant part of rose plants were found to lack nitrogen, sufficient in phosphorus, calcium, magnesium, zinc and mangan, generally sufficient in iron. Even in soils rich in exchangeable potassium, potassium deficiency was detected in the majority of rose plants. In such regions, the use of potassium fertilizers is required (Yalçın et al. 1992).

Although soils in the region are sufficient as far as extractable copper is concerned, copper content in the plants were found to be near critical. Therefore, copper fertilization is necessary in the region. Fertilization must be performed according to soil analysis and especially fertilizers rich in nitrogen, phosphorous, potassium and copper must be used. It is a widespread application to utilize 15 kg/day diammonium suphate (18 % N and 46 % P_2O_5) as base fertilizer in November-December period, and 20 kg/da ammonium sulphate fertilizer (21 % N) as top fertilizer in February-March period in the Lakes District. Furthermore, composite fertilizers such as 20–20–0, 15–15–15 can also be applied. Since rose plant absorbs 3 kg N per 100 kg flower and 1 da rose garden yields *ca.* 500 kg flowers, 15 kg N must be delivered to the soil. In limy and alkaline soils, a sulphur containing fertilizer such as ammonium sulphate must be applied as top fertilizer to decrease the pH. In addition, pH of the soil can be adjusted by humic acid in order to better retain nutrients and to convert them to a more beneficial form. Especially in limy and alkaline soils, iron deficiency is very frequent. In such soils, 5–14 % iron containing Fe EDTA-OH chelate can be applied (Anon. 1987, 2009a; Arslan 1994, 2002; Baydar and Kazaz 2010; Sarıbaş and Aslancan 2011; Singh and Ram 1987).

In an Iranian study, 3 times application of Fe chelate was shown to increase flower and essential oil yield. Liquid fertilizers rich in micronutrients are applied by spraying on the leaves generally late April during the preflowering period. It is very useful to apply organic fertilizer between ditches after rejuvenation pruning (2–4 t/day) in Autumn months. Farmyard manure must not be applied in spring. Otherwise, fragrance value of rose flowers diminish. Soils poor in organic matter must be fertilized with organic fertilizers. While preparing the field 3 t of burnt farmyard manure must be applied and this must be repeated after every 3 years. This must be performed in Autumn or as late as February. Green fertilizers and rose marc (spent roses) can also be recommended next to farmyard manure for organic rose farming. Phosphorous and potassium fertilizers can also be applied in conventional rose farms in February according to the results of soil analysis. Multiple application of nitrogen fertilizers instead of one can be preferred for better performance on the quality and quantity of the crop.

Fertilizers must be well mixed with soil by hoeing. 8–10 kg N and 9–12 kg P_2O_5 per decare is recommended as regional condition. In plantations using drip irrigation it is possible to apply fertilizers continuously *via* fertigation (Anon. 2009a; Baydar and Kazaz 2010; Sarıbaş and Aslancan 2011).

Irrigation

Oil rose can be cultivated without irrigation. However, while crop yield in unirrigated fields is ca. 500 kg/da, it can increase up to 1,000 kg/da in those irrigated. Irrigation of rose plantations help plants to grow stronger and increases the crop yield. In general, rose plantations are not irrigated in Isparta region. This creates problems especially in years of drought and insufficient rainfall resulting in a considerable decrease in crop yield. Only ca. 60 % of the rose fields are possible to irrigate in Isparta region. Irrigation possibilities are more in villages and lands adjacent to Burdur lake and in those of Keçiborlu and Gönen, however, those in Başmakçı and Dinar of Afyon province have more limited irrigation facilities. Therefore, it is recommended to establish irrigation facilities in regions where irrigation is possible. Flood irrigation and sprinkling must be avoided since they make land unproductive and cause spreading of diseases. Ideal method of irrigation for rose plantations is drip irrigation since it keeps the soil moist and productive. It is also necessary for plantations on a slope. Fertilizers dissolved in irrigation water are piped to the plantations and only the sets between ditches are irrigated avoiding the development of weeds. Encouragements given in recent years for drip irrigation have made drip irrigation popular and enlarged the size of irrigated rose plantations.

In irrigable regions irrigation can be performed in summer months. Three times irrigation in July–September period in which rose needs water most is necessary. It is required to apply irrigation during budding in drought stricken spring months.

Irrigation is not recommended during the flowering period since it makes harvesting difficult. Crop yield in non-irrigable fields is 350–600 kg/day with 1,000–1,500 TL/day income while in irrigable fields it is 800–1,000 kg/day with 2,000 TL/day income (Anon. 1987, 2009a; Arslan 2002; Baydar and Kazaz 2010; Lebaschi 2012; Sarıbaş and Aslancan 2011).

Weed Control

Weed control studies for oil rose are insufficient. Weeds compete with cultivated plants for water, nutrients, growth factors, etc. Rose is a deciduous shrub in the first 2–3 years possessing lesser amounts of leaves and branches. Therefore, it is prone to a heavy attack by weeds throughout the year. In the first years of plantation rose plants are less resistant to such attacks. Weeds invade the rose fields even before rose seedlings suppressing roses to emerge regularly. In an Indian study it was revealed that if weeds are not controlled properly reduction in flower yield may go up to 39 % (Singh and Singh 2004). Mechanical and physical weed control measures are meticulous and expensive although effective and environment friendly. Generally, three hoeings a year is necessary. In organic cultivation of roses since herbicides are prohibited mulching is an effective weed control measure. Mulching with organic matters such as sawdust etc. in 5–10 cm length decreases especially annual weed density and facilitates mechanical hoeing. Mulching can be supplemented by mechanical hoeing and the use of herbicides.

Since rose roots do not grow too deep, this must be taken into account in mechanical hoeing and the soil should be hoed no deeper than 20 cm. The choice of herbicides depends on the distribution of weeds in the plantation. Rose is sensitive to herbicides such as triclorpyr, dicamba and 2, 4-dichlorophenoxyacetic acid (2,4-D) used for broad leaved weeds (Karlık and Tjosvold 2003). By applying herbicides such as simazin and atrazine 2 kg/ha in light soils and 5 kg/ha in medium and heavy soils annual weeds can be controlled. Singh and Singh (2004) reported that atrazin, pendimethalin and metribuzin are less effective than rice-straw mulch.

Rryzalin and pendimethalin used before emergence of roses can be used before the germination of weeds. Such herbicides control primarily narrow leaved grasses and broad leaved weeds. However, perennial weeds are not affected. Herbicides such as fluazifop-p-butyl (Fusilade), sethoxydim (Sethoxydim) and clethodim (Envoy) can be used for perennial weeds after emergence. Roses are highly sensitive to glyphosate (Roundup or other commercial brands). Leaves as well as young shoots can also absorb herbicides (Karlik 2008).

Sing and Singh (2004) had conducted a study between 1985 and 1991 by using herbicides and mulch singly or together for weed control in oil rose. Application was made once a year after pruning and first hoeing in the first week of January. Herbicides and organic mulch had decreased weed density and weed dry weight compared to control either singly or in combination. Combined herbicide and organic mulch had given a better response. Herbicide use had given better result than organic mulching. However, 1.5 kg/ha simazin and 10 t/ha citronella marc combination had given a much better response. 39 % decrease in yield was observed in plots with no weed control. Successive herbicide use for 5 year did not damage oil roses, however, slowing in growth was observed after application. If organic mulching is to be carried out any treatment must be avoided in order to keep the soil moisture between the pruning and flowering periods. There is no report on the harmful effects of herbicides on oil rose. Herbicides must be used according to user instructions and expert advice.

In Turkey, weed control is affected by plowing ditches between the plants. First soil treatment kills also the early emerging weeds. In spring months, weeds appear again due to warming and rain. In such cases, ditches must be replowed with equipment such as Rotavator or Crow's feet. Plowing the soil properly and in time is the most effective weed control measure. Weeds growing nearby rose plantations can also be host to diseases and pests. Such weeds must be eradicated by mowing or plucking before shedding seed. After flower harvest, a third soil treatment can be carried out according to the weed formation or creamy layer formation. Although not yet applied for oil rose, covering ditches between roses with black plastic can also be an effective weed control (Anon. 2009a; Baydar and Kazaz 2010; Sarıbaş and Aslancan 2011).

Pest Control

The most harmful pests to oil rose are Rose soft scale (*Rhodococcus perornatus* Cockerell & Parrott), Rose aphid (*Macrosiphum rosae* L.), Spring beetle (*Perotis chlorana* Leporte & Gory), Scion rose bee (*Syrista parreyssi* Spinola), Rose tip infesting sawfly (*Ardis brunniventris* Hartig), Rose plume moth (*Cnaemidophorus rhododactyla* Denis & Schiffermüller), Mussel scale (*Lepidosaphes ulmi* L.), Rose rhychites (*Rhynchites hungaricus* (Herbst), Red spider mite (*Tetranychus urticae* Koch) and Peach thrips (*Thrips meridionalis* Priesner) (Anon. 2009a; Baydar and Kazaz 2010; Demirözer 2008; Flint and Karlık 2008; Karlık and Tjosvold 2003; Karlık 2008).

Rose soft scale (*Rhodococcus perornatus*): Cultural measures include burning of the cut dry and drying branches, avoiding the use of infected branches in establishing new plantations. Chemical measures can be applied before ovulation of the insect end-April and early-May or at the peak of ovulation for the first period larvae. Chemicals must be applied to ensure distribution all parts of the plant during cool hours of the day.

Rose aphid (*Macrosiphum rosae* L.): Leaf aphids feed by sucking the juice via their thin and long proboscis. They have forms with or without wings and establish colonies on shoots, buds and leaves. When overpopulated development of rose buds ceases and can not form flowers. If uncontrolled, great losses may be inflicted. Cultural measures include keeping the plantations clean and free from weeds, timely annealing, irrigation and fertilization. Chemical warfare should commence when 20 out of 100 shoots are infested and colonies are observed in shoots and underneath young leaves. Chemical spraying must be carried out on all parts of shoots, buds and leaves during cool hours of the day.

Rose rhynchites (*Rhynchites hungaricus*): It appears on flower buds generally end April – early May. Females lay eggs in holes bored through the flower buds using their proboscis. During egg laying some buds fall down and some hang off on branches. Eggs open in 1–2 weeks and larvae starts feeding on the bud. Mature larvae abandon the buds and go in the soil. Adults inflict damage on newly opening flower buds. Affected buds stop opening, anomalies on flowers occur and crop yield drops. As cultural measures infested buds must be collected and destroyed in order to reduce the infestation density and to prevent the damage in the following year. Chemical warfare can be performed end-April mid-May. Other pests include Rose Shoot Sawfly (*Syrista parreyssi*), Tip-infesting sawfly (*Ardis brunniventris*), Spring beetle (*Perotis chlorana*), Red spider mite (*Tetranychus urticae*). Similar control measures are carried out for these pests as well.

Integrated Biological Control and cultural and mechanical control are the most important measure for oil rose pests and diseases. Unless absolutely necessary pesticide use must be avoided. Chemical spraying must be stopped during flowering in rose plantations. Otherwise, pesticides go into rose oil spoiling its quality. For the control of leaf aphids, weeds acting as host must be removed. In controlling rose shoot sawfly, dried branches and shoots growing downward should be cut and destroyed. In controlling spring beetle, dry branches in which eggs are laid between mid-May and July and adult insects should be collected and eradicated; for rose soft scale control, spraying the rose plants with pressurized water during the diapose stage in winter months is effective. In pest control, biological life span of insects must be followed carefully and chemicals must be applied when they are most sensitive to them. For example, in controlling rose soft scale the application of Methidathion is very effective only in the 1st nymph period. A study has shown that for spring beetle larvae Azinphos-methyl containing Gusathion (M EC 20) and Carbaryl containing Agrovin (50WP); for adults Methyl parathion containing Folidol (M EC 360) were found the most effective. Chemical controls must be carried out for shell lice and scale insects 2–3 weeks before the flower buds open. Chemical control for shell lice, rose rhynchites, rose shoot sawfly and tip-infesting sawfly can be performed in April. Insecticides can be used for oil rose plants before flowering in the 1st week of May. Control for shell lice and scale insects is continued after the flower harvest is over. A whole rose plantation may be destroyed when adults of spring beetle cut the shoots and larvae open galleries in the roots. However, as mentioned before, cultural and mechanical measures must be considered first. Chemicals must be applied in recommended dosage and expert advice must be followed (Anon. 2009a; Baydar and Kazaz 2010; Demirözer 2008; Sarıbaş and Aslancan 2011).

Disease Control

Most common diseases found in oil rose are powdery mildew, rust, chlorose and root rot. The most important applications in controlling diseases of oil rose are cultural and mechanical controls and integrated biological control. Chemical control is not applied during the flowering period (Anon. 1987, 2009a, 2013f; Baydar and Kazaz 2010; Ikiz 2011; Sarıbaş and Aslancan 2011).

Powdery mildew: *Sphaerotheca pannosa* (Wall.:Fr.) Lév. var. *rosae* Wor. (*Sphaerotheca rosae*) (syn. *Podosphaera pannosa*). Disease is observed in the leaves, shoots and flower buds. White spots turn to dark brown in time. It reveals itself as flour or ash like appearance on the leaves. Diseased leaves curl, grooved and hardened and reddened. It prevents bud opening and keeps buds small. Disease progresses well at 15–17 °C and especially at high night moisture. As cultural measure, overirrigation and especially sprinkling must be avoided. Diseased leaves, if any, remained from the previous year must be gathered and destroyed. For chemical control, first application must be performed during leaf and bud formation and the second application ca. 2 weeks later. Chemical control must be carried out early morning in wind-free and rain-free days, and all parts of the plant surface must be sprayed. Application of powdered sulphur is effective. Protect foliage with fenarimol, myclobutanil, neem oil, propiconazole, stylet oil potassium bicarbonate, or sulphur. Eradicate infections with horticultural oils, neem oil, or jojoba oil.

Rose rust (*Phragmidium mucronatum*): It appears as yellow spots on leaves, branches and bud stalks and the colour turns red in time in spring. Spots are seen as slightly swollen pustules.

Rose rust inflicts damage in two ways directly or indirectly; firstly, it prevents opening of the buds by infesting stalks and calyces of flower buds, secondly reduces the yield by preventing photosynthesis due to spots formed on the leaves. All these spoil physiological balance reducing the number of flowers and retarding flowers to develop normally. Cool and moist weather encourages the spreading of rose rust. Cultural measures for its control include pruning of infested branches in spring, collecting and destroying fallen leaves on which the disease can overwinter, disinfecting pruning shears frequently and avoiding sprinkling. After Whisking type pruning, plants must be treated with 2 % Bordeaux mixture (see Pruning). For chemical control; 1st application 3–4 weeks before flowering; 2nd application after 2–3 weeks during bud formation and 3rd application after the flower harvest must be carried out. Surface area of whole plants must be treated with myclobutanil, triadimefon and mancozeb early morning, in wind-free and rain-free weather.

Rose black spot (*Diplocarpon rosae* F.A. Wolf (Syn. *Marssonina rosae* (Lib.) Died.) common rust in roses does not inflict too much damage. However, root rot which can destroy a whole rose plantation occurs due to the choice of wrong field. It is particularly common in fields with water-retaining loamy soils and the roots of roses cultivated in such fields rot due to fungus grown in moist soil. There is no chemical control for this fungus. It is advisable to hoe the soil frequently and take precautions to reduce ground water in infested rose fields.

Chlorose is mainly due to iron deficiency in rose plantations. It is seen in rose plantations established on limey soil. Disease begins in young leaves and shoots and spread to other leaves. Edges of some leaves are dried and some are fallen. Due to nutritional disbalance growth of rose is disrupted and flower yield drops. In advanced stages rose can completely run dry. Cultural measures include application of farmyard manure and chemical fertilizers. Iron fertilizers must be applied at once.

Fungicides such as Antracol, Korunep and Hexamor can be used for rust and rot before flowering in the first week of May for oil rose. Excess fertilization and irrigation can augment the damage of rust and rot. Chemicals must be used in recommended doses and expert advice must be sought (Anon. 1987, 2009a, 2013f; Baydar and Kazaz 2010; Ikiz 2011; Sarıbaş and Aslancan 2011).

Harvesting

Harvesting of oil roses begins mid-May or early June depending on the climatic conditions in April–May. Harvesting time is much later than that of Morocco but about the same time as Bulgaria. Blooming period and length of blooming time largely depends on weather conditions. In cool, cloudy and rainy conditions blooming period is long (more than a month) and oil yield is good. However, in dry and hot weather conditions, blooming period is short (15–20 days) and the oil yield is poorer. Harvesting period lasts 25–30 days. Accordingly, rose harvest is carried out 3–7 weeks according to the weather conditions. Best flower yield is obtained between the first and the last weeks of harvesting. In the beginning and towards the end harvest is poor due to low flower density. Roses are harvested in early hours of the day for ca. 40 days between mid-May and mid-June. In full bloom the oil yield is at its maximum. 65 % of the flowers consist of pink petals and 35 % green calyces, stalk, leaves and reproductive organs. Full blooming flowers are harvested by plucking right below the calyces every morning between 4 to 10 a.m. Flower harvest must be carried out before the dew dries. It is recommended to use gloves during harvest due to thorns and bees. A worker can harvest *ca.* 5 kg of flowers per hour. Oil content in flowers harvested early morning (5–7 a.m.) is almost twice as much as those harvested at noon. In cool and cloudy days, oil content in flowers collected between 7 and 9 a.m. is high. Oil content drops drastically in late harvested flowers. While in early harvested flowers, oil content is 0.04 % (1 kg rose oil from 2.5 t of flowers), it is 0.02 % (1 kg oil from 5 t of flowers) in flowers harvested late afternoon.

Similarly, north wind affects the oil content positively while the south wind has a negative effect. This is due to the fact that north wind brings about coolness resulting in less evaporation in early hours of the day while hot southernly wind facilitates evaporation. Therefore, roses are harvested in early hours of the day and transported to rose oil factories. As a general rule, rose oil factories should not accept roses after 10 a.m., however, in practice, this rule is not followed. In hot and windy days, harvests may take place between 5 and 7 a.m. while in cool and overcast days it may be carried out between 6 and 9 a.m. Oil yield drops drastically in roses harvested after 10 a.m. This is due to the effect of hot sun expediting evaporation. Roses must be harvested daily. Otherwise, oil loss occurs in old flowers. Petals contain ca. four times more oil than the other parts of the flower. Flower yield of oil rose varies between 250 and 1,000 kg/da depending on factors such as age of the plants, maintenance conditions, soil conditions, altitude, etc. It is 450–500 kg/da in dry conditions and 750–1,000 in moist conditions. Generally, 3.5 t of flowers are harvested from 5 day rose plantation yielding 1 kg rose oil. Therefore, rose oil is a low volume high value product.

A study conducted in Bulgaria has shown that 15 % oil loss occur in roses harvested at 7 a.m. compared to roses harvested at 4 a.m. It is 30–40 % in roses harvested at noon time and 65 % in roses harvested at 8 p.m. It has also been shown that oil yield and quality depends on meteorological conditions and therefore, it is not possible to recommend a fixed time for harvesting roses for the best quality and yield of oil. Farmers is Bulgaria are said to be given advice on the best harvesting times depending on weather conditions.

The oil yield and oil quality of rose flower are not fixed and depend on the factors such as weather conditions, harvest time, variety of the rose, type of the distillation still, distillation method, etc. In wind-free and rain-free hot days, oil yield is low and 7,000–8,000 kg of roses are required for 1 kg of oil. However, in appropriate weather and other conditions, 1 kg of rose oil can be obtained from 2,600 to 2,800 kg of flowers. In Turkey, ca. 3,500–4,000 kg flowers are needed to obtain 1 kg of rose oil; 300–400 kg of flowers for 1 kg of rose concrete (Acatay 1969; Anon. 1987, 1991, 2009a; Arslan 1994; 2002; Baser 1992; Baser et al. 2012; Baydar and Kazaz 2010; Baytop 1963; Dobreva and Kovacheva 2010; Işık 1953; Kurkcuoglu and Baser 2003; Yılmaz 1992).

Production Schedules

Distillation and Extraction

Rose oil is produced by water distillation of fresh *R. damascena* flowers. Commercial oil-bearing rose is a totally cultivated plant. After planting the rose twigs in a rose field, it takes at least 3 years for a rose plant to attain maturity. A mature rose field normally yields 5 t of fresh roses per hectare. However, in a carefully nurtured field the yield may increase to 7–8 t per hectare. It is normal for a field to be productive for up to 20–30 years. Harvest of roses last for about a month between mid-May and mid-June. Roses are hand-picked in the early hours of the day and either transported to factories or the collection sites of various firms. A skilled worker can pick about 40 kg of roses in 8 h. Factories work 24 h a day in 3 shifts for 1 month. When the season is over, the factories are cleaned and closed down until the next season.

In the village-type distillation, freshly picked flowers are loaded into 150–1,000-L copper or galvanized steel open fire stills; most stills have a 300-L capacity and consist of a retort and a head. The removable spherical head is connected to a pipe which leads through a pool filled with lukewarm water to cool the condensate. At the outlet, there is a 9-L glass collecting flask.

Typically, 10 kg of flowers and 60 L of water are loaded into 300-L stills and are distilled for 1–2 h to collect 2 flasks full of the distillate (18 L). Due to the low concentration of oil in the distillate, the oil does not separate. Therefore, about 60 L of the distillate are redistilled, yielding another 18 L of distillate from which the oil that floats to the top is decanted. The aqueous phase is diluted with distilled water and marketed as rose water.

For the industrial production, generally 3,000-L copper or stainless steel stills are employed. Each still has a charge size of 400–500 kg flowers, and 1,500–2,000 L of warm water. The stills are steam-jacketed, *i.e.,* they contain a double-wall inside of which steam is circulated. There may also be provision for the injection of live steam into the still to speed up distillation. The distillation is carried out for 1.5 h. The condenser temperature is kept at 35 °C (95 °F) to avoid the solidification of waxes.

The distillate is collected in 200 L stainless steel Florentine flasks. The oil that separates out is called "crude oil", "first oil", or "direct oil". Distillation is terminated when the distillate no longer has a bitter taste.

The overflow of the Florentine flasks is collected in 500-L tanks. These "bottom waters" or "first waters" are then pumped into 5,000-L stainless still tanks. These are cohobated in 3,000-L stills for 1–1.5 h to obtain what is called the "second oil", "cooked oil" or "indirect oil". The distillate after removal of oil is sold as rose water. The first and second oils are filtered and kept in glass flasks in the dark. When the production season is over, all the first and second oils are mixed to yield the Turkish rose oil and packed in special 2–5-L tinned steel containers called *kumkuma*.

Generally 3.5–4 t of flowers yield 1 kg of rose oil, yielding about 0.02% oil (Baser 1992).

Rose concrete is obtained by extracting fresh roses with n-hexane. Removal of hexane leaves a highly fragrant solid extract appearing somewhat like shoe polish wax. When rose concrete is extracted with ethanol and cold-filtered upon evaporation of ethanol under vacuum, the dark liquid obtained is rose absolute. Annually, about 7,000 t of roses are processes to produce about 1,600 kg of rose oil and 2,400 kg of rose concrete. 400 kg of roses are required to produce 1 kg of rose concrete. The primary constituent of rose concrete is 2-phenylethyl alcohol. It is also the main constituent of the headspace odor of roses and rose water (Baser 1992; Baser et al. 2003).

Chemical Composition

Rose oil, like other essential oils, is analyzed by gas chromatography (GC) and gas chromatography/mass spectrometry (GC/MS).

Main component of rose oil is citronellol (31–44 %). Then comes geraniol (9–15 %), nonadecane (8–15 %), nerol (5–11 %), 1-nonadecene (2–5 %), methyl eugenol (2–4 %), heneicosane (3–4 %), geranyl acetate, phenylethyl alcohol, beta-caryophyllene, citronellyl acetate, germacrene D, linalool, (2E,6E)-farnesol are in the range of 1–2 % each (Baser et al. 2012).

Other typical constituents of rose oil are nonanal, citronellyl formate, eugenol, cis-rose oxide, alpha-terpineol. Damascenones and some sulfur compounds are among the minor components. Stearoptenes (paraffins) are natural constituents of rose oil (the major one being nonadecane) and due to their presence rose oil solidifies at room temperature or when refrigerated. Their content in village oils is lower (Baser 1992).

In order to simplify the comparison of Gas Chromatographic results citronellol/geraniol ratios of each oils are taken. Village oils give a ratio of 0.83–1.92 % while in factory oils the ratio is 2.30–4.84 %. In Bulgarian rose oil the citronellol/geraniol ratio is around 1 (Baser 1992).

According to ISO9842:2003, standards of rose oil have been established (Anon. 2003). The monograph gives a table of comparison of the rose oils of Bulgarian, Turkish (factory and peasant type) and Moroccan origin as shown below. We compared the Taif rose oil results with those given in the ISO rose oil monograph in the following table (Kurkcuoglu et al. 2013) (Table 4).

Table 4 Comparison of main components in rose oils

	Bulgaria		Turkey		Morocco		Turkey "peasant" type		Taif	
	Min.	Max.	Min.	Max.	Min.	Max.	Min.	Max.	Min.	Max.
Ethanol	–	2	–	7	–	3	–	2	–	–
Citronellol	20	34	34	49	30	47	26	40	22.8	27.5
Nerol	5	12	3	11	3	11	6	12	6.4	10.5
Geraniol	15	22	8	20	6	23	12	29	13.5	19.9
Phenyletanol	–	3.5	–	3	–	3	–	3	1	2.9
Heptadecane	1	2.5	0.8	3	0.6	4	0.7	3	0	2.2
Nonadecane	8	15	6	13	7	16	6	8.5	10.9	15.7
Heneicosane	3	5.5	2	4	2	5.5	1.5	4	2.7	7.2

Kurkcuoglu et al. (2013)

Utilization

Rose water (distilled water of roses), rose confection (Rose paste – a thick jam made of blending roses with sugar or honey) and rose oil (made by steeping roses in sesame seed oil or olive oil left under the sun) are frequently mentioned in old Islamic texts.

The arab physician Al-Kindi (ninth century CE) prescribed rose products in medicines for stomach pain, ulcers, liver diseases, sore throat and mouth diseases. Rose oil was used by him for burns, ulcerated wounds and as an ingredient of haemorroid salves. Al-Dinawari (ninth century CE) recommended rose water for fever and rose oil for alleviating fever by applying it to the head. Abu Bakr Mohammad ibn Zakariya Al-Razi (Rhazes), the ninth century (CE) Arab physician, noted that "the rose diminishes drunkennes". The World famous physician Ibn Sina (Avicenna) (eleventh century CE) was the first scientist to emphasize rose fragrance's beneficial effects on the heart and the brain. He said "Because of its exquisite fragrance, the rose addresses the soul." "It has a calming effect and is highly beneficial for fainting and for rapid heart beats." He praised the rose water's effects on mind and spirit, and its beneficial effects on functioning of the brain and its cognitive power by the statement that "It enhances comprehension and strengthens memory." Ibn-Al-Baitar also noted rose water's beneficial effects on the brain by saying "Rose water strengthens the mind and the brain, sharpens the senses, increases the life force; It is beneficial for rapid heart beats due to anxiety; because of its beneficial fragrance it empowers the body." Ibn-Al-Baitar also stated that boiling rose water and exposing the head to its steam had healing effects. It was especially beneficial for eye diseases. He also recommended this inhalation to alleviate drunkenness and to relieve headache. Mahmud of Shirvan (fifteenth century CE) mentions a powder prepared by crushing dried rose petals in a mortar for applying it to neck, breast, and armpits after bathing while the skin is still moist to enable body to smell favorably and to "treat the spirit." He claimed that this scent empowers spirituality and purifies the heart. The same powder is also mentioned by Ishak bin Murat (fourteenth century CE) for use in Turkish Baths. He also recommended it for scabies. If rubbed on pimples, it reportedly clears pimples. If sprinkled on the skin with smallpox or measles lesions is said to be beneficial. Ibrahim Hakki of Erzurum (eighteenth century) recommended rose water for headaches due to fever and as beneficial in fainting in his famous book Marifetname (Book of Talents) (Baser et al. 2012).

Dried rose flower buds are used as herbal tea especially in Central Asian Countries. Rose jam is prepared from fresh rose flowers. Petals are washed, sprinkled with sugar and lesser amount of citric acid and rubbed with fingers. This is called rose jam stock (gul mayasi). It can be kept for longer periods. When needed, it is boiled with water and sugar to the consistency of a jam. Rose water is often sprinkled on the hands of participants in religious ceremonies and it is also added to some deserts such as *güllach* and *su muhallebisi* [a thick rice flour pudding] for flavoring. Rose syrup is also made from fresh rose flowers. *Gülbeşeker* is a rose-flavored confectionary which was particularly popular during the Ottoman period (Baser et al. 2012).

References

Açatay A (1969) Gül ve Gülyağı. Özaydın Matbaası, İstanbul
Anon. (1987) Gülcülük.T arım Orman ve Köy işleri Bakanlığı Isparta İl Müdürlüğü
Anon. (1991) Gülçiçeği Raporu.Gülbirlik Isparta
Anon. (2003) Oil of rose (Rosa x damascena Miller). International Standard ISO 9842
Anon. (2009a) http://gular.sdu.edu.tr/. Accessed 10 Oct 2009
Anon. (2009b) Symrise Geschäftsberichte
Anon. (2011) 2010. Yılı Gülçiçeği Raporu. Gümrük ve Ticaret Bakanlığı, Kooperatifçilik Genel Müdürlüğü
Anon. (2012) 2011 Yılı Gülçiçeği Raporu. Gümrük ve Ticaret Bakanlığı, Kooperatifçilik Genel Müdürlüğü
Anon. (2013a) 2012 Yılı Gülçiçeği Raporu. Gümrük ve Ticaret Bakanlığı, Kooperatifçilik Genel Müdürlüğü
Anon. (2013b) http://www.meteoroloji.gov.tr. Accessed 20 June 2013
Anon. (2013c) http://www.dr.hauschka.com/en_DE/legal. Accessed 20 June 2013
Anon. (2013d) http://1000fragrances.blogspot.com/2011/01/afghanistan-starts-to-produce-special.html. Accessed 20 June 2013
Anon. (2013e) http://www.kasbahdesroses.com/. Accessed 20 June 2013
Anon. (2013f) Rose (*Rosa* spp.) disease control outlines (Reviewed 3/09, updated 3/09). http://www.ipm.ucdavis.edu/PMG/menu.homegarden.html. Accessed 12 July 2013
Anon. (2013g) http://en.wikipedia.org/wiki/List_of_Rosa_species. Accessed 27 Sept 2013
Arslan N (1994) Tütün, ilaç ve baharat bitkileri ders notları, Ankara. (Not printed)
Arslan N (2002) Kokulu bitkiler yetiştirme ve ıslahı ders notları, Ankara. (Not printed)
Baser KHC (1992) Turkish rose oil. Perfum Flavour 17(3):45–52
Baser KHC, Kurkcuoglu M, Ozek T (2003) Turkish rose oil: Recent results. Perfum Flavour 28(2):34–42
Baser KHC, Altintas A, Kurkcuoglu M (2012) Turkish rose: A review of the history, ethnobotany and modern uses of rose petals, rose oil, rose water and other rose products. HerbalGram 96:40–53
Baydar H (2006) Oil- bearing rose (*Rosa damascena* Mill.) cultivation and rose oil industry in Turkey. Euro Cosmet 14(6):13–17
Baydar H, Baydar GN (2005) Isparta gülünde (*Rosa damascena* Mill.) genetik ilişkilerin AFLP ve mikrosatellit markırları ile belirlenmesi. Türkiye VI. Tarla Bitkileri Kongresi, 5–9 Eylül 2005, Antalya. Bildiriler Kitabı Cilt II:1123–1126
Baydar H, Kazaz S (2010) Organik gülcülük. SDÜ. Gülcülük Araştırma Merkezi yayınları no. 1.50 pages, Isparta
Baydar GN, Baydar H, Debener T (2004) Analysis of genetic relationships among *Rosa damascena* plants grown in Turkey by using AFLP and microsatellite markers. J Biotechnol 111:263–267
Baytop T (1963) Türkiye'nin Tıbbi ve Zehirli Bitkileri, İsmail Akgün Matbaası, İstanbul, s 197
Baytop T (2001) Türkiye'de Eski Bahçe Gülleri. T.C. Kültür Bakanlığı Yayınları 2593, Yayın Dairesi Başkanlığı, Sanat Eserleri Dizisi 319, Ankara
Bilir S (2010) Isparta ilinde gülcülük ve ekonomik önemi. İstanbul Üniversitesi Sosyal Bilimler Enstitüsü Yüksek Lisans Tezi, 65 pages
Cheers G (ed) (1999) Rosen Enzyklopaedie. Könemann, Köln. (704 pages)
Dehghan P, Lebaschi MH, Abbaszadeh B, Zakerian A (2012) The main compounds in essential oil composition of Damask Rose genotypes under Mulched Rainfed condition. Ann Biol Res 3(9):4470–4473
Demirözer O (2008) Isparta yağ gülü (*Rosa damascena* Mill.) üretim alanlarında bulunan zararlılar, yayılışları, doğal düşmanları ve önemlilerinin populasyon değişimleri. Süleyman Demirel Üniversitesi Fen Bilimleri Enstitüsü, Doktora Tezi. Isparta

Dobreva A, Kovacheva N (2010) Production Systems Daily dynamics of the essential oils of *Rosa damascena* Mill. and *Rosa alba* L. Agric Sci Technol 2(2):71–74

Farooq A (2011) Analysis of genetic diversity for oil content, morphological characters and microsatellites in some *Rosa damascena* landraces and selected scented rose species. Institute of Horticultural Sciences, University of Agriculture, Faisalabad, p 162

Flint ML, Karlik JF (2008) Roses: insect and mite pests and beneficials. http://www.ipm.ucdavis.edu/PDF/PESTNOTES/pnrosesinsect.pdf. Accessed 10 July 2013

Frederick G, Wagner A, Morvillo N (2002) Randomly amplified polymorphic DNA (RAPD) analysis of the musk roses (*Rosa moschata*). Proc Fla State Hortic Soc 115:117–119

Ginova A, Tsvetkov I, Kondakova V (2012) *Rosa damascena* Mill. – an overview for evaluation of propagation methods. Bulg J Agric Sci 18:545–556

Gora J, Lis A, Kalemba D (1995) Chemical composition of the essential oil of *Rosa centifolia* L. petals. J Essent Oil Res 7(1):89–90

Hussain A, Khan M (2004) Effect of Growth Regulators on Stem Cutting of *Rosa bourboniana* and *Rosa gruss-an-teplitz*. Int J Agric Biol 6(5):931–932

İkiz M (2011) Göller yöresinde organik ve konvonsiyonel yağ gülü (*Rosa damascena*) yetiştiriciliğinin karşılaştırmalı. Isparta Ekonomik Analizi. Süleyman Demirel Üniversitesi, Fen Bilimleri Enstitüsü, Yüksek Lisans Tezi

Işık H (1953) Isparta'da Gülcülük ve Gülyağcılığı. Tarım Bakanlığı, Ankara

Javani M, Taher M, Arslan N, Valizadeh N (2013) İran'da Gülcülük. V. Süs Bitkileri Kongresi 6–9 May 2013, Yalova

Karabulut AR (1993) Tıbb-ı Nebevi Ansiklopedisi, iki cilt, Ankara

Karlik JF (2008) Roses in the garden and landscape. Cultural practices and weed control. http://www.ipm.ucdavis.edu/PMG/PESTNOTES/pn7465.html. Accessed 10 July 2013

Karlik JF, Tjosvold SA (2003) Integrated Pest Management (IPM) for Roses. In: Roberts AV (ed) Encyclopedia of Rose Science. Elsevier Science, Amsterdam

Kaul K, Karthigeyan S, Dhyani D, Kaur N, Sharma RK, Ahuja PS (2009) Morphological and molecular analyses of *Rosa damascena* × *R. bourboniana* interspecific hybrids. Sci Hortic 122:258–263

Kovacheva N (2011) Selection of oil-bearing rose in Bulgaria – tendencies and perspective. Agric Sci Technol 3(3):189–192

Kovacheva N, Rusanov K, Atanassov I (2010) Industrial cultivation of oil bearing rose oil production in Bulgaria 21st Century, directions and challenges. Biotechnol Biotechnol Equip 24(2):1793–1798

Kürkçüoğlu M (1988) Türk Gül Yağının Üretimi ve Özelliklerinin Tespiti. Anadolu Üniversitesi Sağlık Bilimleri Enstitüsü, Yüksek Lisans Tezi, Eskişehir, 128 pages

Kurkcuoglu M, Baser KHC (2003) Studies on Turkish rose concrete, absolute and hydrosol. Chem Nat Comp 39(5):457–464

Kürkçüoglu M, Abdel-Megeed A, Başer KHC (2013) The composition of Taif rose oil. J Essent Oil Res 5(5):364–367

Lebaschi HM (2012) Yield and Quality of Damask Rose (*Rosa damascena* Mill.) Genotypes under Irrigated Conditions. Ann Biol Res 3(5):2148–2152

Nedkov N, Dobreva A, Kovacheva N, Bardarov V, Velcheva A (2009) Bulgarian Rose Oil of White Oil-bearing Rose. Bulg J Agric Sci 15:318–322

Nilsson O (1972) Rosa L. In: Davis PH (ed) Flora of Turkey and the East Aegean Islands, vol 4. Edinburgh University Press, Edinburgh, pp 106–128

Özcan Ö (1998) Isparta ili ve çevresinde gül tarımı ve Türkiye Ekonomisine katkısı. Marmara Üniversitesi Sosyal Bilimler Enstitüsü, Yüksek Lisans Tezi, 154 s

Özçelik H, Gül A, Özgökçe F, Ünal M, Özcan G, Fakir H, Orhan H, Sakçalı H, Korkmaz M (2009) Türkiye Rosa L. (Gül) taksonlarının genetik çeşitliliğinin tespiti, ekonomiye kazandırılma olanaklarının araştırılması. TÜBİTAK-105O627 proje kesin raporu, 634 pages

Pati KP, Rath PS, Sharma M, Sood A, Ahuja SP (2012) In vitro propagation of rose – a review. Biotechnol Adv 24:94–114

Riaz S, Sadia B, Awan FS, Khan IA, Sadaqat HA, Khan IA (2012) Development of a species – specific sequence-characterized amplified region marker for roses. Genet Mol Res 11(1):440–447

Ritz C (2005) Evolution von Hundsrosen (*Rosa* L. sect. *Caninae*) Dissertation Biologisch-Pharmazeutischen Fakultät der Friedrich Schiller Universität Jena, 95 pages

Roberts AV, Debener T, Gudin S (eds) (2003) Encyclopedia of Rose Science, vol III. Elsevier Academic Press, Amsterdam

Rusanov K, Kovacheva N, Vosman B, Zhang L, Rajapakse S, Atanassov A, Atanassov I (2005) Microsatellite analysis of *Rosa damascena* Mill. accessions reveals genetic similarity between genotypes used for rose oil production and old Damask rose varieties. Theor Appl Genet 111:804–809

Rusanov K, Kovacheva N, Stefanova K, Atanassov A, Atanassov I (2009) Rosa damascena genetic resources and capacity building for molecular breeding. Biotechnol Biotechnol Equip 23(4):1436–1439

Saffari VR, Kalighi A, Lesani H, Babalar M, Obermeier JF (2004) Effects of Different Plant Growth Regulators and Time of Pruning on Yield Components of *Rosa damascena* Mill. Int J Agric Biol 6(6):1040–1042

Sarıbaş R, Aslancan H (2011) Yağ Gülü Yetiştiriciliği. Eğirdir Meyvecilik Araştırma İstasyonu Müdürlüğü Yayın no. 45

Singh DV, Ram M (1987) Effect of spacing, extent of pruning, growth hormone and nutrients on flower yield of essential oil bearing rose (*Rosa damascena*) in subtropical India. Acta Horticult 208:83–86

Singh A, Singh M (2004) Weed management in essential oil yielding Damask rose (*Rosa damascena*). J Med Aromat Plant Sci 26:721–726

Temurçin K (2004) Isparta ili ekonomik coğrafyası. Ankara Üniversitesi Sosyal Bilimler Enstitüsü, Doktora Tezi, 377 pages

Yalçın R, Usta S, Yüksel M, Topçuoğlu B (1992) Gül tarımı yapılan Isparta yöresi topraklarının bazı fiziksel ve kimyasal özellikleri üzerine bir araştırma. TÜBİTAK TOAG-GÜLAR 2

Yılmaz Ö (1992) Isparta ve yöresinde yetiştirilen güllerin (*Rosa damascena* Mill.) bazı özellikleri Üzerine bir İnceleme. Uludağ Üniversitesi Fen Bilimleri Enstitüsü, Yüksek Lisans Tezi, Bursa

Zeinali H, Aghdaei SRT, Arzani A (2009) A Study of Morphological Variations and Their Relationship with Flower Yield and Yield Components in *Rosa damascena*. J Agric Sci Technol 11:439–448

Opium Poppy (*Papaver somniferum*)

Kemal Hüsnü Can Baser and Neset Arslan

Abstract Opium poppy which is an indispensible source of opium (dried exudate of fresh fruits of opium poppy after lancing) and poppy seeds is a cultivated species of *Papaver somniferum*. Due to narcotic properties of morphine, the major alkaloid in opium, Opium poppy cultivation is controlled by the States where it is cultivated and the UN worldwide. Major opium poppy cultivating countries are Turkey, India, Australia, France, Spain, Hungary, Czech Republic and China. However, illicit opium poppy cultivation is widespead in Afghanistan, Burma, Mexico, Laos, Pakistan and Columbia for opium production. Illicit opium is used to isolate mophine which is then converted to heroine in makeshift production plants.

India is the world's largest producer of raw opium and Turkey, having banned opium production in 1971, is the largest producer of poppy straw (crushed dried seedless poppy fruits). Both countries are recognized as "traditional supplier countries" in the world.

The chapter covers cultivation aspects, pest and weed control, production, pharmaceutical, food and cosmetic uses of opium poppy comprehensively.

Keywords Opium • Poppy • Opium poppy • *Papaver somniferum* • Poppy straw • Poppy seed • Poppy cultivation • Morphine • Heroine • Illicit opium

K.H.C. Baser (✉)
College of Science, Department of Botany and Microbiology, King Saud University, Riyadh, Saudi Arabia
e-mail: Khcbaser@gmail.com

N. Arslan
Faculty of Agriculture, Ankara University, Ankara, Turkey
e-mail: Neset.Arslan@agri.ankara.edu.tr

Z. Yaniv, N. Dudai (eds.), *Medicinal and Aromatic Plants of the Middle-East*, Medicinal and Aromatic Plants of the World 2, DOI 10.1007/978-94-017-9276-9_17

Part I: General Aspects

Description of the Plant

Root: Opium poppy has a tap root with weak lateral roots. They may grow down to 20–40 cm according to the time of cultivation, development of the plant and condition of the soil. The root system is considerably weaker than the above ground parts. Extreme rain or irrigation followed by strong wind may give rise to downing and tilting. Stem and branches: Branches of opium poppy are green, glaucous while developing and off white, yellowish in maturity. Stems erect up to 40–175 cm. Stem length differs according to climatic and cultivation conditions. Due to our long observational experience; length of winter range poppies ranges between 90 and 110 cm while those of summer range between 70 and 90 cm. Rainy years and irrigation result in taller poppies. Opium poppy shows branching in middle and upper parts of the stem. Number of branches is generally 2–3. However, in some opium poppies lateral branches also branch more and the number of branches may be 8–10 and in some extreme cases 20–30. Branching is from top to bottom. Generally, lateral branches are longer than the main stalk. Stalks of opium poppy are round, hollow inside and 1–2 cm thickened at the base. Branches are glabrous, however upper parts may have white hirsute cover and such hairs drop during maturity (Arslan et al. 2011; Bernath 1998; Broszat 1992; Honermeier 2006; Incekara 1949, 1964; Işıkan 1955, 1957).

Stem and Branches

Leaf; Opium poppy develops rosulate leaves in the first vegetative stage. These are with short repand petiole sometimes crenate, dentate or smooth edges. Rosette stage is longer in winter range poppies than the summer range poppies. Strength of the plant depends on its ability to develop good rosette leaves and such plants are resistant to cold weather. Development of leaves on the stem is from bottom to top and phyllotactic arrangement is 2/5. Opium poppy leaves are rough, thick, serrate-dentate and with prominent median vein. Form and size of the leaf depends on its location on the plant. Lower leaves have a short petiole with ca. 10–30 cm length and 4–13 cm width. These are the narrowest leaves. Median leaves are sessile and linguiform with ca. 15–35 cm length and 7–20 cm width. Such leaves contribute best to photosynthesis. Upper leaves are cordate, sessile and amplexicaul with ca. 9–18 cm lenth and 7–14 cm width. Additionally, there are cymbiform protective leaves surrounding the buds. Leaves of opium poppy are serrate-crenate, uneven towards the apex, upper leaves generally with double toothed edges. Leaves are covered more or less with a waxy layer and are greyish green, glaucous or green. Main stalk contains ca. 7–9 leaves (Arslan et al. 2011; Bernath 1998; Broszat 1992; Honermeier 2006; Incekara 1949, 1964; Işıkan 1955, 1957).

Flower; Flower buds first appear from among the surrounding leaves. Branching starts with budding and flower formation occurs on the tips of branches. Flower buds are inclined in the beginning. Towards the evening of bud formation, stalks of buds become erect and flowers appear next morning. Flowers have two large sepals which drop during flowering. Four large and showy white, violet or red petals are with smooth or undulate margins and dark or White spot at the bottom. Petal and spot colours are genetical and specific to the cultivar. Numerous stamens envelop stigmas. Anthers are yellow or bluish green. They mature from inside to outside.

Ovary is superior and has many stigmas. These can be seen in immature fruits of different shapes and sizes. The period between budding and flowering ranges from 7 to 18 days according to the type and climatic conditions. Flowering period, however, ranges between 1 and 8 days depending on the number of flowers. Pollens mature later than the pistil and keeps its fertilizing ability up to 8 days. Since ovary and pollens mature before flowering self pollination is more abundant (Arslan et al. 2011; Bernath 1998; Broszat 1992; Honermeier 2006; Incekara 1949, 1964; Işıkan 1955, 1957).

Fruit; Opium poppy fruit is a glabrous capsule. Fruit may be ovoid, depressed or long. Variation of forms can be seen even on the same plant. Fruits contain inside symmetrically located numerous cells. Seeds develop on these cells and their number is related to the number of stigmas on placenta. This number affects the number of seeds. In cultivated varieties, those with high number of stigmas and placenta membranes are preferred (Arslan et al. 2011; Bernath 1998; Broszat 1992; Honermeier 2006; Incekara 1949, 1964; Işıkan 1955, 1957).

Seed; Opium poppy fruit contains ca. 1,000–5,000 seeds. Seeds are very small and 1,000 seeds weigh 0.2–0.7 g. Seeds have a thin coat which is reticulated. It stores fat in endosperm; seed colour may be dark blue, black, grey, light grey, brown, white, yellow, dark green or pink (Arslan et al. 2011; Bernath 1998; Broszat 1992; Honermeier 2006; Incekara 1949, 1964; Işıkan 1955, 1957).

Classification

Order: Ranunculales (Papaverales)
Family: *Papaveraceae*

The family Papaveraceae is cosmopolitan occurring in temperate and subtropical climates. Its main distribution is around the Mediterranean region and the Middle East. Most genera occur in Asia, Europe and North Africa; and a few in North America, South Africa and Australia. The family Papaveraceae is represented by 41 genera and *ca.* 250 species which are annual, biennial, perennial and seldom bushy. The family is represented by 7 genera in Turkey. However, the number of genera is shown as 3 in the Taxonomic Database of Turkey while the other four are included in the family Fumariaceae. Nevertheless, these genera are included in Papaveraceae in the International Plant Names Index (IPNI) (Anon. 2010a, b, c, 2011; Seçmen et al. 1995).

The number of species belonging to *Papaver* L. in the World is controversial. According to Kapoor (1995) 110; Carolan et al. (2006), and Mihalik (1998) 80; and Kadereit et al. (1997) 70 species exist. Turkey is quite rich in the number of Papaver species. Flora of Turkey listed 39 species including 19 annual and 20 perennial. However, the recent reports indicate 31 species and 46 taxa (Cullen 1965; Davis 1965, 1988, http://turkherb.ibu.edu.tr/; Güner et al. 2000). The genus Papaver has been divided into 5–11 sections by different researchers. These are *Argemonidium, Californicum, Carinatae, Horrida, Macrantha (Oxytona), Meconella, Meconidium (Milthantha), Papaver, Pilosa, Pseodopilosa* ve *Rhoeadium* sections. The sections *Horrida, Meconella* and *Californicum* do not exist in Turkey (Arslan et al. 2009; Goldblatt 1974; Kadereit 1988; Mihalik 1998; Sarıyar 2002).

Among the Papaver species, the most important and economically valuable is *Papaver somniferum* L. Furthermore, *P. bracteatum* Lindl., *P. orientale* L. and *Papaver nudicale* L. have partial importance. *Papaver rhoeas* L. (Corn poppy) is a widespread wild growing species in Europe, Asia, North Africa, North America and Australia.

Cultivated opium poppy has been classified differently by various researchers according to geographical distribution, seed and flower colour, fruit shape and other morphological properties. There are over 60 classifications in subspecies level. According to latest findings cultivated opium poppy is represented by three subspecies as *Papaver somniferum* L. subsp. *setigerum* (DC.) Corb., *Papaver somniferum* L. subsp. *songaricum* Basil ve *Papaver somniferum* L. subsp. *somniferum.*

Origin and Distribution

Opium poppy has been cultivated in Turkey and other parts of the World since ancient times. Turkey is considered as a gene centre for opium poppy among many other cultivated plants. Anatolia is important as the homeland and cultural source of opium poppy. Most ancient sources of written information on opium poppy and opium are Sumerian cuneiform tablets. Sumerians who reigned in 5000 BC in Mesopotamia used words for opium and Assyrian reliefs depicted opium poppies. It is known that opium poppy was cultivated by the Sumerians in 3300 BC. Sumerians called it as "the joy plant". Hittites (2000 BC) also cultivated opium poppy in Anatolia. Hittites named it as "Hassikka" and this is believed to be a prenom of today's haşhaş in Turkish. Furthermore, the same name also relates to "to sleep" and "to tranquilize" in Hittite language.

Among the many breads of the Hittites exists a bread with honey and poppy seeds (Ertem 1974). A city coin minted in 330 BC by the Romans in Şuhut, Turkey depicts a wheatear and poppy fruit. Furthermore, a city goddess holding a bunch of opium poppy, wheatear and flax is depicted in Emirdağ and Şuhut coins (Akçiçek 1997). Dioscorides of Anazarba (Kozan, Adana) describes the cultivation of opium poppy and production of opium in his Materia Medica printed in 1 AD.

Opium poppy is also considered as one of the most ancient plants of Europe. Archeological data reveal that opium poppy had been cultivated and used in Western and Central Europe since 4600–3800 BC. Cultivation areas were concentrated in the Alpine region. Opium poppy seeds and fruits were discovered in excavations in Switzerland (Incekara 1964). Furthermore, opium poppy seeds were also discovered in Spain, the Netherlands, Greece and Cyprus. This goes to show the widespread cultivation and use of opium poppy in ancient times. It is also claimed that earliest cultivation of opium poppy had occurred in Europe (Schultze-Motel 1979).

French researcher Belon who had visited Turkey in 1516–1519 said "Turks grow opium poppy as good as French grow wheat". Evliya Chelebi mentioned in his famous Book of Travels that the Inner Western Anatolia was described as the "land of opium" and opium poppy was very important as the main source of income of the inhabitants of Afyonkarahisar and its environs. Opium produced in Turkey used to be exported from the Port of Izmir to India, China and Indonesia by British and Dutch companies (Mat 2009; Matsui 1995; Schmidt 1998; Gevenkiriş 2011). Akbar shah of the Moghul Empire (1543–1605) initiated the cultivation of opium poppy in India, however, its widespread cultivation was achieved by the Eastern Indian Company established by Great Britain. In early eighteenth century, Portuguese merchants traded India-grown opium in China.

China banned the sale and consumption of opium in 1729. Eastern India Company purchased the trade concessions of opium in India. This company started illegally selling opium to Chinese pirates and smugglers. Chinese precautions to stop illegal opium trading failed. Chinese police had destroyed 21,000 boxes of opium confiscated in the warehouses of the Eastern India Company in Canton. This event gave rise to the 1st and 2nd Opium Wars, and the loser China had to pay compensation. Finally, China had almost surrendered to opium and started consuming 75 % of the opium produced in India. Hong Kong became an opium port and had been considered as the Opium State. In this period, these countries imported opium from Turkey as well (Babaoğlu 1997; Bhattacharji 2007; Booth 1996; Gevenkiriş 2011; Matsui 1995; Schmidt 1998).

Unlimited opium poppy cultivation which had been affected until 1933 in Anatolia was stopped and opium fields were restricted and opium production was controlled after the signing of the International Opium Convention by Turkey in 1933.

Soil Requirements

Opium poppy is not selective for soil requirements and grows in every kind of soil. Ideal development conditions and high crop yield is achieved in reasonably alluvial-loamy soils. Such soils are ideal for keeping the moisture for better development of the plant. Sandy soils, heavy soils, coarse soils and pressed or hard pressed soils are not suitable. Heavily loamy soils may create problems for root development, and due to the forming of a creamy layer hinders the emergence of embryotic shoot from seeds to the ground. Sandy soils do not retain water and prevents the plant to acquire the water needed regularly. In sandy soils, the plant has the high risk of downing during wind and rain. Unfavourable soil conditions lead to diseased or dwarf plants, narrowing of the leaves, discoloration of the leaves, lack of branching and eventually cropping losses.

Good conditioned soils with humus and without extreme humidity leads to favourable germination and first stage development. Opium poppy prefers neutral to weakly acidic soils. In heavily calciferous soils alkaline pH increases and the acquisition of nitrogen and potassium by the plant becomes complicated (Anon. 1950, 2005, 2010d, 2012a, 2013a, b; Bernath 1998; Broszat 1992; Honermeier 2006; İncekara 1964; Kadıoğlu 2007).

Climatic Requirements

Transient regions between the marine climate and the continental climate are considered as ideal for opium poppy. In Turkey, opium poppy is cultivated in western transitionary region and inner black sea region. It adapts very well to the climatic conditions of Central Europe. Primary development of opium poppy is very slow. Therefore, opium poppy cultivation in moist climates is difficult due to weed strains and disease risk. Windy places are not suitable due to breaking or downing of the stems. Fast warming of the soil is good for poppy cultivation. Opium poppy is relatively more tolerant to the cold during germination and primary development stages. Minimum germination temperature is +3 °C, and opium poppy seeds can germinate at +4 °C when optimum moisture is attained in the soil. At lower temperatures germination is not achieved. The plant shows good development above +4 °C at fall. Plants with a good root system and 6–8 rosette leaves can pass a normal winter without damage. Opium poppy is relatively more tolerant to cold during primary development.

Depending on the strength and period of coldness, foreign opium poppy types are tolerant to −5 °C while native types and populations are resistant to −10 °C and even to −20 °C under cover of snow. However, harsh winters and frozen soils may cause first leaves to wither, and even to total death of the plant at later stages. Plants late sowed in fall and not emerged due to unsufficient rain, and those whose cotyledons exposed to lower temperatures may be damaged from frost heaving and frost shock in spring. For better vernalization and healthy development, opium poppy requires 2–3 degrees of temperature for 30–35 days. Plant without a good rosette formation and good vernalization becomes weak and thin. Opium poppy likes the sun and hot weather especially during the bolting period. During the flowering and maturing stages relatively high temperature is required for seed development and maturing.

Opium poppy is a long day plant and especially where long day light persists and during the vegetation period clear weather exists morphine content in the fruits increase.

During the development period, total temperature requirement of opium poppy is 23–27 °C. Temperature during seed development and maturing period is crucial for the yield and oil content. In this period, sunny, dry and warm weather (18–20 °C) is suitable. However, extreme hotness and low relative humidity during the flowering period retard pollination. As a consequence, seed yield drops. Ideal weather temperature during the maturing period is above 20 °C, and sunny weathers. These information imply that regions with very harsh winters and cool summers are not so suitable for opium poppy cultivation.

Annual rain requirement of opium poppy is 600–700 mm. It is ideal for 300–400 mm to persist during the primary vegetation period till flowering. In provinces where opium poppy cultivation is affected annual rain fall is between 321.7 and 694.4 mm. A good germination is not possible if the soil is too dry. Therefore, seeding is preferred after the autumn or spring rains. Rain during the flowering period affects pollination negatively and seed yield drops. Furthermore, rain causes fungal infections, especially, the mildew (*Peronospora arborescens*) to spread. Overcast days and high humidity during the vegetation period also may lead to diseases. It may also cause opium to ooze out. In full blooming period, early morning rain retards opening of the petals and sticking of petals to the capsule on top hinders pollination leading to slightly poorer seed yields. In countries where lancing is practiced extreme precipitation right after lancing leads to a considerable loss of opium. Nevertheless, it should be remembered that water demand of opium poppy increases during the flowering period. However, rain is not preferred during the maturing period.

Freezing losses are high due to strong winds in winter without snow cover. Although opium poppy has a taproot, its root system is weak. Hail and heavy rains riddle the leaves and strong winds following a rain or irrigation may down the plants especially during the development stage of the capsules.

In countries where lancing is allowed, strong winds after lancing frictions between the plants lead to a considerable loss of opium. Likewise, strong winds also cause seed losses for dehiscent capsule poppies. Wind damage can be avoided by applying appropriate cultivation techniques (Anon. 1950, 2005, 2010d, 2012a, 2013a, b; Bernath 1998; Broszat 1992; Erdurmuş and Öneş 1990; Honermeier 2006; İncekara 1964; Kadıoğlu 2007; Kahraman 2011; Marquard and Malko 2006; Öğretir 1985).

Production Levels

The cultivation, production, control and trade of opium poppy in the World is carried out according to the protocol of 1972 related to Single Convention on Narcotic Drugs of 1961. Single Convention of 1961 mandates the global control of narcotics single-handedly by the United Nations. This convention regulates the cultivation of narcotic drugs, production, import and export of narcotics. Cultivation of opium poppy has been affected in the following countries under UN control, Turkey, India, Australia, France, Spain, Hungary, Czech Republic and China (Table 1).

Table 1 Legal opium poppy cultivation areas (hectare) based on main cultivating countries

Years	Turkey	India	Australia	France	Spain	Hungary	Total
2003	99,430	12,320	9,811	7,919	5,732	2,937	138,149
2004	30,343	18,591	6,644	8,312	5,986	7,084	76,960
2005	25,335	7,833	6,599	8,841	4,802	5,106	58,516
2006	42,023	6,976	3,457	6,632	2,146	4,322	65,556
2007	24,603	5,913	4,661	3,198	5,606	3,269	47,250
2008	20,042	2,653	4,108	3,683	5,507	2,262	38,255
2009	48,893	8,853	4,598	6,750	6,865	1,114	77,073
2010	51,897	12,237	9,127	9,400	6,439	7,308	96,408
2011	54,911	16,518	10,973	8,592	9,488	6,025	106,507
2012	13,510	12,092	10,158	8,680	9,308	3,755	57,503

Source: INCB Narcotic Drugs Report (2012a,b,c)

United Nations has declared Turkey and India as "Traditional Producers of Opium Poppy". Turkey, as a traditional producer of opium poppy holds about half the opium poppy cultivation areas of the World (Table 1). Opium poppy cultivation areas for alkaloid production in the World range between 138,146 ha (2003) and 38,255 ha (2008). Cultivation areas show a fluctuating trend in all the producing countries.

Opium poppy cultivation in Turkey is regulated according to law no. 3298 and related regulations on Narcotic Products based on certified controlled and unlanced opium poppy capsule production. Opium poppy cultivation and unlanced poppy capsule production is allowed in areas declared by the state at a limit of 70,000 ha quota given by the UN according to the planning realized by the Soil Products Office (TMO) General Directorate. TMO is responsible for the cultivation and control of opium poppy and the production of morphine and its derivatives in Turkey. In Turkey, due to security concerns and heavy labour use, opium poppy is cultivated near towns and villages. Opium poppy cultivation is mainly a family business and cultivation is affected in small plots of 1–10 da (0.1–1.0 ha).

At present, opium poppy is cultivated in the whole of Afyon, Amasya, Burdur, Çorum, Denizli, Isparta, Kütahya, Tokat and Uşak provinces and in some towns of Konya, Balıkesir, Eskişehir and Manisa provinces (Fig. 1). Previously, opium poppy had been cultivated in 52–70 provinces, however, during 1971–1974 its cultivation was banned due to international pressure. By years, based on the 70,000 ha legal cultivation limit, cultivation areas are determined according to the requirements of the Opium Alkaloids Factory (annual processing capacity 20,000 t) and critical stock (10,000 t), and it shows variations year by year. Therefore, opium poppy cultivation fields follows uneven development. Producers number between 35,000 and 103,000. Excess stock is depleted or deficient stock is completed with production planning to normal levels (Anon. 2010d, 2013a; Kamminga 2011).

Some countries cultivate opium poppy only for its seeds (Tables 2 and 3).

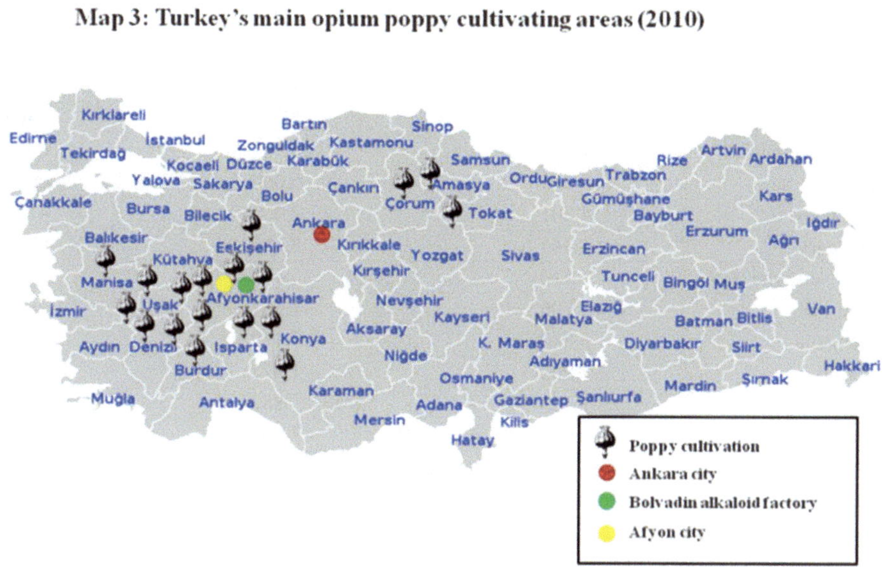

Fig. 1 Opium poppy cultivation fields in Turkey (Kamminga 2011)

Table 2 Poppy seed production in the World (FAOSTAT 2013)

Years	Area harvested (ha)	Production quantity (ton)	Yield (kg/ha)
2006	130,559.00	81,508.00	624.3
2007	116,997.00	66,583.00	569.1
2008	125,754.00	89,254.00	709.8
2009	140,878.00	98,872.00	701.8
2010	147,412.00	97,188.00	659.3
2011	128,601.00	106,419.00	827.5

World production of poppy seeds is around 100,000 t and major producers are Turkey and Czech Republic. These two countries meet 68 % of the World demand (Table 3).

Illicit opium poppy cultivation is as important as the legal cultivation (Table 4). Illegal cultivation areas far exceed the legal cultivation. No figure exists for illegal poppy seed production.

No poppy cultivation exists in Africa.

Cultivars

As mentioned above, cultivated opium poppy *P. somniferum* has been classified according to geographical distribution, seed and flowers colours, other morphological characteristics especially fruit characteristics. Over 60 subspecies level

Table 3 Poppy seed producing countries in 2011 (FAOSTAT 2013)

Countries	Area harvested (ha)	Crop yield (ton)	Yield
Austria	1,740	1,614	927.6
Croatia	3,492	2,256	646.0
Czech Republic	31,495	26,918	854.7
France	10,000	6,000	600.0
Germany	4,801	2,030	422.8
Hungary	8,344	8,156	977.5
The Netherlands	510	492	964.7
Palestine	206	2,561	1,243.2
Romania	3,880	2,173	560.0
Serbia	900	1,022	113.6
Slovakia	1,115	887	795.5
Spain	6,923	7,000	1,011.1
Macedonia	284	233	820.4
Turkey	54,911	45,077	820.9
World	128,601.0	106,419.0	827.5

Table 4 Illicit opium production in 2009 (http://chartsbin.com/view/h0z)

Countries	Cultivation areas (ha)	Opium production (ton)
Afghanistan	123,000	6,900
Columbia	356	9
Laos	1,900	11
Mexico	15,000	325
Burma	31,700	330
Pakistan	1,779	44
World	137,735	7,619

classifications exist. Major opium poppy taxonomists are Russian scientists Basilevskaya, (1928 (According to Hammer 1981)) and Veselovskaya (1933 (translation 1976)) furthermore, Danert 1958 ve Hammer 1981 had been interested in the topic and Mihalik (1998) has reviewed their work (Table 5).

Veselovskaya has classified cultivated poppies to seven subspecies and further to strains and 64 varieties. The first scientific study on Anatolian opium poppies has been made by Zhukovski 1933 (translated in 1951); later Incekara (1949) has carried out a detailed study on Anatolian poppies. Incekara (1949, 1964) mentioned that one of the seven subspecies of Veselovskaya is found widespread and the others rare or little cultivated in Anatolia, and white and purple flowering (White, yellow and blue coloured seeds) varieties of *Papaver somniferum* subsp. *anatolicum* are cultivated most. According to latest studies, cultivated opium poppy is divided to three subspecies one being wild. These are ***Papaver somniferum* L. subsp.**

Table 5 Classification of cultivated opium poppy (Hammer 1981; Mihalik, Bernath 1998)

Papaver somniferum				
Seed colour	*Subsp. somniferum*		*Subsp. songaricum*	
	Indehiscent fruit	Dehiscent fruit	Indehiscent fruit	Dehiscent fruit
White	var. *somniferum*	var. *somniferum*	var. *albescens* Vess	var. *orientale* Danert
	var. *candidum* Vess	var. *candidum* Vess	var. *rubicundum* Vess	var. *apertum* Danert
Yellowish	var. *roseolum* Vess	var. *roseolum* Vess	var. *rhodanthum* Vess	var. *foratum* Danert
Pink	var. *paeonifolium* Alef	var. *paeonifolium* Alef	var. *igneum* Danert	var. *fulgidum* Danert
	var. *macrocarpum* Coss	var. *macrocarpum* Coss	var. *parmulatum* Danert	var. *maculosum* Danert
	var. *papyrinum* Danert	var. *papyrinum* Danert	var. *apiatum* Vess	var. *hapalanthum* Danert
	var. *clausum* Danert	var. *clausum* Danert	var. *limboflorum* Danert	var. *palleolum* Danert
	var. *baageanum* Alef	var. *baageanum* Alef	var. *mundum* Danert	var. *leucomelum* Danert
	var. *coerulescens* Rothm	var. *coerulescens* Rothm	var. *livens* Vess	var. *gaucescens* Danert
Light grey	var. *oculatum* Danert	var. *oculatum* Danert	var. *holonatum* Danert	var. *pratextum* Danert
Light blue	var. *nigrum* Hayne	var. *nigrum* Hayne	var. *rubidum* Vess	var. *rotundilobum* Danert
Dark blue	var. *serenum* Danert	var. *serenum* Danert	var. *sigillatum* Danert	var. *ocellatum* Danert
	var. *subgriseum* Vess	var. *subgriseum* Vess	var. *nubeculosum* Danert	var. *poriferum* Danert

setigerum* (DC.) Corb.** (Syn; *Papaver setigerum* DC), ***Papaver somniferum* L. *subsp. songaricum* Basil (Syn;** *Papaver somniferum* L. *subsp. indicum*) and ***Papaver somniferum* L. *subsp. somniferum (Syn; *Papaver somniferum* L. *convar. nigrum* (Hayne) Alef., *Papaver somniferum* L. *convar. somniferum*, *Papaver somniferum* L. *subsp. hortense* (Hussenot) Lange).

At present, various seed coloured populations and improved types have been cultivated in Turkey. By 2013, 24 opium poppy cultivars have been registered in Turkey. Most of them have been developed through hybridization breeding and some by selection breeding.

Part II: Cultivation Practices

Propagation

Opium poppy is a cultivated plant propagated by seeds. Fruits contain numerous seeds. Therefore, it is one of the cultivated plants with highest propagation rate (Table 6).

Seed purity of opium poppy must be 98 % and germination rate minimum 80 %. Min. moisture content should not exceed 10 % and the number of other plant and weed seeds should not be more than 25 in a kg (Dejnega et al. 2003).

Soil Preparation

Opium poppy seeds are minute. 1,000 seeds weigh 0.5 g. Due to small size of the seeds, soil preparation of the cultivation field necessitates great care. First of all, seeds must be enveloped with moist soil in order to germinate. Opium poppy likes deeply tilled field being a taproot plant.

In order to meet these requirements, the following procedures have to be implemented. After harvesting the crop, farm manure is applied, deep plowing is carried out and crop residue and manure is mixed with the soil. This must be performed immediately after the harvest. In fallowing field, farm manure should be applied in the fall followed by deep plowing. For winter range sowing, all of phosphorous fertilizer and half of nitrogenous fertilizer must be applied at end-September or early October to till the field. Cultivator (crow's feet) must be used for soil tilling.

If moisture content is not enough, the soil must be brought to the right condition by irrigation if possible. During tilling the soil should not be pulverized and normal structure must be maintained in opium poppy cultivation like in other plants. Keeping the soil slightly lumpy prevents creaming and keeps winter range opium poppies from frost.

For summer range sowing, the soil must be tilled in Autumn. In Spring, base fertilizers must be applied with deep plowing. Soil tilling in Spring and preparation of the seed bed is similar to that of sugar beet. In order to eradicate weeds, soil must be tilled superficially, moisture must be maintained and an efficient seed bed must

Table 6 Maximum cultivation field where one plant can propagate in successive years (Becker 1993 modified by Arslan 2002)

Years	Potato	Wheat	Rape	Opium poppy
0	1 plant	1 plant	1 plant	1 plant
1	3 m^2	1 m^2	20 m^2	120 m^2
2	30 m^2	30 m^2	1.2 ha	7.2 ha
3	300 m^2	900 m^2	720 ha	4,320 ha

be prepared after slightly roller pressed. Annealing the soil is important in soil tilling. In annealed soil, the soil maintains its structure and makes it easy to ward off the weeds (Anon. 1950, 2005, 2010d, 2012a, 2013a, b; Bernath 1998; Broszat 1992; Erdurmuş and Öneş 1990; Honermeier 2006; Incekara 1964; Kadıoğlu 2007; Marquard and Malko 2006).

Planting

Opium poppy must be sown as early as possible in order to avoid the loss of moisture during soil tilling and for hardiness to cold and frost during the primary development stage. In Turkey, opium poppy is planted as summer sown and winter sown. Their planting may show some variations, nevertheless, winter range sowings take place in October while summer range sowings are affected in February, March or early April. A good germination is not possible if the soil is too dry. Therefore, planting is carried out after autumn or spring rains.

Early planting for summer range poppy is highly advantageous since the long day plant opium poppy exhibits a good primary development. Early planting also leads to better crop yields. When lancing was allowed, 60–70 % of planting used to be winter range and 30–40 % summer range. Presently, winter range planting is over 90 %. Summer range planting is affected as replanting especially following frost damage. In areas where winters are harsh and irrigation is possible summer range planting can also be made.

In Turkey, sowing is generally by hand spreading, however, the use of sowing machine is increasing in recent years. In hand sowing, seeds are spread as such or after mixing with sand half and half, and the seed sown per decare is 1–2.5 kg. In machine sowing, the amount of seed per hectare drops to 300 g. Seed economy cannot be overlooked when shifting from hand sowing to machine sowing.

In machine sowing, emergence is uniform and considerable economy is achieved in labour use for hoeing and thinning. It may be difficult to achieve uniform sowing in a unit area even when using a sowing machine due to minute size of the seeds. Therefore, sensitive sowing machines must be utilized. In order to achieve uniform distribution seeds may be mixed with sand, fertilizer, etc. Sowing depth should not exceed 1 cm irrespective of hand or machine sowing since poppy seeds are very small. In hand spreading, the soil must be treated with wood harrow, disc harrow or rake after sowing. It is not necessary in machine sowing. Especially during the use of planter, planting direction is important. As mentioned before, opium poppy suffers from heavy winds. Therefore, in such areas, poppies should be planted in the direction of the wind.

Optimum plant density is arranged after the maintenance. Generally, additional thinning is not necessary with the use of sowing machine. Arrangements must be made to maintain minimum 50 and maximum 100 plants per square meter. Optimum plant density of 50–80 per square meter plants are recommended in practice. In such a case, each plant develops 1–4 fruits. However, the number of fruits per plant depends on plant density as well as the poppy type, manuring, soil structure and climatic conditions.

Row spacing depends upon the preferred maintenance procedures and weed control. In mechanical sowing, maintenance is also mechanical with row spacing of 40–50 cm and 20 cm between plants. In the case of hand spreading type sowing, it is enough to leave 30 cm between plants. Narrow row spacing is advantageous in order to reduce competition among the plants like in grains. In the case of machine hoeing, wider than 30 cm row spacing is necessary (Anon. 1950, 2005, 2010d, 2011, 2012a, 2013b; Bernath 1998; Broszat 1992; Erdurmuş and Öneş 1990; Honermeier 2006; Incekara 1964; Kadıoğlu 2007; Marquard and Malko 2006).

Fertilisation

According to a study, the amount of nutrients taken up by opium poppy from the soil depend on the amount of dry matter (seed, fruit, stem, stalk, etc.) on the surface of the soil.

Another researcher reported that opium poppy takes up 71.4 kg N, 26.6, kg P, 92.7 kg K, 81.7 kg Ca, 15.3 kg Mg, 347 g Mn, 35 g Cu, 205 Zn and 113 g Borax per hectare based on 270,000 plant density and 3,864 kg superficial dry matter. Of the dry matter, 41 % consists of leaves and stems; 14.8 % fruits and 26.7 % seeds. Harvest index is 0.15 for fruits and 0.27 for seeds (Edelbauer and Stangl 1993). Other studies confirmed these findings. As evident, nutrient requirement of opium poppy is not high. However, for better yields fertilizing is necessary. Better bioavailability of nutrients from the soil and enriched content of nutrients is important for higher yield.

Opium poppy likes farmyard manure. Farmyard manure not only provides nutrients but also adjusts the water retention and character of the soil. Since opium poppy cultivation fields are closer to settlements, some farmers prefer barnyard manure. In Autumn, before the first ploughing, the soil must be enriched with refuted farmyard manure and be mixed well with the soil by deep plowing. Unrefuted barnyard manure increases the risk of weeding; encourages the multiplication of soil pests and causes them to inflict damage on opium poppy especially during the young period. Indian farmers reportedly use farmyard manure (10–20 t per ha^{-1}) in opium poppy cultivation (Kapoor 1995).

Application of chemical fertilizers is important depending on the soil structure and nutrient content for a good development and crop yield. 1/2–1/3 of nitrogenous fertilizers, and whole of phosphorous and potassium fertilizers must be applied and mixed with the soil before or during autumn or spring plantings. It has been reported that during machine sowing application of nitrogen to the rows affects the yield positively. Laughlin and Chung (1992) has reported to obtain 197 kg seed and 2.84 kg morphine per decare after applying 6 kg/da nitrogen on the rows.

Nitrogen uptake of opium poppy is the highest during bolting and budding. Therefore, half of the nitrogenous fertilizer must be given during bolting. When deciding on the nitrogen dosage, expected yield as well as pre-plant, nitrogen content in the soil and resistance to tilting must be taken into account. Application of excess nitrogen retardingly prolongs vegetative growth and retards maturing. Several researchers have reported the positive effect of nitrogenous fertilizing on morphine synthesis and morphine content. Nitrogenous fertililizer recommendation is 7–14 kg per decare depending on the circumstances (Anon. 1950; 2005, 2010d, 2011, 2012a, 2013a, b; Bernath 1998; Broszat 1992; Engin 1995; Honermeier 2006; Marquard and Malko 2006; Ramanathan et al. 1975).

Studies carried out in Turkey have revealed that the amount of nitrogenous fertilizer needed for optimum fruit and seed yield vary between 8 (Eyüpoğlu 1995) and 10 kg da^{-1} (Engin 1995). Turkish Grain Board (TMO) recommends 8–10 kg nitrogenous fertilizer per hectare for a good development and yield depending on soil structure and nutrient content (Anon. 2005, 2010d, 2013a, b). Aytekin and Önder (2006) reported the highest seed yield (180.7 kg/da), crude oil yield (93.71 kg/da), fruit yield (140.99 kg/da), morphine percentage (0.74 %) and crude oil percentage (50.86 %) from 12 kg/da N and P_2O_5 plots.

Like nitrogen, P, K, Ca and trace elements are also important for opium poppy. Phosphorous is crucial for the synthesis of proteins and phospholipids. According to our observations, it also increases the resistance of the plants to lower temperatures. Potassium is necessary for the activity of several enzymes. Potassium also affects alkaloid synthesis at different levels. As soils in Turkey are rich in potassium, generally potassium fertilizers are not used. Phosphorous fertilizers are recommended 3 kg per decare. Boron (B), a minor element, is particularly important for the development of secretory ducts. Boron deficiency leads to abnormal vegetative development, and to death of sprouts and buds of the plant. Boron fertilizing is particularly important in high pH soils since boron uptake in such soils is decreased (Anon. 2005; Honermeier 2006). Oktay and Akademir (1997) has reported that zinc is also important and zinc sulphate fertilizing increases crop yield by 15.5–20.5 %.

In conclusion, appropriate and timely fertilization is necessary for hardiness to winter conditions, fast development in spring, bountiful and healthy crop.

Irrigation

Winter range opium poppy can be cultivated in dry conditions. In Turkey, opium poppy cultivation is affected 35 % in fields with ground water and 65 % in arid and dry land. Necessary moisture must be present at seed bed for the germination of seeds. Timely emergence is necessary for plants to enter winter in rosette period. Due to rain delay, plants cannot enter winter at this period and are damaged from the cold. Fields must be in proper condition during sowing or the soil must be irrigated if possible after sowing. In some areas, flood irrigation is resorted to in order to till the soil before sowing. After the sowing, sprinkling may be necessary for better emergence. Both methods can be recommended to capable farmers.

It must be remembered that flood irrigation may lead to tilting in fields where throat filling is not applied; sprinkling may affect pollination negatively in flowering period and may lead to loss of alkaloids. In a study, water consumption of opium poppy was determined as 752 mm for winter range poppies and 425 mm for summer range poppies. For winter range poppy one irrigation (111 mm) during the budding period, and for summer range poppy two irrigations each (75–80 mm) one in budding and one in flowering periods are necessary. Such irrigation increases the seed and fruit yield. An irrigation trial in Australia has shown that controlled irrigation increases the yield and 50 mm each of 1–2 irrigation before flowering, 50 % during flowering and at the end of flowering were found useful. If possible, opium poppy plants must be irrigated especially in years when spring period is dry (Anon. 2005, 2010d, 2012a, 2013a, b, Bernath 1998, Broszat 1992; Chung 1987,1992; Erdurmuş and Öneş 1990; Honermeier 2006; Incekara 1964; Laughlin and Chung 1992; Öğretir 1985).

Pest Control

There are many pests and diseases of Opium poppy but their economical significance is not high. However, they inflict important damage in crop yield in the years of epidemics. Therefore, it is useful to know important diseases and pests and fighting methods for opium poppy farmers and workers.

Poppy Root Weevil *Ceuthorrhynchus denticulatus* Schr.

Adults hybernate in winter in the soil. In spring, especially in March when poppy plants are at 4–6 cm in length and with 3–4 pairs of leaves, they migrate to poppy. They feed on leaves. After 1–2 weeks of feeding they start mating and the females oviposite under the leaves along the midrib in batches of 1–4. They reproduce once a year. Adults inflict damage on the leaves by boring small holes. Most damage is caused at the larval stage. Larvae descend to the roots on which they feed, form galeries on the surface causing considerable damage. When uprooted in mid-May pearl like larvae formations can be seen on the roots. Inflicted roots darken, leaves turn yellow and fall, plant length stay short. Since boring does not go to the centre of the roots, plants can recover however, produce less and unhealthy fruits compared to healthy plants. Alkaloid yield of such capsules is also less. Control measures include early sowing and crop rotation. Chemical control includes the use of Azinphos methyl, Fenitrothion, Diazinon, Fenthion. During heavy infestation (20 adults in sq. m.) pest control recommended to be done twice in 7 day intervals.

Tettigometra *Tettigometra hexaspina* (Kolenati)

Adults are seen end-April or early-May in poppy fields. Females oviposite eggs in packs along root node to the roots sequently. It is possible to see the eggs from seedling stage on until green fruit stage in poppy roots. It is widely distributed in Afyon and Uşak provinces. Both nymphs and adults inflict damage by sucking juice from roots and the stems. Pest control for underground pests is made by using Trichlorphon, Endsulfan or Chlorpyrifos. It must be ensured that the chemicals reach the roots.

In addition, polyfagous insects such as aphids (*Aphis fabae* Scopoli), black cutworm (*Agrotis ipsilon* Hufn.), onion thrips (*Thrips tabaci* Lindl.), European mole cricket (*Gryliotalpa gryllotalpa* L.) also cause damage. Birds can also inflict damage in poppy fields. Furthermore, rodents (mice and rats) inflict damage on stored poppies. Cultural measures, physical and chemical methods can be affected in pest management (Erdurmuş and Öneş 1990; Giray 1985; Zümreoğlu and Akbulut 1984).

Disease Control

Fungal diseases are common in opium poppy. Two of them are important for Turkey.

Poppy Root Stem Blight Collar Rot Disease (*Dendryphion papaveris* Sawada)

Its synonyms are *Dendryphium penicillatum*, *Helminthosporium papaveris.* This fungus can inflict disease in all vegetative stages of the plant. Damping off makes the most damage during seedling stage. In older plants root collar is damped off and stains and rotting is seen in the stem and the capsules and leaves are withered. Capsule rot can cause up to 50 % reduction in crop yield. This is also common in other countries where poppy is cultivated. According to a study conducted in 1976 in Afyon, it was discovered that the disease was mainly transmitted by the infected seeds and had caused collar rot disease in poppy fields. The fungus can maintain viability on spent plant parts and the seeds for 1 year. The utilization of clean seeds is important to curb spreading of the disease.

Cultural measures include;

1. Seed should not be gathered from diseased plants.
2. Seeds for cultivation should be dried very well and stored under dry conditions.
3. Since the fungus spend winter on spent plants, fields should be cleaned after the harvest.
4. At least 2 year alternation must be applied.

Applying chemicals on seeds give good results. Before sowing 100 kg seeds must be mixed well with 1 kg of the mixture of tetramethyl thiuram disulfide 37.5 + Carbendazime 37.5 (Thiram + Vitavax) (Karahan and Maden 1978; O'Neill et al. 2000; Honermeier 2006).

Downy Mildew (*Peronospora arborescens* (Berk) de Bary)

Disease agent fungus spend the winter on diseased spent plant parts on the soil. It is transported by seeds and diseased plant parts infected with the fungus. Disease is spread by the wind in rainy and moist periods. It is mainly seen in rainy and moist years. It occurs as primary and secondary infections. Primary infection is observed while the plant is at seedling stage in spring. First indications of the disease are seen on lower leaves close to the soil. Leaves turn light green, lower leaves develop yellowish spots, leaves are curled and damaged. In moist conditions a grey rust develops underneat the leaves. This layer is crucial multiplication and spreading of the disease. In dry and warm conditions, diseased leaves develop spots but the grey rust does not occur. Plants infected by primary infection can not develop and stay dwarf. Such plants can be observed in poppy fields. Cultural measure is alternation. Dense sowing should be avoided. Diseased plants must be removed immediately. Although there is no licenced product against this fungus in Turkey, preparations used against tobacco mildew can be utilized. Since disease in transmitted by the seeds, seeds must be disinfected.

Although not so well known in Turkey, beet mosaic virus (BMV) (observed in Turkey), beet yellow virus (BYV), bacterial blight (*Xanthomonas papavericola),* bacterial soft rot (*Erwinia carotovora)* are also known to occur in Turkey. Some other fungal diseases can also inflict damage on opium poppy plants (Erdurmuş and Öneş 1990, Honermeier 2006; Türkoğlu and Fidan 1984).

Weed Control

Opium poppy seeds germinate and emerge within 1–2 weeks after planting in prepared soil. Cotyledons are 3–4 mm long and 1 mm wide, according to the cultivars they can be green or with anthocyanins. Primary development stage of opium poppy is very slow, therefore its challenge against weeds is weak. For the preparation of the field and for better yield weed control is very important. Weed flora of opium poppy fields consist of typical annual summer weeds. Problematic weeds may vary according to cultivation areas. Weeds can be cleared mechanically or by hand hoeing after the emergence of poppy seedlings. Extreme caution should be taken during mechanical hoeing not to damage the seedlings. Two to three hoeing is necessary for opium poppy cultivation. There are studies to eradicate weeds by the use of herbicides. However, herbicides are not used in Turkey for poppy cultivation.

First thinning and hoeing is carried out when opium poppy plants develop 7–10 leaves following winter. Interrow hoeing is applied 4–6 weeks after the emergence. If weeds have a tendency of early development, it is necessary to add marker plants like false flax, buckwheat, etc. to the poppy seeds in order to predetermine the rows. Hoeing is particularly important for heavy and hard soils. If mechanical sowing is applied interrow must be 40–50 cm and 10–20 cm between plants in a row must be left after thinning. In fields where hand sowing is applied, it is necessary to leave 30 cm between plants. In practical terms, it is preferred to plant 50–80 plants in square meter. In hand sowing by spreading, especially the first hoeing is difficult and labour intensive. In row-wise sowing, interrow weeds are cleared and thinning is carried out on the rows. This results in at least one third savings in hoer per decare.

Especially in spring sowing, soil must be prepared during mechanical hoeing. Otherwise, clods can damage the plants. 15–20 days after thinning and 1st hoeing, 2nd hoeing and throat filling are carried out. Although opium poppy has a tap root, since its lateral roots are not well developed and above ground parts are large, it falls over especially after heavy winds accompanied with rain. Therefore, throat filling is important.

After the 2nd hoeing, opium poppy shows fast improvement and the plants cast shade on the soil preventing the development of weeds. As a result, 3rd hoeing may not be necessary (Anon. 2005, 2010d, 2012a, 2013a; Broszat 1992; Bernath 1998; İncekara 1964; Erdurmuş ve and Öneş 1990: Honermeier 2006; Marquard and Malko 2006).

Harvesting

Opium poppy harvest is performed when poppy capsules dry completely. At that time, ripened seeds stuck on the placenta drop to the bottom of the capsule. Ripened capsules make a sound when shaken. Not all capsules in a field or on a plant ripen at the same time. Since capsules formed of last flowering blossoms ripen later, lowest capsules should be looked at before deciding on the harvest. Although it is a type characteristics to have dehiscent or indehiscent capsule, opening can be observed in over exposed to sun and hence over dried capsules which had not been harvested though ripened. As such situations cause loss of seeds due to spillage, harvest time should not be delayed.

Poppy capsules are harvested by hand breaking them at the point of attachment to the stalk. If the farmer wishes to keep some seeds for the following year's cultivation, necessary number of capsules are selected and cut by a knife to collect the same colour seeds. Rest of the crop is passed through poppy crushing machine or crushed by wooden mallet and screened. This is carried our using special equipment. During harvest, capsules must be crushed in such a way to allow inspection of probable lancing and dusting should not be allowed. In optimum conditions, crop yield from a decare is 100 kg capsule or 100–125 kg seed. Average crop yield of Turkey in 2011 was 75 kg/da for fruits and 82 kg/da for seeds. Stalk yield was 300–500 kg per da. Moisture content for seeds and fruits during storage must not exceed 8–9 %.

Part III: Post-Harvest Handling

Part Harvested and Harvesting Techniques Used

Grading

Poppy seeds are cleaned, sorted according to white, blue and yellow colours and offered to the market. According to Spice Decree, in poppy seeds, foreign matter content should not exceed 1.5 %, moisture content should not exceed 11 %, ash content should not be over 10 %, damaged seed content should not exceed 0.5 %. TMO (Turkish Grain Board) specifications for the purchase of poppy straw (broken capsules) is as follows:

1. Should be in natural colour and scent.
2. Moisture content should not exceed 12 %.
3. Should not contain green fruit parts.
4. Should not have been burnt, rotten, moldy or fermented.
5. Seeds deliberately adulterated by moistening and redrying to change colour, shriveled, shrunkened will not be purchased but confiscated.
6. Foreign matter should not exceed 8 %. Part exceeding it will be confiscated and necessary procedure will be applied.
7. Lanced capsules are not allowed. Such capsuled will be confiscated and necessary legal procedures will apply.
8. Foreign matter consists of not only fruit parts remain over the sieve, but also unripe burnt capsules, organic and inorganic matter as well as those organic and inorganic matter remaining on 2.2 mm round hole American sieve.
9. Capsule powder in the capsule will be regarded as foreign matter and its content should not exceed 7 %.
10. Seeds in capsule will be treated as foreign matter and its content should not exceed 1 %. Those exceeding will have to be cleaned by the farmer before purchase. During cultivation only 0.5 % of other types than the certified seeds are allowed. For other seeds it should not exceed 2 % and the seed packs should not exceed 25 kg (Dejnega et al. 2003).

Storage

Opium poppy capsules are stored in warehouses after purchase by TMO before worked up at the Opium Alkaloids Factory. A separate safety stock is also kept. As capsules which had been roughly crushed to allow inspection of possible lancing are voluminous requiring large storage places. Stored capsules, in some years, exceed 30,000 t per season. Poppy capsules are purchased in standard 30–35 kg gunny sacks and kept as such. Majority of poppy capsules are stored in factory warehouses and some in warehouses of regional TMO offices.

Table 7 Export of morphine and derivatives and poppy seeds from Turkey

	Alkaloid (Morphine and derivatives)		Seed	
Year	Amount (kg)	Value ($)	Amount (Ton)	Value (1,000$)
2006	87,687	29,870,797	22,951	33,200
2007	123,219	38,573,288	14,934	35,266
2008	124,282	42,521,262	9,523	50,504
2009	96,004	37,840,602	14,008	47,493
2010	96,747	44,691,867	16,228	52,312

Source: 2010 Yılı Haşhaş Faaliyet Raporu (2010 Activity Report for Poppy) and FAOSTAT/Trade (03.06.2013)

Since poppy seed is an oil seed, its storage is important. If moisture content at harvest is high, it should be decreased below 10 %. Moisture content of poppy seeds at harvest in Turkey hardly exceeds 10 %. Seeds are stored by farmers or the merchants.

Packaging

Poppy capsules are transported in 30–35 kg standard gunny sacks and delivered to TMO and then to the factory as such.

Marketing

Sole buyer of poppy capsules in Turkey is TMO. There is no other buyer and the farmers are required to sell unlanced capsules to TMO. Capsule prices are declared by the Government every year before the season.

Trade of poppy seeds is free. A small portion is kept as agroseed, some used as local food ingredient and the rest is either sold to merchants, oil factories or exported. There are small size oil extraction plants in the cultivation areas, however, its production and sale is highly limited. The following table gives export figures of alkaloids and poppy seeds (Table 7).

Part IV: Production Schedules

India is the world's largest producer of raw opium, producing 1,075 mt of opium in 1999, or 118.3 mt in morphine equivalent. It is the only country to export raw opium. Turkey, having banned opium production in 1971, began to cultivate opium poppy for the production of poppy straw in 1974. Today, it is the largest producer of poppy straw in the world, producing 31,332 mt in 1999 or 97.1 mt in morphine equivalent. Both countries are recognised as 'traditional supplier countries' by the United Nations and by some member states, including the US and Japan (Mansfield 2001).

Utilization

Cosmetic

Poppy oil nourishes the skin and used as massage oil in aromatherapy.

Pharmaceutical and Therapeutic

Poppy is mainly used for pharmaceutical and therapeutical purposes. Opium which is obtained by collecting the milky exudate that appear after lancing of unripe live fruits has been used in therapy for the last 4,000 (to some 6,000) years and it has been the most widely used drug in history. It has kept its importance over the centuries. Its medicinal and narcotic properties have always been highlighted and other uses have not been so popular. Durability of opium to years long storage without losing its properties under normal conditions has increased its commercial importance. Opium has analgesic, antidiarrheic, narcotic and antitussive properties. Isolation of morphine from opium in 1804 by the German scientist Sertürner opened an era and research into opium alkaloids has intensified over the years but interest in opium has continued till today. It is monographed in many pharmacopoeias. Since Sertürner, over 40 alkaloids have been isolated from opium. Research has shown that the activity of opium is due to its alkaloids. Five major alkaloids morphine, codeine, thebaine, noscapine and papaverine has been most widely utilized. Information on major alkaloids of opium can be found in the following table (Arslan 2012; Gümüşçü et al. 2008; Kapoor 1995; Önmez 2007; Tanker and Tanker 2003) (Table 8).

Morphine: This is the first isolated and most frequently used opium alkaloid. It is the most effective analgesic ever found. However, it has high potential for addiction, tolerance and psychological dependence. Therefore, its actual use is limited.

Table 8 Major alkaloids of opium (Kapoor 1995)

Alkaloid	Chemical formula	Content (%)	Inventer	Year	Properties
Morphine	$C_{17}H_{19}O_3N$	3.0–23.0	Sertürner	1816	Pungent, alkali and strong toxic structure
Codeine	$C_{18}H_{21}O_3N$	0.3–3.0	Robiquet	1832	
Thebaine	$C_{19}H_{21}O_3N$	0.3–1.0	Thioumery	1835	
Noscapine (Narcotine)	$C_{22}H_{23}O_7N$	2.0–12.0	Derosne	1803	Mild, light toxic structure
Narceine	$C_{23}H_{22}O_8N$	0.1–0.2	Delletier	1842	
Papaverine	$C_{22}H_{21}O_4N$	0.8–1.5	Merck	1848	

Table 9 Distribution of important alkaloids in opium poppy capsules (Gümüşçü et al. 2008)

Alkaloid	Minimum content (%)	Minimum content in total alkaloids (%)	Maximum content (%)	Maximum content in total alkaloids (%)	Mean alkaloid content (%)	Mean alkaloid content in total alkaloids (%)
Morphine	0.110	42.6	1.140	87.6	0.475	67.9
Codeine	0.005	0.5	0.270	25.5	0.049	7.69
Thebaine	0.005	1.9	0.134	18.1	0.041	5.66
Papaverine	0.001	0.3	0.440	29.7	0.029	3.70
Noscapine	0.006	1.1	0.418	41.2	0.099	15.05
Total	**0.141**		**1.643**		**0.693**	

Codeine: It is a native alkaloid of opium but is also obtained semi-synthetically from morphine by methylation. It has analgesic, antitussive and dormitive properties. Addiction to codeine is 6 times less than that of morphine. **Noscapine (Narcotine)**: Its antitussive effect is almost equal to that of codeine. It is also used to treat cancer cells. Its activity has been proven against colon cancer, prostate cancer, brain cancer and lymphoma. It is not addictive, narcotic and analgesic. No toxicity with noscapine is reported, however, it should not be used by patients with bronchial asthma. **Thebaine**: Thebaine is the most toxic opium alkaloid and not used as such in therapy. However, it can easily be converted to codeine, hydrocodone, acetyldihydrocodeine and oxycodone. It is also used in the synthesis of super-analgesic oripavine derivatives. **Papaverine**: Papaverine is used in the treatment of gynacological diseases. **Oripavine**: It is a metabolite of thebaine used to treat cocaine dependence (Arslan 2012; Kapoor 1995; Önmez 2007; Tanker and Tanker 2003) (Table 9).

Alkaloid synthesis in opium begins soon after germination; alkaloids accumulate in vascular bundles and laticifers in different parts of the plant. After budding and flowering opium latex starts accumulating in capsules (seed pods) and exudes out after lancing that is slightly cutting the longitudinally oriented laticifers. Several genes and enzymes affecting the synthesis of opium alkaloids have been discovered and synthesis mechanism has been disclosed (Kapoor 1995; Kempe 2008; Weid 2004).

Food and Flavouring

Poppy seeds are used as appetizer after roasting or as filling in pastries after roasting and mashing, or as such in cakes and pastries. Poppy seeds contain 45–54 % fixed oil which comprises linoleic acid (omega 6) (62–72 %), oleic acid (omega 9) (15–20 %), palmitic acid (4.8–9.5 %) and stearic acid (2.0–2.9 %). Poppy oil is an edible oil but is also used in paint dyes like other semi-drying oils, soap and other related industries. Gradual reduction of cultivation fields after 1965 and total ban in cultivation in 1972 has forced many small oil plants to close down. In the same years, fast developments in refined oil industries have led refined sunflower oil to take the lead and high price of poppy seed as an export commodity decreased the use of poppy oil as edible oil. Traditional use of haşhaş ezmesi (Poppy seed mash) as filling in pastries is continuing (Arslan 2012; Arslan et al. 2000; Broszat 1992; Incekara 1964; Ipek and Arslan 2012; Rahimi et al. 2011a, b).

Industrial

Dried opium poppy fruits (capsules) are worked up to isolate morphine at the Opium Alkaloids Factory in Bolvadin, Afyon. Morphine consists of 68–70 % of total alkaloids in the fruits. The remaining 30–32 % of alkaloids are just wasted. The factory has the capacity to work up annually 20,000 t of poppy capsules. The factory consists of two main units. In the Alkaloid Extraction Unit only base morphine is produced. In the adjacent Derivatives Unit half of the morphine extracted is converted to morphine hydrochloride, codeine, codeine phosphate, codeine sulphate, codeine hydrochloride, morphine sulphate, ethylmorphine hydrochloride. These alkaloids are sold after pharmacopoeial compliancy is achieved (Anon. 2005, 2010d, Gümüşçü et al. 2008).

Other

Oil cake is a good animal feed and it positively affects the production of milk and milk fat in cows. Being rich in protein, oil, and non-nitrogenous basic materials it is a valuable animal feed especially for beef cattle and water buffalo. Famous Afyon kaymağı is made from the milk of water buffalo (Arslan and Camcı 1986).

Showy poppy flowers used to be popular in the gardens of Europe and USA and present in the seed catalogues. Plantlets and seeds used to be sold for horticulture. Double varieties of poppy flowers used to be grown in gardens in the sixteenth century. Hortus Lusatie of Johannes Franke published in 1594 mentions double varieties of opium poppy. Eight forms of ornamental poppies were illustrated in Hortus Eystettennis published in 1613. Due to ban on the cultivation of poppies even for ornamental reasons in horticulture under the International Narcotics Control measures there are cases of prosecution of gardeners. Therefore, cultivation of poppies is confined only to botanic gardens, museum gardens and some public gardens. In short, horticultural cultivation of opium poppy requires permission from relevant authorities. International trade of poppy seeds was addressed by a UN resolution “to fight the international trade in illicit opium poppy seeds” in 1998. In this resolution, reference is made to the United Nations Single Convention on Narcotic Drugs of 1961 for poppy cultivation and member countries are reminded and warned that majority of poppy seeds in international trade is from countries where illegal poppy cultivation is carried out. However, seeds of ornamental poppies are freely traded for horticultural purposes in various countries.

Opium poppy has no value as cut flower since they do not keep and petals fall out quickly. However, double varieties are more suitable for cut flower trade. Dried poppy capsules are used in dry flower arrangements. Large fruits, long fruits lighter in colour are more favoured. Growing opium poppies for dry flower trade also requires permission. Therefore, many countries have legal obstacles for its cultivation. This can be seen as an opportunity for Turkey's export potential (Arslan and Sarıhan 2011).

Young fresh leaves of opium poppy during rosette period are consumed as salad vegetable where it is grown.

References

Akçiçek E (1997) Anadolu Tıp Tarihinde ve Halk Hekimliğinde Haşhaş. Eren'ce, Halk Bilim Yazıları, İzmir, s.142–143

Anon. (1950) The cultivation of the opium poppy in Turkey. Bull Narcotics 01:13–25

Anon. (2005) Haşhaş Raporu (2004 yılı). Toprak Mahsulleri Ofisi Genel Müdürlüğü, 40s

Anon. (2010a) http://bioces.tubitak.gov.tr/. Accessed 4 Oct 2010

Anon. (2010b) http://www.ipni.org/. Accessed 4 Oct 2010

Anon. (2010c) http://wwweski.tubitak.gov.tr/tubives/. Accessed 4 Oct 2010

Anon. (2010d) Haşhaş Raporu (2009 yılı). Toprak Mahsulleri Ofisi Genel Müdürlüğü, 25s

Anon. (2011) http://www.turkherb.ibu.edu.tr/. Accessed 12 Sept 2011

Anon. (2012a) Haşhaş Raporu (2011 yılı). Toprak Mahsulleri Ofisi Genel Müdürlüğü, 15s

Anon. (2012b) Toprak Mahsulleri Ofisi Genel Müdürlüğü, Mali İşler Dairesi Başkanlığı 74. Hesap dönemi faaliyet raporu 2011, 112s

Anon. (2012c) Narcotic drugs. United Nations/International Narcotics Control Board, New York, 444s

Anon. (2013a) Haşhaş Raporu (2012 yılı). Toprak Mahsulleri Ofisi Genel Müdürlüğü, S: 5

Anon. (2013b) http://www.tmo.gov.tr/Main.aspx?ID=147. Accessed 15 May 2013

Anon. (2013c) http://www.faostat.fao.org. Accessed 15 May 2013

Anon. (2013d) Illicit opium production in 2009 (http://chartsbin.com/view/h0z)

Arslan N (2002) Haşhaş Islahı Kurs notları. Ocak 2002, Bolvadin (Not printed)

Arslan N (2012) Haşhaş ve Alkaloidleri. Türkiye'de Tarım Dergisi, İzmir

Arslan N, Er C, Camcı H (1986) Haşhaş ekimi yasağının kaldırılmasından beri haşhaş tarımı ve problemleri, VI. Bitkisel İlaç Hammaddeleri Toplantısı, 10–16 Mayıs, Bildiler Kitabı, pp 99–118, Ankara

Arslan N, Sarıhan EO (2011) *Papaver* (Haşhaş) Türlerinin Süs Bitkisi Olarak Değerlendirilme İmkanları. IV. Süs Bitkileri Kongresi 20–23 Ekim 2010, Mersin Bildiri Kitabı: pp 546–552

Arslan N, Büyükgöçmen R, Gümüşçü A (2000) Türk haşhaş populasyonlarının yağ ve morfin muhtevaları. Tarla Bitkileri Merkez Araştırma Enstitüsü Dergisi 9(1–2):56–60

Arslan N, İpek A, Gümüşçü A, Sarıhan EO, Parmaksız I (2009) *Oxytona* Seksiyonundaki yabani haşhaşların Tebain Yönünden araştırılması. Türkiye VII. Tarla bitkileri kongresi, Hatay

Arslan N, İpek A, Rahimi A, İpek G (2011) Ankara Üniversitesi Ziraat Fakültesi Tarla Bitkileri Bölümü haşhaş (*Papaver somniferum* L.) genetik kaynaklarının morfin ve bazı özellikler yönünden değerlendirilmesi, Bursa, Tarım Kongresi, II. Cilt: 1185–1190

Aytekin M, Önder M (2006) Azot ve fosfor dozlarının haşhaşta (*Papaver somniferum* L.) verim ve bazı verim unsurları ile kalite üzerine etkileri. Selçuk Üniversitesi Ziraat Fakültesi Dergisi 20(38):68–75

Babaoğlu AN (1997) Uyuşturucu ve Tarihi Bağımlılık Yapan Maddeler *Kaynak Yayınları*

Becker H (1993) Pflanzenzüchtung. UNI Taschenbuch, Ulmer Verlag, Stuttgart

Bernath J (ed) (1998) Poppy. Medicinal and aromatic plants: industrial profiles, vol 3. Harwood Academic Publishers, Amsterdam

Bhattacharji R (2007) Case study India's experiences in licensing poppy cultivation for the production of essential medicines. (http://www.poppyformedicine.net/modules/case_studies)
Booth M (1996) Opium, a history. Simon & Schuster, London, 381 p
Broszat W (1992) Der Mohn (*Papaver somniferum* L.). Anbau und Markt einer wiederentdeckten Kulturpflanze. Der Tropenlandwirt, Beiheft Nr. 47
Carolan J, Hook I, Chase M, Kadereit JW, Hodkinson TR (2006) Phylogenetics of *Papaver* and related genera based on DNA sequences from ITS nuclear ribosomal DNA and plastid trnL intron and trnL–F intergenic spacers. Ann Bot 98(1):141–155
Chung B (1987) The effect of irrigation on growth and yield components of poppies (*Papaver somniferum* L.). J Agric Sci 108:389–394
Chung B (1992) The effects of plant density and irrigation on the lodging, yield and yield components of poppies (*Papaver somniferum* L.). Acta Horticulturae 306:458–465
Cullen J (1965) *Papaver* L. In: Davis PH (ed) Flora of Turkey and the East Aegean Islands, vol 1. Edinburg University Press, Edinburgh, pp 219–236
Danert S (1958) Zur systematik von *Papaver somniferum* L. Kulturpflanze 6:61–88
Davis PH (1988) Flora of Turkey and the East Aegean Islands, vol 1 and 10. Edinburgh University Press, Edinburgh
Dejnega D, Girsch L, Kramberger I (2003) Betrachtungen zur Saatgutproduktion und zu Saatgutregelungen bei MOHN (*Papaver somniferum* L.) 10s. www.ages.at/uploads/.../mohn_endf_12_02.pd. Accessed 20 Sept 2012
Edelbauer A, Stangl J (1993) Nährstoffentzug durch den Waldviertler Graumohn (*Papaver somniferum* L.) im Verlaufder Vegetationszeit. Bodenkultur, Band1/;15–27
Engin D (1995) Azot Dozları ve Hasat Zamanının Haşhaşta (*Papaver somniferum* L.) Verim ve Kalite Üzerine Etkisi. Doktora Tezi. Uludağ Üniversitesi Fen bilimleri Enstitüsü, Tarla Bitkileri Bölümü – Bursa
Erdurmuş A, Öneş A (1990) Haşhaş. Mahsulleri Ofisi Genel Müdürlüğü Ankara
Ertem H (1974) Boğazköy Metinlerine Göre Hititler Devri Anadolu'sunun Florası, TTKY 7/65, Ankara
Eyüpoğlu F (1995) Göller Bölgesinde yetiştirilen haşhaşın azotlu ve fosforlu gübre isteği. Toprak ve Gübre Araştırma Enstitüsü. Yayın No:205, Ankara
Gevenkiriş A (2011) Determination of morphine and total phenolic content in poppy seed of Turkish origin. Thesis, The Graduate School of Natural and Applied Sciences of Middle East Technical University
Giray H (1985) Türkiye haşhaş (*Papaver somniferum* L.) zararlılarına ait liste ve önemlilerinin zarar şekilleri hakkında notlar. Türk Bitki Kor Derg 9:109–124
Goldblatt P (1974) Biosystematic studies in Papaver section Oxytona. Ann Missouri Bot Gard 61(2):264–296
Gümüşçü A, Arslan N, Sarıhan EO (2008) Evaluation of selected poppy (*Papaver somniferum* L.) lines by their morphine and other alkaloids contents. Eur Food Res Technol 226:1213–1220
Güner A, Özhatay N, Ekim T, Başer KHC (2000) Flora of Turkey and the East Aegean Islands (supplement 2), vol 11. Edinburgh University Press, Edinburgh
Hammer K (1981) Problems of *Papaver somniferum*-classification and some remarks on recently collected European poppy land-races. Kulturpflanze 29:287–296
Honermeier B (2006) Mohn. In: Heyland K-U, Hanus H, Keller ER (eds) Handbuch des Pflanzenbaues, vol 4. Ulmer, Stuttgart, pp 262–270
Incekara F (1949) Türkiye haşhaş çeşitleri ve bunların tohum ve afyon verimi bakımından değerleri. Toprak Mahsulleri Ofisi Yayınları, Ankara
Incekara F (1964) Endüstri Bitkileri ve Islahı Cilt II, Yağ Bitkileri. Ege Üniversitesi Yayınları; No. 83, İzmir
Ipek G, Arslan N (2012) Gıda Maddesi Olarak Haşhaş (*Papaver somniferum* L.) Tohumunun Değerlendirilmesi. Türk Bilimsel Derlemeler Dergisi 5(2):99–101
Işıkan M (1955) Haşhaş Çeşitleri Üzerinde Biyolojik Araştırmalar. A.Ü.Z.F Yayınları No:50, Ankara

Işıkan M (1957) Anadolu haşhaşlarının tohum renkleri üzerinde genetik araştırmalar. Ankara Üniversitesi Ziraat Fakültesi Yayınları No: 128, Ankara

Kadereit JW (1988) Sectional affinities and geographical distribution in the genus *Papaver* L. (*Papaveraceae*). Beitraege zur Biologie der Pflanzen 63:139–156

Kadereit JW, Schwarzbach AE, Jork KB (1997) The phylogeny of *Papaver* s. l. (*Papaveraceae*): polyphyly or monophyly? Plant Syst Evol 204:75–98

Kadıoğlu Y. 2007. Uşak'ta haşhaş tarımının coğrafi özellikleri. Eastern Geographical Review 18:165–185.

Kahraman M (2011) Türkiye'de haşhaşta uygulanan politikalar ve haşhaşın Türkiye için önemi. Namık Kemal Üniversitesi Fen Bilimleri Enstitüsü, Yüksek Lisans Tezi, Tekirdağ

Kamminga J (2011) Opium poppy licensing in Turkey a model to solve Afghanistan's illegal opium economy? Research Paper International Council on Security and Development (ICOS), 63s

Kapoor LD (1995) Opium poppy: botany, chemistry and pharmacology. Food Products Press, New York

Karahan O, Maden S (1978) Haşhaşta kök boğazı yanıklığı hastalığı (*Dendirophion papaveris* (Saw.) Sawada) tohumla taşınma durumu ve tohum ilaçlarının bu etmene etkisi üzerinde çalışmalar. Bitki Koruma Bülteni 18(1–4):1–9

Kempe K (2008) Analyse der Benzylisochinolinalkaloid-Biosynthese des *Papaver somniferum* L. mit Hilfe der RNA*i*-Technik. Dissertation, Mathematisch-Naturwissenschaftlich-Technischen Fakultätder Martin-Luther-Universität Halle-Wittenberg

Laughlin JC, Chung B (1992) Nitrogen and irrigation effects on the yield of poppies (*Papaver somniferum* L.). Acta Hort 306:466–473

Mansfield D (2001) An analysis of licit opium poppy cultivation: India and Turkey. Foreign and Commonwealth Office of the Government of the United Kingdom

Marquard RA, Malko A (2006) Mohn. In: Heyland K-U, Hanus H, Keller ER (eds) Handbuch des Pflanzenbaues, vol 4. Ulmer, Stuttgart, pp 465–468

Mat A (2009) Osmanlı İmparatorluğu'nda afyonun tarihi. Osmanlı Bilimi Araştırmaları XI/1–2:285–290

Matsui M (1995) Production and trade of opium in the Ottaman empire 1828–1838. Master thesis, Boğaziçi University İstanbul

Mihalik E (1998) Biology of poppy taxonomy. In: Bernath J (ed) Poppy. Medicinal and aromatic plants: industrial profiles, vol 3. Harwood Academic Publishers, Amsterdam

O'Neill NR, Jennnings JC, Bailey BA, Farr DF (2000) *Dendryphion penicillatum* and *Pleospora papaveracea*, destructive seedborne pathogens and potential mycoherbicides for *Papaver somniferum*. Phytopathology V 90(1):691–698

Öğretir K (1985) Afyon (Çay) koşullarında haşhaşın Su tüketimi. Eskişehir Toprak-Su Araştırma Enstitüsü Yayınları No:188

Oktay M, Akdemir H (1997) Haşhaş Yetiştiriciliğinde Çinko Sülfat Gübrelemesinin Ürün Miktarı ve Bazı Kalite Kriterlerine Etkisi. I. Ulusal Çinko Kongresi, pp 243–249

Önmez H (2007) *Papaver somnferum* bitkisinden elde edilen alkaloidlerin ekstraksiyonunda kullanılan çözücü ve metodların karşılaştırılması. Selçuk Üniversitesi Fen Bilimleri Enstitüsü/ Yüksek Lisans Tezi, Konya

Rahimi A, Arslan N, İpek A (2011a) Düşük Morfinli Haşhaş (*Papaver somniferum* L.) Hatlarından Çeşit Geliştirilmesi Yönünde Yapılan Çalışmalar. IV. Tohumculuk Kongresi, 14–17; Haziran 2011 *Samsun.* Bildiriler kitabı1: 115–120

Rahimi A, Kıralan M, Arslan N, Bayrak A, Doğramacı S (2011b) Variation in fatty acid composition of registered poppy (*Papaver sumniferum* L.) seed in Turkey. Acad Food J 9(3):22–25

Ramanathan VS, Ramachandran C, Prakash V (1975) Effect of organic manure and chemical fertilizers on opium poppy. III. Role of phosphate on the yield of opium, morphine and poppy seeds. Indian J Agron 20(1):65

Sarıyar G (2002) Biodiversity in the alkaloids of Turkish *Papaver* species. Pure Appl Chem 74(4):557–574

Schmidt J (1998) From Anatolia to Indonesia, opium trade and The Dutch community of Izmir (1820–1940), PIHANS, 82, Istanbul

Schultze-Motel J (1979) Die urgeschichtlichen Reste des Schlafmohns (*Papaver somniferum* L.) und die. Entstehung der Art Die Kulturpflanze 27(2):207–215

Seçmen Ö, Gemici Y, Görk G, Bekar L, Leblebici E (1995) Tohumlu Bitkiler Sistematiği. Ege Üniversitesi Fen Fakültesi Ders Kitapları Serisi No: 116, Sayfa: 396, İzmir

Tanker M, Tanker N (2003) Farmakognozi Cilt I. Ankara Üni., Eczacılık Fak., Yayınları No: 66, Ankara

Türkoğlu T, Fidan Ü (1984) Ege Bölgesi Haşhaş (*Papaver somniferum* L.) ekim alanlarında görülen virus hastalıkları üzerinde ön çalışmalar. Bitki Koruma Bülteni 24(2):88–93

Veselovskaya MA (1976) The poppy: its classification and importance as an Oleiferous crop. Translated from a Russian edition dated 1933, New Delhi

Weid M (2004) Gewebe- und zellspezifische Lokalisation der Alkaloidbiosynthese in *Papaver somniferum* L. Dissertation, Mathematisch-Naturwissenschaftlich-Technischen Fakultät der Martin-Luther-Universität Halle-Wittenberg, 106S

Zhukovski P (1951) Türkiye'nin Zirai Bünyesi (Anadolu) [Çev: C. Kıpçak, H. Nouruzhan, S. Türkistanlı]. Türkiye Şeker Fabrikaları A. Ş. Neşriyatı No.: 20, 887S

Zümreoğlu S, Akbulut N (1984) Uşak ili ve çevresindeki haşhaş (*Papaver somniferum* L.) ekim alanlarında görülen zararlılar üzerinde ön çalışmalar. Bitki Koruma Bülteni 24(3):159–172

Index

Z. Yaniv, N. Dudai (eds.), *Medicinal and Aromatic Plants of the Middle-East*,
Medicinal and Aromatic Plants of the World 2, DOI 10.1007/978-94-017-9276-9

MIX
Papier aus verantwortungsvollen Quellen
Paper from responsible sources
FSC® C105338

If you have any concerns about our products,
you can contact us on
ProductSafety@springernature.com

In case Publisher is established outside the EU,
the EU authorized representative is:
Springer Nature Customer Service Center GmbH
Europaplatz 3, 69115 Heidelberg, Germany

Printed by Libri Plureos GmbH
in Hamburg, Germany